Food Engineering Series

Springer's Food Engineering Series is essential to the Food Engineering profession, providing exceptional texts in areas that are necessary for the understanding and development of this constantly evolving discipline. The titles are primarily reference-oriented, targeted to a wide audience including food, mechanical, chemical, and electrical engineers, as well as food scientists and technologists working in the food industry, academia, regulatory industry, or in the design of food manufacturing plants or specialized equipment.

More information about this series at http://www.springer.com/series/5996

Colm P. O'Donnell • Colette Fagan • P.J. Cullen
Editors

Process Analytical Technology for the Food Industry

Springer

Editors
Colm P. O'Donnell
University College Dublin
Belfield
Ireland

Colette Fagan
University of Reading
Reading
United Kingdom

P.J. Cullen
Dublin Institute of Technology
Dublin
Ireland

ISSN 1571-0297
ISBN 978-1-4939-0310-8 ISBN 978-1-4939-0311-5 (eBook)
DOI 10.1007/978-1-4939-0311-5
Springer New York Heidelberg Dordrecht London

Library of Congress Control Number: 2014940790

Printed on acid-free paper

Springer is part of Springer Science+Business Media (www.springer.com)

Contents

Contributors

J. Burger BurgerMetrics SIA, Jelgava, Latvia

Fabien Chauchard Indatech, Clapiers, France

Qiaofen Cheng Department of Food and Nutritional Sciences, Whiteknights, Reading, UK

P.J. Cullen School of Food Science and Environmental Health, Dublin Institute of Technology, Dublin 1, Ireland

School of Chemical Engineering, University of New South Wales, Sydney, Australia

Cheng-Jin Du Warwick Systems Biology Centre, University of Warwick, Coventry, UK

Colette C. Fagan Department of Food and Nutritional Sciences, University of Reading, Reading, UK

Food and Nutritional Science, Department of Food and Nutritional Sciences, University of Reading, Reading, UK

Tat Hean Gan Brunel University, Middlesex, UK

E. Gaston IRIS-Innovació i Recerca Industrial i Sostenible, Castelldefels, Barcelona, Spain

Jarka Glassey School of Chemical Engineering and Advanced Materials, Newcastle University, Newcastle upon Tyne, UK

A. A. Gowen School of Food Science and Environmental Health, Dublin Institute of Technology, Dublin 1, Ireland

Joseph Irudayaraj Agricultural & Biological Engineering, Purdue University, West Lafayette, IN, USA

Digvir S. Jayas Department of Biosystems Engineering, University of Manitoba, Winnipeg, MB, Canada

Julie Lundtoft Johnsen Arla Strategic Innovation Centre, Arla Foods amba, Aarhus, Denmark

Ramazan Kizil Chemical Engineering Department, College of Chemical and Metallurgical Engineering, Istanbul Technical University, Maslak, Istanbul, Turkey

Jordane Lallemand Ondalys, Clapiers, France

Kathryn L. McCarthy Department of Food Science and Technology, University of California, Davis, CA, USA

Michael J. McCarthy Department of Food Science and Technology, University of California, Davis, CA, USA

Colm P. O'Donnell School of Biosystems Engineering, University College Dublin, Dublin 4, Ireland

Sébastien Preys Ondalys, Clapiers, France

Sylvie Roussel Ondalys, Clapiers, France

R. Vadivambal Department of Biosystems Engineering, University of Manitoba, Winnipeg, MB, Canada

Chapter 1
Benefits and Challenges of Adopting PAT for the Food Industry

P.J. Cullen, Colm P. O'Donnell and Colette C. Fagan

1.1 Introduction

Process analytical technology (PAT) is a framework for innovative process manufacturing and quality assurance. The concept is to design, analyse and control manufacturing processes through the measurement of identified critical control parameters which govern product variability. The identified benefits of the framework include increased process efficiency, reduced operating costs, increased process validation and ultimately improved final product quality and safety.

1.1.1 Evolution of PAT

Process analytical chemistry (PAC) is a term which developed during the 1940s to describe the application of analytical chemistry with techniques, algorithms and sampling equipment to solve developing problems related to various chemical processes. Although industrial process analysers have been in use for more than 60 years, the modern period of PAC essentially began with the formation of the Centre for Process Analytical Chemistry (CPAC) in 1984. The goal of PAC was to "supply quantitative and qualitative information about a chemical process" for monitoring, control and optimization. They went on to define five "eras" of PAC: (1) off-line, (2) at line, (3) on-line (4) in-line and (5) non-invasive, which describe the evolution

P.J. Cullen (✉)
School of Chemical Engineering, University of New South Wales,
Sydney, Australia
e-mail: patrick.j.cullen@dit.ie

C. P. O'Donnell
School of Biosystems Engineering, University College Dublin, Dublin 4, Ireland

C. C. Fagan
Department of Food and Nutritional Sciences, University of Reading,
P.O. Box 226, Reading RG6 6AP, UK

C. P. O'Donnell et al. (eds.), *Process Analytical Technology for the Food Industry*,
Food Engineering Series, DOI 10.1007/978-1-4939-0311-5_1,

of sensor technologies (Mishra et al 2008). Its definition has evolved over the years to encompass analytical measurements and understating of chemical, physical and microbiological parameters governing processing. Changing the term "chemistry" to "technology" allowed a broader scope of the approach to other processes. The pharmaceutical industry, in particular, has adopted the approach as a strategy to understand and control variability within the sector. The broad definition given by the US Food and Drug Administration (FDA): "A system for designing, analyzing, and controlling manufacturing through timely measurements (i.e., during processing) of critical quality and performance attributes of raw and in-process materials and processes with the goal of ensuring final product quality" covers the requirements and desires of manufacturing within the food industry.

Since 1987, PAT has had a dedicated international conference (International Forum Process Analytical Chemistry, IFPAC, which brings together instrumentation manufacturers, researchers and industry users.

1.1.2 Learning From Other Process Industries

The food industry has always been to the fore with regard to adoption of sensors and the use of risk analysis strategies. By comparison, the pharmaceutical industry has been more restricted in the adoption of advanced control strategies due to validated batch production processes, high-value-added products and lack of specialised technologies. Nevertheless, it was identified that an improved production process was required and PAT emerged as a platform of future good manufacturing practice (GMP). Industrial adoption has still remained relatively low; however, the ideology and desire for the approach is evident between regulatory agencies and industry alike.

The pharmaceutical industry also recognised that PAT could provide additional benefits such as continuous validation. Q7A GMP Guidance for Active Pharmaceutical Ingredients (API) defines validation as: *a documented program that provides a high degree of assurance that a specific process, method, or system will consistently produce a result meeting predetermined acceptance criteria.* PAT tools are capable of continuously measuring product *acceptance criteria* and critical control points (CPPs), thereby continuously evaluating if the process is behaving *consistently.*

Food treatment processes such as pasteurization need validation; the emergence of novel treatment processes such as high-pressure processing, pulsed electric fields, etc. necessitates process validation to ensure adequate treatment. A PAT strategy may facilitate such validation. Also, similar complementary benefits as found in the pharmaceutical industry may occur within the food industry with increased PAT implementation. For example, Hazard Analysis Critical Control Point (HACCP) is used in the food industry to identify potential food safety hazards so that key actions can be taken to reduce or eliminate the risk of the hazards being realized. HACCP is *a systematic preventive approach to food safety that addresses physical, chemical, and biological hazards as a means of prevention rather than finished product inspection.* Integrating a PAT strategy within HACCP may support the overall goal of ensuring food safety through the use of process monitoring.

Comparing both industries, we can see similarities and differences which influence the drivers to PAT adoption. The pharmaceutical industry is highly regulated and is risk averse with actives synthesised via both chemical and biological routes; many processes are operated under sterile conditions in highly controlled environments. Batch production continues to dominate with low levels of automation. Product quality is typically monitored off-line using laboratory-based methods of analysis. Food production within developed countries is increasingly regulated and is risk adverse to microbial or chemical contamination. Production is typically non-sterile.

1.1.3 PAT Drivers in the Food Industry

Consumer: For the food industry, the consumer is a key driver for food production methodologies. Produce taste, nutrition, appearance, cost and shelf life are important parameters influencing consumer purchase and thereby the choice of production methodology. Technologies which can optimize the process will ultimately lead to consumer loyalty and repeat purchase. Consequently, the food industry has traditionally employed technologies to monitor food produce particularly as end-point quality control strategies.

Regulators Pharmaceutical regulators have played a fundamental role in the push to adopt PAT strategies within the pharmaceutical industries. For pharmaceutical current GMP (cGMP), both the FDA and the European Medicines Agency (EMA) openly encourages the use of PAT. The FDA formed a PAT advisory committee that includes industry participation which is intended to facilitate dialogue between the regulators and industry. They have issued a PAT guidance document and created a PAT training program for industry. By comparison, there has been little involvement from food regulators in such a strategy. However, as PAT technologies begin to show promise as food safety prevention techniques this may change. Apart from production optimization, many of the PAT tools discussed within this book such as hyperspectral imaging have showed potential for contamination identification of foods within production facilities (Chap. 9). Similarly, PAT data may be useful in food traceability or potentially have a role to play in the development of proactive hazard alert system in ensuring food security throughout the food chain. If the approach can show reduced risk to consumers, food regulators may become increasingly interested in the approach.

Business One of the common reasons forwarded as to explain the limited uptake of PAT tools by the pharmaceutical industry has been the lack of a business case to improve current production processes given the high margins that drug companies traditionally operated within, coupled with the validation implications with adopting new production approaches. By comparison, the food manufacturing industry has typically been governed by lower profit margins and production efficiency. PAT tools may offer improved production cost and energy efficiency through process optimization along with increased quality control.

Sustainability The concept of environmental sustainability in food processing has become a key issue in recent times as awareness of the importance of environmental protection, and the possible impacts associated with the manufacture and consumption of food products, has increased. Environmental sustainability can be achieved by developing and implementing alternative environmental best-practice technologies and products which maximise the efficient use of resources and achieve cost savings, while minimising negative human and environmental impacts (Clark 2011). Recently, PAT has been linked to green production strategies:

> The Process Analytical Technology (PAT) initiative has made its name as a mechanism for monitoring processes in real time, facilitating process understanding and, in some cases, real-time release of product. With PAT, the focus has been on monitoring Critical Quality Attributes and controlling Quality Critical Process Parameters, and to a lesser degree manufacturing efficiency. It may be time to add sustainability to the list of PAT's objectives. (Thomas 2009)

1.1.4 Technology Advances

Originally, PAC measurements were performed off-line; however, they moved continually closer to production to result in: at line, on-line and in-line measurements. The principal benefit from such a paradigm shift was a significant decrease in the time delay between sampling and analysis along with monitoring of more representative samples. The past two decades have seen significant progress in the incorporation of on- and in-line process monitoring using advanced instrumentation. Optical and spectroscopic technologies have been to the fore in this advancement including computer vision, ultraviolet–visible (UV–Vis), near- and mid-infrared (NIR and MIR) and Raman spectroscopy. This rise was facilitated by related technological advances in solid-state detectors, fibre optics and instrumentation innovations for in situ sampling (Chew and Sharrat 2010) along with a parallel evolution of computer processing power.

This use of increasingly sophisticated process analysers resulted in increasingly large data sets that require appropriate numerical strategies to unravel chemical information (or process signatures) and associated process states encoded within the analytical data. With increasing know-how and affordability of spectroscopic and chromatographic instrumentation for on-line and in-line process analysis, PAT data sets are often intrinsically multivariate in nature (Chew and Sharrat 2010). Chemometrics has emerged and is widely embraced as a useful tool to unravel the data obtained by PAT.

1.1.5 Challenges

To facilitate widespread adoption of PAT within the food industry, a number of challenges need to be overcome. Technologies must meet the challenges posed by the food manufacturing environment including: cleaning-in-place (CIP) compatibility, harsh environments, real-time analysis, low-cost and ease of use. Technologies which provide food safety information directly from the production environment would be advantageous, with identification of microbial or chemical contamination. Suitable data management systems need to be developed and integrated with production to ensure that the benefits offered by PAT are achieved.

Unlike the pharmaceutical industry, variability with food production is sometimes welcomed. Indeed, one of the admirable aspects of artisan or "home-made" food is the variable nature of product. However, adoption of PAT does not mean that we have to produce foods which appear perfectly similar to some defined optimum. It is up to the manufacturers themselves to decide on what parameters they wish to optimize; indeed, PAT could be used to produce foods which display more artisan-like features.

Finally, the food industry needs to be made more aware of PAT as a framework for innovative process manufacturing and quality assurance. More collaboration between industry, academia and regulators is required to unify the disperse efforts currently underway. Adoption of PAT as a strategy would bring together process engineers, food scientists, technologists and microbiologists under one umbrella with the goal of providing the industry with a manufacturing framework for the twenty-first century. Here, we can learn from and cooperate with other industries such as the pharmaceutical and petrochemical to further develop the strategy.

References

Chew W, Sharrat P (2010) Trends in process analytical technology. Anal Methods 2:1412–1438

Clark J (2011) Introduction to green chemistry. In: Proctor A (ed) Alternatives to conventional food processing. RSC Publishing, London, p 1–10. http://pubs.rsc.org/en/content/ebook/978-1-84973-037-2

Mishra A, Banerjee S, Bhatwadekar N, Mahajan P, Karode P (2008) Process analytical technology (PAT): boon to pharmaceutical industry. Pharm Rev 6:3

Thomas P (2009) Is it time for PAT to go green? Pharma Manufacturing. http://www.pharmamanufacturing.com/articles/2009/148.html. Accessed 20 Oct 2011

Chapter 2
Multivariate Data Analysis (Chemometrics)

Sylvie Roussel, Sébastien Preys, Fabien Chauchard and Jordane Lallemand

2.1 Introduction

2.1.1 Definition of Chemometrics

Chemometrics, or multivariate data analysis, is the science which applies optimal mathematical and statistical methods to process data. Chemometrics includes the design of experiments upstream and the analysis of data to get valuable information after measurements have been taken. The need for chemometrics tools mainly comes from the development of analytical instruments providing large amounts of increasingly complex data.

This scientific arena consists of a large variety of mathematical methods, aiming at processing numerous data sets to achieve diverse objectives. The scheme below is an overview of the chemometrics approach any scientist should follow when facing a multivariate data analysis issue (Fig. 2.1).

Even though the principles of chemometrics are based in mathematics and statistics, one does not need to have deep knowledge of either of these disciplines to analyse multivariate data. However, thorough knowledge of the application as well as common sense are required in order to analyse the outputs of the chemometrics software packages and avoid pitfalls and misinterpretations.

2.1.2 PAT and Chemometrics

Process analytical technology (PAT), as defined in Chap. 1, includes appropriate measurement devices, that can be placed at-, in- or on-line, combined with mul-

S. Roussel (✉) · S. Preys · J. Lallemand
Ondalys, ZA La Plaine, 4 rue Georges Besse, 34830 Clapiers, France
e-mail: contact@ondalys.fr

F. Chauchard
Indatech, ZA La Plaine, 4 rue Georges Besse, 34830 Clapiers, France

C. P. O'Donnell et al. (eds.), *Process Analytical Technology for the Food Industry*,
Food Engineering Series, DOI 10.1007/978-1-4939-0311-5_2,

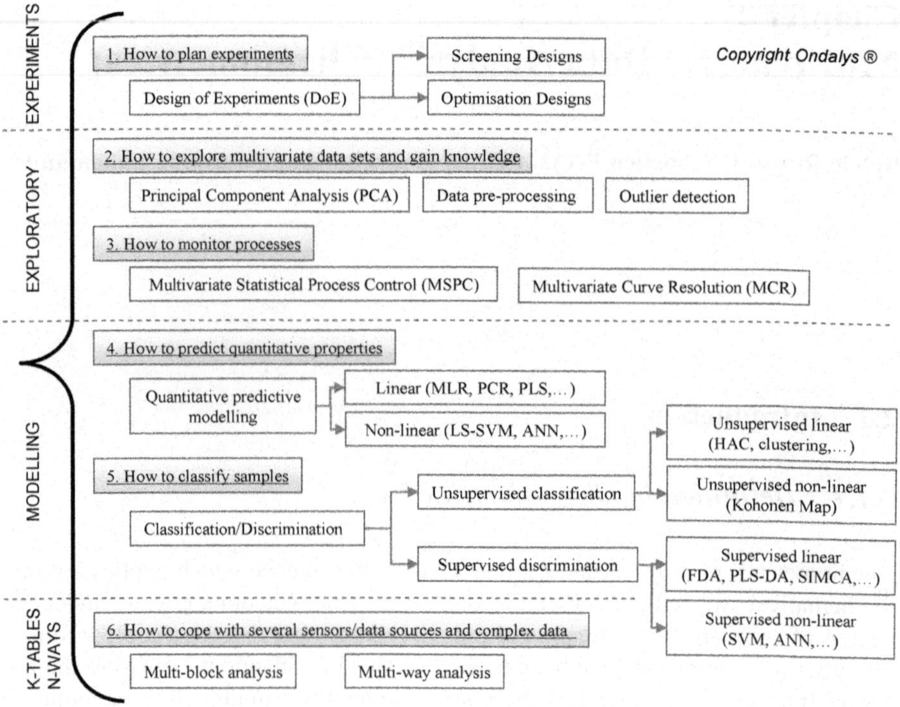

Fig. 2.1 The multivariate data analysis approach: classification of chemometrics methods

tivariate statistical (chemometrics) tools to analyse data and monitor and control processes. Chemometrics is therefore essential to understanding and diagnosing real-time processes, and keeping them under multivariate statistical control. PAT is also strongly linked to the quality by design (QbD) concept, which implies quality integration from the product development stage. Within the PAT framework, Wold et al. identified five levels of chemometrics analysis corresponding to different data and objective complexity levels (Wold et al. 2006):

- PAT-1: calculating critical quality attributes (CQA), such as concentrations, from rapid and real-time multivariate measurements, such as spectra, by multivariate calibration (predictive modelling).
- PAT-2: sorting samples (raw materials, intermediate or final products) as acceptable or not, based on multivariate measurements, such as spectra or property profiles, using multivariate statistical process control (MSPC).
- PAT-3: monitoring and classifying batch processes as acceptable or not from real-time multivariate measurements, such as process data, raw material data and spectra, using batch SPC (BSPC).
- PAT-4: combining data from all the critical process steps and raw materials to assess the final product quality using multi-block analysis.
- PAT-5: including feedback control to the process settings from the multivariate models, using process dynamic identification and time series modelling among others. This last level is not discussed further in this chapter.

For each of these levels, the modelling approach works well if the training set used to build the model is representative of the acceptable (in-control) samples. This is ensured by covering the desired variability, either by using design of experiments (DoE) during the process development at laboratory and pilot scales or by using huge historical laboratory or production databases with sufficient variability. Finally, the robustness of the model has to be regularly evaluated during its life cycle through maintenance and updating.

2.2 Design of Experiments

Using historical databases to model processes requires a very large amount of observations to ensure a minimum of variability. When it is possible, a more rational way consists in choosing the observations or experiments to span the whole desired operating conditions, i.e. the design space, with a maximum of variability. DoE (experimental designs) corresponds to that part of chemometrics which aims at planning the relevant experiments, minimising the cost without decreasing information quality, quantifying the different factor effects, modelling and optimising the processes (Gacula and Jagbir Singh 1984; Box and Draper 1987; Lundstedt et al. 1998; Leardi 2009). Different designs corresponding to different objectives are discussed in the following sections, such as screening and optimisation designs.

2.2.1 Problem Formulation

Understanding and modelling a process requires first to determine its multivariate inputs and outputs. On the one hand, the inputs represent the different factors or parameters which may have an influence on the outputs, such as temperature, pressure or the type of catalyst for a chemical reaction. They correspond to the independent measurements or variables which can be set independently of one another. The user's expertise and some tools, such as the Ishikawa diagram, are needed to determine an exhaustive list of the potential variability sources. Factors which cannot be precisely set by the user, i.e. uncontrolled factors, cannot be considered as inputs. On the other hand, outputs correspond to the response measurements (or dependent variables) which have to be optimised, such as the yield of a chemical reaction.

2.2.2 Screening Designs

The DoE methodology often includes a first step which consists in implementing a screening design (Araujo and Brereton 1996a). The experiments are chosen in order to quantify the influential factors among a large number of factors.

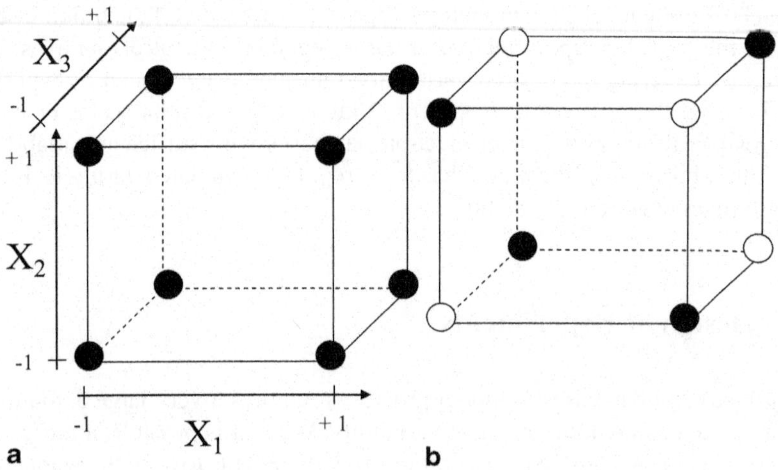

Fig. 2.2 **a** Full factorial design. **b** Fractional factorial design with three factors

Table 2.1 Example of an experimental design for a full factorial design with three factors

Experiment	Average I	X_1	X_2	X_3	X_{12}	X_{13}	X_{23}	X_{123}	Response Y
1	+1	-1	-1	-1	+1	+1	+1	-1	60
2	+1	+1	-1	-1	-1	-1	+1	+1	72
3	+1	-1	+1	-1	-1	+1	-1	+1	54
4	+1	+1	+1	-1	+ 1	-1	-1	-1	68
5	+1	-1	-1	+1	+1	-1	-1	+1	52
6	+1	+1	-1	+ 1	-1	+1	-1	-1	83
7	+1	-1	+1	+ 1	-1	-1	+1	-1	45
8	+1	+1	+1	+ 1	+1	+1	+1	+1	80
Effects	64.25	11.5	-2.5	0.75	0.75	5	0	0.25	

2.2.2.1 Full Factorial Designs (2^k)

Full factorial designs are the basic designs which carry out all possible experiments with *k* two-level factors, low and high levels. All experiments at the boundaries of the design space are planned, as illustrated for three factors in Fig. 2.2a. The corresponding experimental matrix with its encoding system is shown in Table 2.1.

The main effects for each factor are calculated as the semi-difference between the high-level average and the low-level average. They represent the average direct impact of each factor on the response when increasing the encoded factor level from 0 to 1.

First-degree interaction effects between two factors are then processed as the semi-difference between the effect of factor 1 at factor 2 high level and the effect of factor 1 at factor 2 low level. The second-degree interaction corresponds to the interaction between three factors. Interactions with a degree higher than 1 are however often small and difficult to interpret.

Table 2.2 Experimental design for the fractional factorial design with three factors coming from the full factorial design in Table 2.1

Experiment	Average I	X_1	X_2	X_3	X_{12}	X_{13}	X_{23}	X_{123}
2	+1	+1	−1	−1	−1	−1	+1	+1
3	+1	−1	+1	−1	−1	+1	−1	+1
5	+1	−1	−1	+1	+1	−1	−1	+1
8	+1	+1	+1	+1	+1	+1	+1	+1

The resulting model equation for the experimental matrix in Table 2.1 is illustrated in Eq. 2.1:

$$y = 64.25 + 11.5x_1 - 2.5x_2 + 0.75x_3 + 0.75x_{12} + 5x_{13} + 0.25x_{123} + \varepsilon. \quad (2.1)$$

The significance of each effect has then to be determined by means of statistical tests. Estimating the experimental uncertainty s_y from m replicate measurements, the uncertainty of each calculated effect value is $\sigma_E = s_y / \sqrt{n}$, where n is the number of designed experiments. An effect is thus significant at a risk α if the calculated effect value (model parameter, e.g. for factor 1) is greater than $t_{(1-\alpha,\nu)} \cdot \sigma_E$, where $t_{(1-\alpha,\nu)}$ is found in the Student's law table and $\nu = m-1$ is the degree of freedom. Another way to assess the significance of each effect is to run an analysis of variance (ANOVA) and observe the calculated p values associated with each effect. The resulting significant effects, meaning that these factors are statistically influent on the response, have to be maintained for the remainder of the model development.

2.2.2.2 Fractional Factorial Designs (2^{k-p})

Fractional factorial designs are used to screen factors when the number of experiments has to be lowered (Fig. 2.2b). The aliase principle allows selection of which experiments from the full factorial design must be run without losing significant information. The idea is to choose the experiments which lead to confound important effects, such as main and first-degree interaction effects, with smaller and less interpretable effects, such as second (and more)-degree interaction effects. For example, with three factors, Eq. 2.2 shows the aliase generator. The resulting experimental matrix is in Table 2.2. The number of experiments is reduced to 2^{k-p}, where p is the number of aliase generators. With three factors, only one aliase generator is allowed, dividing the number of experiments by two for similar model accuracy. The experimental matrix shows that main effects are confounded with first-order effects as the encoding is the same two by two. This only allows the interpretation of principal effects:

$$\mathbf{I} = \mathbf{X}_{123}. \quad (2.2)$$

Randomisation of the experiment order is usually needed to correct eventual systematic response errors. When experimental blocks are clearly identified, such as analysis days, aliase generators are used to confound the block effects with high-order interaction effects.

Some food applications were developed by Ellekjaer et al., who studied the effects of process variables and ingredients on sensory variables for processed cheese (Ellekjær et al. 1996), and Christiansen et al., who implemented a fractional factorial design to model food dressings (Christiansen et al. 2004).

2.2.2.3 Other Screening Designs

Other screening designs using linear models are also commonly used to identify the few significant factors among many.

The Plackett–Burman designs are two-level saturated designs where all interaction effects are neglected (Plackett and Burman 1946). The number of experiments is a multiple of four, and "saturated" means that this number is equal to the number of model parameters, i.e. the number of factors plus one (model constant), without any degree of freedom left. For example, a six-factor Plackett–Burman design requires theoretically a minimum of seven experiments, running finally eight experiments (multiple of four). The total number of experiments is hence drastically reduced.

The Rechtschaffner screening designs correspond to saturated two-level fractional factorial designs to estimate main and first-order interaction effects (Rechtschaffner 1967). For example, a six-factor Rechtschaffner design requires 22 experiments ($1+k+k(k-1)/2$, with k the number of factors).

2.2.3 Optimisation Designs: Response Surface Methodology

When two-level factorial designs have difficulties to model a process, showing a significant lack-of-fit when observing ANOVA results or when using validation experiments at the centre of the experimental domain, second-order designs, also called optimisation designs, are used. These designs propose to carry out experiments at more than two levels, allowing curvature modelling. Non-linear response surfaces can thus be drawn to achieve the main goal of these designs, i.e. estimating the experimental area corresponding to the response optimum (Araujo and Brereton 1996b).

2.2.3.1 Central Composite Designs

Central composite designs are widely used, since they can complete an existing full factorial screening design with $2k$ additional experiments, designing a "star" around the existing hyper-cube. Additional experiments to the centre can be required (Fig. 2.3 and Table 2.3). Five levels for each factor are thus investigated to model non-linearity.

Fig. 2.3 Central composite design with three factors

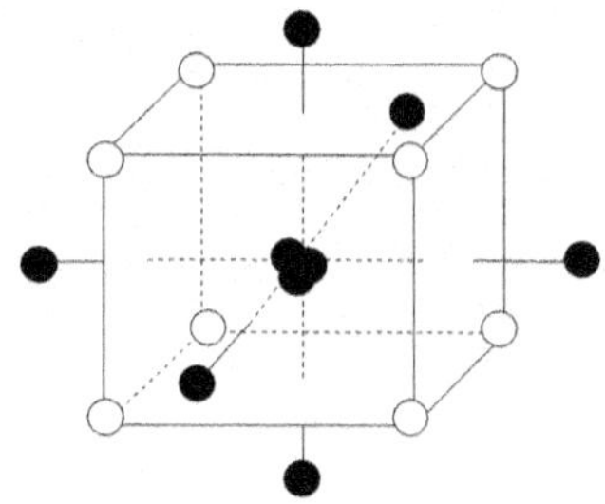

Table 2.3 Number of experiments for central composite designs

k	Full factorial design 2^k	Star points Number 2k	α	Centre	Total
2	4	4	1.414	3	11
3	8	6	1.682	3	17
4	16	8	2	3	27
5	32	10	2	4	46

The model equation for a second-degree design is shown in Eq. 2.3:

$$y = \beta_0 + \beta_1 x_1 + \beta_2 x_2 + \beta_3 x_3 + \beta_{12} x_{12} + \beta_{13} x_{13} + \beta_{23} x_{23} + \beta_{11} x_1^2 + \beta_{22} x_2^2 + \beta_{33} x_3^2 + \varepsilon. \quad (2.3)$$

2.2.3.2 Other Optimisation Designs

Central composite designs are the most popular optimisation designs. However, when the number of factors increases, the number of experiments rapidly becomes very large. Thus other second-degree designs are also commonplace.

The 3^k three-level full factorial designs are an extension of the two-level full factorial designs seen in Sect. 2.2.2.1.

The Rechtschaffner optimisation designs, similar to the screening Rechtschaffner designs in Sect. 2.2.2.3, correspond to saturated three-level fractional factorial designs to estimate main and first-order interaction effects. For example, a six-factor Rechtschaffner design requires 28 experiments ($1+k+k+k(k-1)/2$, with k the number of factors). Additional experiments in the centre are always recommended.

The Box–Behnken designs are incomplete three-level full factorial designs, without experiments in the corners of the experimental domain (Ferreira et al. 2007). Application on several responses related to bread-making quality is illustrated in Rouillé et al. (2000).

Doehlert designs allow the estimation of all main effects, first-order interactions and quadratic effects without any confounding effects (Ferreira et al. 2004). Their geometric shape is polyhedronic based on hyper-triangles (simplexes). The specificity of the Doehlert designs is related to the ability to extend them to contiguous experimental domains for one or more factors in a sequential way. The number of levels is finally not the same for each factor.

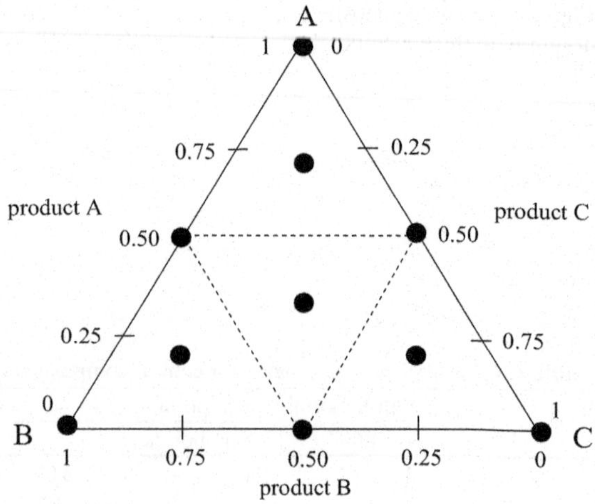

Fig. 2.4 Example of mixture design for three products (augmented simplex-centroid design for quadratic models)

2.2.3.3 Mixture Designs

When dealing with formulation optimisation, the closure constraint in the mixture (Eq. 2.4) has to be taken into account (Cornell 1990; Eriksson et al. 1998):

$$\sum_{i=1}^{c} x_i = 1, \tag{2.4}$$

where x_i is a compound (factor) of the mixture and c the total number of the compounds.

This mixture constraint leads to an interdependency between all factors and hence has several consequences. First, the representation of the experiments does not imply hyper-cubes but hyper-tetrahedrons (Fig. 2.4). Second, the underlying models are simplified. The linear model loses its constant term and the second-order model loses its constant and quadratic terms.

2.3 Exploratory Analysis

The first step of chemometrics analysis consists in performing an exploratory analysis in the multivariate space, also called descriptive analysis or unsupervised analysis which occurs without prior knowledge concerning neither the nature nor the group membership of the samples. Initially, data have to be preprocessed or "cleaned" before the exploratory treatment. This is often performed using principal component analysis (PCA).

2.3.1 *Data Preprocessing*

Data preprocessing techniques are often used prior to modelling in order to reduce noise and undesired perturbations in the signal. Several preprocessing methods have been initially developed in near-infrared spectroscopy, due to its sensitivity to the external environment (temperature changes, humidity, etc.). The most suitable preprocessing technique will depend on the conditions; these must be compared to find the optimal combination on a given data set.

2.3.1.1 Classical Preprocessing Methods

The most widely used preprocessing methods consist in mean centring or scaling the data. They can be used for all types of multivariate data: continuous, discrete, spectroscopic or process data.

- *Mean centring* is the most common preprocessing. The principle is to subtract the variable mean to each value. Mean centring is quasi-systematic in projection methods such as PCA or PLS. It is used in order to centre the subspace to the barycentre of the original data set, for a better data visualisation (see Sect. 2.3.2). When building a predictive model, mean centring **X** data set implies that the constant term (b_0) of the equation is not equal to zero (see Sect. 2.4.2.1, Eq. 2.13). Thus, in the cases where the intercept is expected to be null, the data should not be centred.
- *Scaling* is used to make the different variables comparable when included in a global multivariate analysis. The most common scaling technique is the *unit-variance scaling* which divides each variable by its standard deviation, like a columnwise normalisation. The method must be systematically applied to data sets containing variables of different scales (e.g. pH, temperature) in order to give them equal weights in further processing. Scaling should not be applied to spectroscopic data because each variable is comparable and the intensity variations between wavelengths constitute the important information (e.g. spectral peaks). Other kinds of scaling are possible for this data, for instance, to stress the importance of specific variables by giving them higher weights.
- *Auto-scaling* is the combination of mean centring and unit-variance scaling.

2.3.1.2 Signal Correction Methods

The signal correction methods aim at correcting the influence of different perturbations and/or enhancing information. In spectroscopic data, perturbations can be additive, i.e. a constant, which can be wavelength dependent, is added to the spectrum, or multiplicative, where each element of the spectrum is multiplied by a constant. These phenomena are typical of light scattering effects, which induce a photon loss (additive effect) and an increased path length (multiplicative effect), among others. Scatter correction must not be applied if the parameter of interest is physical in nature (e.g. particle size, turbidity).

Almost all the methods cited below are "rowwise" methods, i.e. the preprocessing is carried out sample by sample. It is not the case for mean centring and scaling, where all (calibration) samples are required in order to preprocess the data set, i.e. they are "columnwise" treatments.

A recent review of some of the mentioned signal correction preprocessing methods can be found in Rinnan et al. (2009).

- *Baseline correction* subtracts the undesired spectral background. The classic way is to subtract the lowest value of each spectrum from all the variables. *Detrending* removes curvilinear baseline by approximating it with a wavelength-dependent second-degree polynomial fit.
- *Normalisation* is used rowwise when there is a non-desired intensity variation between objects due to multiplicative effects. This allows focus on the data profile rather than the global intensity. Normalisation is done by dividing each spectrum by an estimation of its spectral intensity. This can be done using the following properties: its area (area normalisation), its maximal peak (maximum normalisation), a specific spectral point (peak normalisation), its length (unit vector normalisation), or the sum of the spectral values.
- *Standard Normal Variate (SNV)* is a path-length variation correction method used, like normalisation, to limit the spectral intensity variation problem (Fig. 2.5b). It is a rowwise auto-scaling, thus removing the spectrum mean value to all the spectrum variables and dividing them by the spectrum standard deviation (Barnes et al. 1989).
- *Multiplicative Signal Correction* (MSC) is also a very common method for correcting multiplicative scattering effects (Geladi et al. 1985). The principle is to fit each spectrum to a reference spectrum (generally, the average calibration database spectrum, Eq. 2.5), and then to correct them as shown in Eq. 2.6. The reference spectrum must be representative to avoid an ill-fitting model. Different versions have been derived. For instance, the extended MSC (EMSC) is based on a polynomial baseline correction depending on the wavelength; it can also allow for the introduction of prior information in the spectra (Martens and Stark 2001):

$$\mathbf{x}_i = a * \overline{\mathbf{x}} + b \tag{2.5}$$

$$\mathbf{x}_i^{\mathrm{MSC}} = \frac{\mathbf{x}_i - b}{a}, \tag{2.6}$$

where $\mathbf{x}_i$ is the measured spectrum, $\overline{\mathbf{x}}$ is the mean spectrum (or a reference spectrum), a is the intercept, b the slope and $\mathbf{x}_i^{\mathrm{MSC}}$ the corrected spectrum.

- *Smoothing* is used to remove random noise. The principle is to use an average of neighbouring points. For example, the moving average method uses the average of a neighbouring window to calculate the new value. The Savitzky–Golay (SG) algorithm uses a polynomial fit (Savitzky and Golay 1964). The latter is the

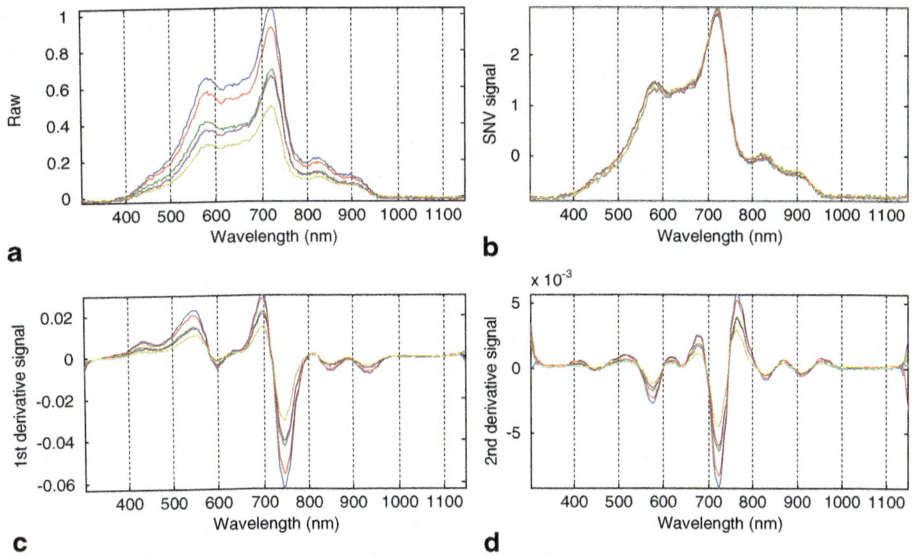

Fig. 2.5 Raw spectra (**a**), SNV (**b**), first derivative (**c**) and second derivative (**d**) spectra for a visible-NIR spectra data set

most widespread algorithm in chemometrics. The wider the window the higher the smoothing while a polynomial degree increase will better fit tiny spectral features and enhancing noise. Thus, a balance between these parameters must be found.

- *Derivatives* are generally used in spectroscopy to enhance spectral features; however, they also correct for additive effect, as a constant baseline (first derivatives, see Fig. 2.5c) or both the offset and the slope of the baseline (second derivatives, see Fig. 2.5d). The most classical algorithms used are the SG (Savitzky and Golay 1964) and the Norris–Williams (Norris and Williams 1984) algorithms. Since computing derivatives enhances the noise, these two methods also smooth the spectrum.

2.3.1.3 Dimensionality Reduction Methods

Dimensionality reduction methods aim at eliminating uninformative signal, thus enhancing information and reducing collinearity:

- *Variable selection* is often used to remove uninformative or noisy variables and keep the relevant ones. For data with only a few variables (e.g. process data), the classical method is the stepwise-multiple linear regression (stepwise-MLR) method, used to select the most informative variables during the model development. For data with a higher number of variables and containing more collinearities (e.g. spectroscopic data), other methods are preferred. A common method in spectroscopy is the interval-partial least squares (i-PLS), which selects the spectral regions in testing all possible combinations with one or several moving window(s) of a fixed size (Nørgaard et al. 2000). A plethora of different methods have been developed. For uninformative variable elimination (UVE), the addition of dummy random variables allows the identification of spectral variables that are as uninformative as noise (Centner et al. 1996). The genetic algorithm (GA) is a stochastic method using the principle of evolution theory to select a subset of variables (Leardi and González 1998). A large number of parameters must be tuned to apply this method. Furthermore, being a stochastic technique, each run provides a different result.

Some comparison studies can be found in Abrahamsson et al. (2003).

- For predictive modelling purposes, orthogonal pretreatments, such as *orthogonal signal correction* (OSC), remove variations which are not linked to **Y** (Wold et al. 1998a). Some variants deriving from the classical OSC and differing in the way the non-relevant part is modelled, can be cited: direct orthogonalisation (Andersson 1999), direct OSC (Westerhuis et al. 2001), piecewise OSC (Feudale et al. 2002a) and orthogonal-PLS (Trygg and Wold 2002). These methods have for objective to extract the net analyte signal (NAS), i.e. the part of the spectra related to the quantity of interest and which is orthogonal to the other compounds. It has been shown that these pretreatments do not always provide better model performances than partial least-squares (PLS) models (see Sect. 2.4.2.4) based on raw data. However, the models based on these preprocessed data provide a better understanding of the model. In this way, Svesson et al. propose some discussions and a comparison on different "OSC" algorithms (Svensson et al. 2002).
- *Data compression* methods aim at reducing the dimensionality of large data sets. Latent structures are extracted and used to rebuild a "cleaned" signal without noise. PCA is the most common method (see Sect. 2.4.2). Other methods are also widely used, like Fourier transformation (FT) working on the frequency domain (McClure et al. 1977; Wu et al. 1996) or wavelet transform (WT) working on time and frequency domains (Daubechies 1990; Alsberg et al. 1997). Both methods present the advantage of working on one spectrum at a time, whereas PCA needs all spectra. These methods are useful to reconstruct the corrected signal or to extract the latent structures (e.g. PCA scores, FT or WT coefficients) and use them as inputs to derive predictive or discrimination models.

A comparison of several preprocessing methods for improving the determination of moisture and protein contents of forage samples is given by Azzouz et al. (Azzouz et al. 2003).

2.3.2 Principal Component Analysis

2.3.2.1 Introduction—Objective of PCA

The PCA is the major workhorse of the chemometrics tools. The PCA method can be used for the following goals:

- Visualisation of X in the multivariate space
- Outlier detection
- Variable selection
- Data compression, when reducing X dimensionality by removing noise
- Be the basics for other multivariate methods, such as unsupervised classification or MSPC

2.3.2.2 Geometrical Interpretation

The PCA can be seen as a better way to visualise samples represented by numerous variables, by projecting their original coordinates into a new set of axes, called principal components (PC). These axes satisfy a number of properties, which make the sample visualisation easier.

The following graphs explain how PCA works for a simple **X**-matrix, composed of only three variables (Fig. 2.6a):

- Each sample is located in the original space with its three coordinates (Fig. 2.6b). The X-matrix can then be visualised as a cloud of points in the three-dimensional (3D) space (Fig. 2.6c).
- The coordinate system is translated to the barycentre of the sample cloud (star in Fig. 2.6d), for a better visualisation, by mean centring the variables.
- A new axis, called first PC, is built following the direction of the maximum spread of the samples (Fig. 2.6d); this helps better visualise the maximum variability of the sample set. The new coordinates of the samples are called Scores.
- A second axis, orthogonal to the first one, is then searched, to represent the maximum of the remaining sample set variance (Fig. 2.6g). This axis can be visualised by placing the eye facing to the first PC (Fig. 2.6e–f).
- This process is done iteratively for the number of PCs equal to the number of original variables (three in this graphical example) (Fig. 2.6h).

Thus, PCA can be seen as a change of axes, designed to better visualise the sample variability, but maintaining the distances and scales between samples. For more convenience, the samples are usually visualised on a 2D plane, corresponding to the projection of the samples on this set of two axes (see Fig. 2.7, an example of plane PC1 × PC3).

2.3.2.3 Mathematical Computation

The spread, inertia or variance of the cloud of samples seen in the previous section is expressed mathematically by the variance–covariance matrix of **X**. Thus, the PCA decomposes the sample set space in the direction of the maximum of **X**-variance. The matrix of variance–covariance ($\mathbf{V}_{(p,\,p)}$) is computed as shown in Eq. 2.7, with $\tilde{\mathbf{X}}$ being the centred matrix of **X**:

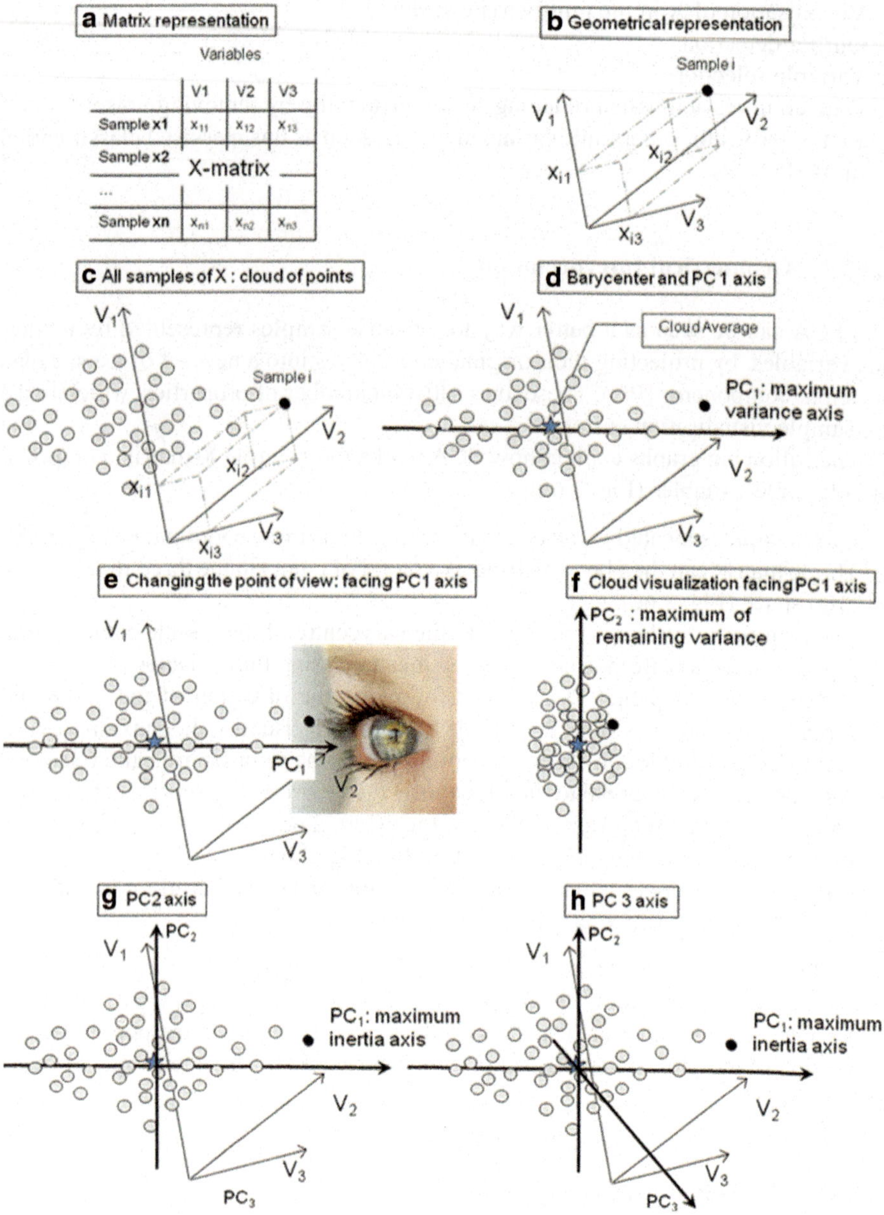

Fig. 2.6 Principal component analysis (PCA) geometrical visualisation

$$\mathbf{V} = \frac{\tilde{\mathbf{X}}^T \tilde{\mathbf{X}}}{n-1}. \tag{2.7}$$

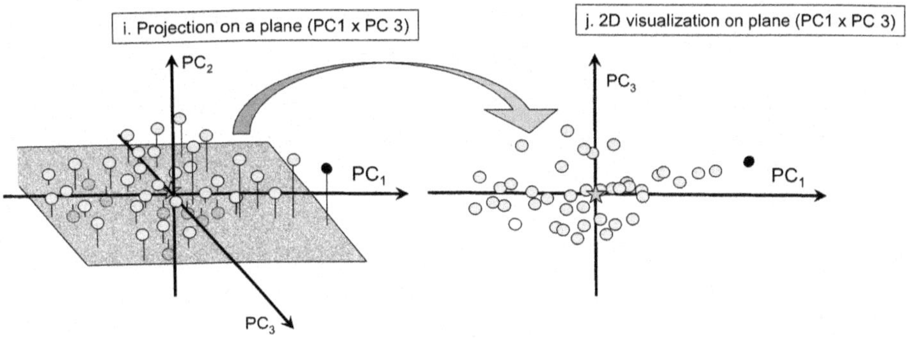

Fig. 2.7 PCA visualisation on a 2D plane

To find the maximum inertia axes, the PCA algorithm diagonalises the matrix of variance–covariance, computing the eigenvalues and eigenvectors of **V**. Then, the eigenvalues are sorted in a descending order, since they are directly related to the variance explained by each axis.

The eigenvectors are called loadings (**P**) and correspond to the weights of each original variable to build the PC: the PC is thus a linear combination of the original variables.

Then, the initial data set **X** can be decomposed with the following expression:

$$\mathbf{X}_{(n,p)} = \mathbf{T}_{(n,p)}\mathbf{P}^{T}_{(p,p)}, \tag{2.8}$$

where **P** is the loading matrix and **T** the score matrix of the new coordinates in the PC subspace. Matrix dimensions are noted in parenthesis.

Since data contains a part of the information and a part of the noise (error), we can express **X** with the first *k* components, the *p*–*k* last components remaining in the error matrix (see Eq. 2.9 and Fig. 2.8a):

$$\mathbf{X}_{(n,p)} = \mathbf{T}_{(n,k)}\mathbf{P}^{T}_{(k,p)} + \mathbf{E}_{(n,p)}, \tag{2.9}$$

where **P** is the loading matrix for *k* components and **T** the corresponding scores. Matrix dimensions are noted in parenthesis.

Thus, when a centred PCA is performed, each sample can be decomposed as shown in Fig. 2.8b. The variables are generally centred (see Sect. 2.3.1.1), to translate the coordinate system origin to the barycentre of the sample cloud, and provide a better visualisation.

2.3.2.4 Interpretation of PCA

When a PCA is performed, scores and loadings are visualised to understand:

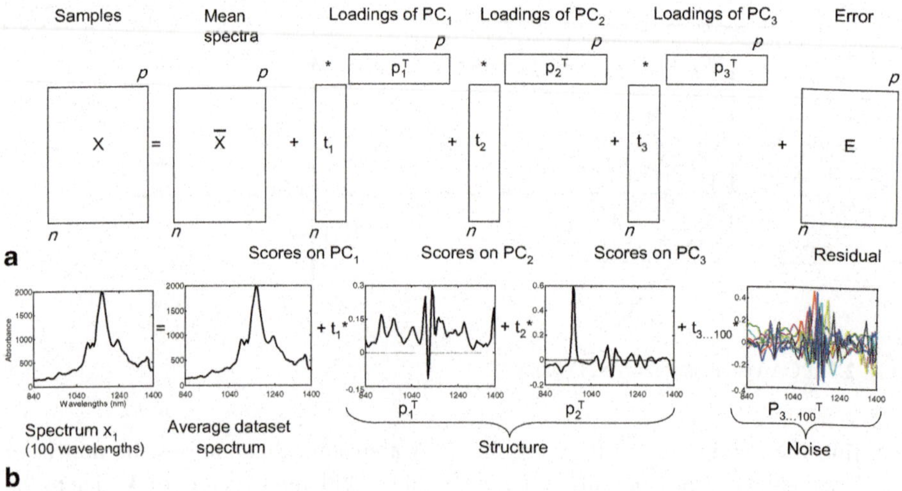

Fig. 2.8 Principal component decomposition: **a** matrix-like decomposition, **b** spectral-like decomposition

- The relationship between samples (Score plot—Fig. 2.9a). If two samples have close scores (neighbouring coordinates in the PCs), it means that they have the same behaviour, and thus have similar levels in the original variables contributing mainly to the construction of these PCs.
- The correlation between variables and their contribution to the PCs (Loading plot—Fig. 2.9b). Variables are considered as vectors, building with different weights the PC directions. Therefore, if two variables show close directions (vectors from frame origin), it means that they are strongly correlated. Additionally, if they have higher values on the axis (high loading weights), it means that they contribute a lot to the calculation of the PC axis. This tool can thus be used for variable selection.
- The relationship between samples and variables (biplot—Fig. 2.9c). Biplot helps to interpret the relationship between samples and variables. Samples with high coordinates (far from origin) in a direction of a variable means that this sample has a high value for this variable. Biplot can be seen on a "correlation circle" plot, normalising each PC to one, thus helping the interpretation of the variable significance.
- The variance expressed by each axis (explained variance plot—Fig. 2.9d). The cumulated explained variance (or residual variance) shows the variability explained by each PC. Thus, the interpretation of trends and neighbouring samples or variables must be weighted by this explained variance. When the explained variance comes close to 0, the remaining PCs only express noise, and not structured information. If PCA is used to compress the information (see Sect. 2.3.1.3), determining this optimum number of PCs is very important. It is also crucial for classification purposes (see clustering, Sect. 2.5.1; SIMCA, Sect. 2.5.2.2) or process control (see MSPC in Sect. 2.6.1).

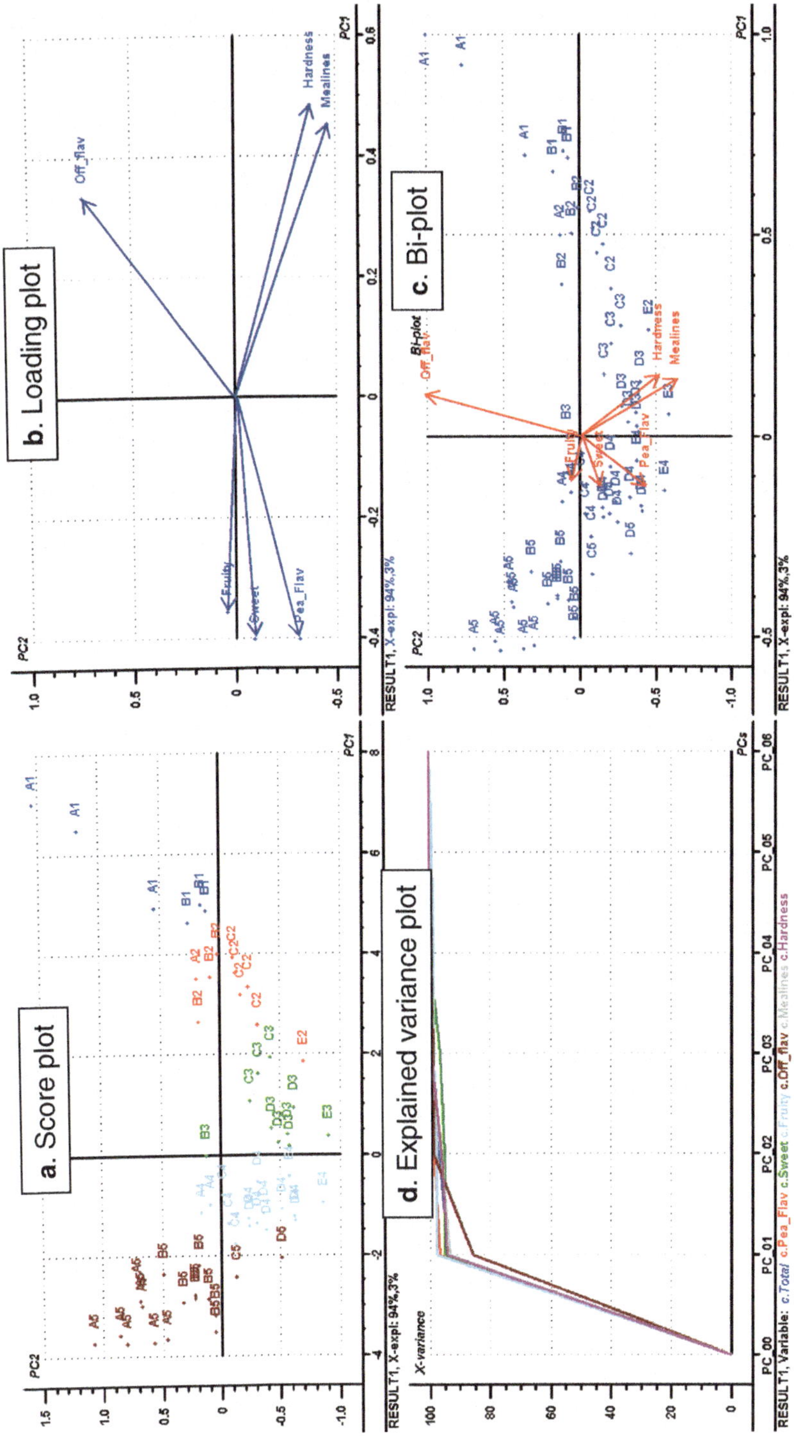

Fig. 2.9 Main plots for PCA interpretation

N.B.: for discrete variables, loading plots are generally shown in 2D scatter plots (see Fig. 2.9b); however, for continuous variables, such as spectra, the loading plots are generally drawn one PC at a time (see Fig. 2.8b) for a clearer view. Furthermore, in spectroscopy, the first PC is generally very high, expressing systematic noise usually due to the light scattering effect.

2.3.3 *Outlier Detection and Handling*

Classical multivariate models for exploratory and predictive analyses are very sensitive to outliers, i.e. extreme, atypical samples that are very different from the other samples of the population. The main reason is that the mean is a non-robust estimator of the data location. Hence, it is of great importance to identify outliers and decide what to do with these samples, i.e. keeping them if they are representative of the variability which has to be modelled or removing them. There are indeed many causes for outliers leading to different decisions: wrong labelling, measurement error, deviating process, etc. Only multivariate outliers and multivariate tools will be investigated here.

2.3.3.1 Outlier Detection in Exploratory Analysis

Outliers can be detected during an exploratory analysis considering the **X** database of the samples.

X-leverage or T^2

The most common diagnostic tool used to identify outliers is the X-leverage (h) or Hotelling's T^2 distance calculated for each sample i:

$$h_i = \mathbf{t}_i \left(\mathbf{T}_0^{\mathrm{T}}\mathbf{T}_0\right)^{-1} \mathbf{t}_i^{\mathrm{T}}, \tag{2.10}$$

where $\mathbf{T}_0$ is the training (calibration) score matrix for a certain number of components and $\mathbf{t}_i$ the score vector for sample i.

The leverage, actually similar to the Mahalanobis distance (De Maesschalck et al. 2000), corresponds to the distance to the centre of the model. Outliers with a high leverage have a large influence on the model and have hence to be well interpreted (Fig. 2.10). This kind of outlier is also called "strong" outlier. A critical limit can be calculated according to a Fisher test (F-test) for a given significance level α (e.g. 5 %) to detect outliers.

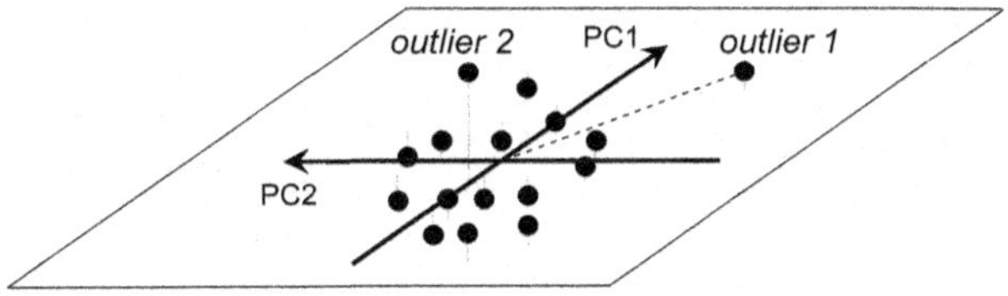

Fig. 2.10 Two kinds of outliers in exploratory analysis: outlier with a high leverage (*outlier 1*) and outlier with a large residual variance (*outlier 2*). *Dashed line:* distance to the centre of the model; *dotted line:* distance to the model

X-residuals or Q

The other most associated diagnostic tool to identify outliers is the X-residuals or Q distance calculated for each sample. This is the non-modelled part of the variance. It corresponds to the distance to the model, i.e. the distance to the model hyperplane. Outliers with a high Q are not well modelled (Fig. 2.10). They often come from an atypical phenomenon in the process. This kind of outlier is also called "weak" or "moderate" outlier. In a similar way to the leverage, a critical limit can be computed using an F-test to detect outliers toward this distance.

Plotting the Q value against the leverage gives the so-called influence plot, which is a very useful diagnostic tool to identify simultaneously all the kinds of outliers.

2.3.3.2 Outlier Detection in Predictive Analysis

In the same way, outliers can be detected during a predictive analysis, with the same tools and using the PLS scores, as in Sect. 2.3.3.1, considering either **X** or **Y** databases for the training then test samples.

The **X**–**Y** relationship can be additionally used in a predictive analysis to diagnose outliers. Plotting for all samples and first components the **t** (**X**-scores) against the **u** (**Y**-scores) is another way to identify outliers, which deviate from the correlation straight line.

2.3.3.3 Robust Statistics

When using the above techniques to detect outliers, some problems can occur if too many outliers are present. Since the mean is a non-robust estimator, some outliers may not be detected (masking effect) or regular samples may be wrongly detected as outliers (swamping effect). Another way to handle outliers is thus to perform robust statistics, i.e. variants of the classical tools using more robust location and covariance estimators (Liang and Kvalheim 1996; Fernández Pierna et al. 2002; Daszykowski et al. 2007). The data majority is hence modelled, neglecting the outliers. Many attempts led to more robust estimators, such as the multivariate trimming (MVT; Gnanadesikan and Kettenring 1972), the minimum volume estimator

(MVE; Rousseeuw 1984) or the minimum covariance determinant (MCD; Rousseeuw 1984).

Robust variants of PCA and PLS exist, coming from either the use of these robust estimators or the projection pursuit (PP) algorithm.

2.4 Quantitative Predictive Modelling

2.4.1 Introduction

The objective of predictive modelling methods is to correlate a set of explanatory variables **X** to one (**y**) or several properties of interest (**Y**).

All the models obey the following equation:

$$\hat{\mathbf{y}} = f(\mathbf{X}), \tag{2.11}$$

where $\hat{\mathbf{y}}$ is the predicted value of the real **y** and f the function relating **X** and **y**.

f is found by minimising the error between the real **y** values and the predicted $\hat{\mathbf{y}}$ values, with diverse algorithms of error minimisation.

The models are developed and optimised on a selected sample set for which **X** and **y** are well known, called the calibration or training set. The ultimate goal is to apply these models to predict **y** value(s) of unknown samples.

We can distinguish linear regression methods (see Sect. 2.4.2), which work well of course when the **X–y** link is linear, but even most of the time when the **X–y** link is non-linear. However, if the **X–y** relationship is more complex, non-linear or local correlation algorithms are available (see Sect. 2.4.3).

2.4.2 Linear Modelling

2.4.2.1 Linear Regression Principle

The basic regression model is the simple (univariate) linear model which relates one quantity of interest **y** (response) to one explicative variable $\mathbf{x}_1$ through the equation:

$$\hat{\mathbf{y}}_{(n,1)} = b_0 + b_1 * \mathbf{x}_{1(n,1)}, \tag{2.12}$$

where **b** is called the regression vector, composed here of two constants: b_0 the offset (i.e. the value of y when x is equal to zero), and b_1 the slope.

When **X** is multivariate, i.e. when each sample is described by p several variables, then a multivariate linear model relates **X** to **y** according to the following equation:

$$\hat{\mathbf{y}}_{(n,1)} = b_0 + b_1 * \mathbf{x}_{1(n,1)} + b_2 * \mathbf{x}_{2(n,1)} + \ldots + b_p * \mathbf{x}_{p(n,1)} = b_0 + \mathbf{X}_{(n,p)} \mathbf{b}_{(p,1)}, \tag{2.13}$$

where $b_1 \ldots b_p$ are the regression coefficients for each variable and refer to the regression vector **b** in the right part of the equation and b_0 the offset. Matrix dimensions are noted in parenthesis.

The modelling goal is to find the optimal approach for calculating the regression coefficients **b** based on a set of experimental data. The following sections show the various ways for finding **b** from an **X** matrix and **y** reference values based on Eq. 2.13.

2.4.2.2 Multiple Linear Regression (MLR)

Multiple linear regression (MLR) is the simplest multivariate modelling method, as an extent of the simple linear regression to the multivariate case. To calculate **b,** the pseudo-inverse calculation is applied on the calibration set:

$$\mathbf{b} = (\mathbf{X}^T \mathbf{X})^{-1} \mathbf{X}^T \mathbf{y}. \tag{2.14}$$

This method has the advantage of being straightforward. However, if the explanatory variables are correlated (collinearity), the pseudo-inverse matrix calculation leads to unstable models. Another important limitation is that MLR cannot be calibrated if the number of samples is smaller than the number of variables (i.e. if $n<p$), because $\mathbf{X}^T\mathbf{X}$ is no more invertible. These two limitations are very often encountered in spectroscopy or imaging where the variables are numerous and strongly collinear. Thus, other modelling methods must be adopted for this kind of data.

2.4.2.3 Principal Component Regression (PCR)

In order to overcome the issue of variable correlation, MLR can be calculated on k scores of a PCA (see Sect. 2.3.2) instead of on original variables **X**. Equation 2.14 is then replaced by Eq. 2.15. Thus, this method is called principal component regression (PCR):

$$\mathbf{b} = (\mathbf{T}^T \mathbf{T})^{-1} \mathbf{T}^T \mathbf{y}. \tag{2.15}$$

Since scores are orthogonal, there is no more correlation between variables. Furthermore, only the first few k informative scores are kept ($k<<p$), as explained in Sect. 2.3.2.3 (see Eq. 2.8) overcoming the high dimensionality problem ($n>k$).

However, the main drawback of this method is that the scores are calculated in order to represent the maximum variance in **X**, which is not necessarily linked with **y**.

2.4.2.4 PLS Regression

PLS regression has then been proposed to improve the **y** modelling (Wold et al. 1983). Rather than calculating scores using only **X**-covariance with PCA as in PCR, the PLS algorithm takes into account the covariance between **X** scores and **y**, i.e. the variances of **X** and **y** and the correlation between **X** and **y**. Scores, referred here as latent variables (LV), are then built in order to capture as much information related to **y** as possible.

PLS can also predict several responses at a time; in this case, a latent subspace is also created for **Y**, and the covariance maximisation is made between **X** scores and **Y** scores. This modelling technique is generally called the PLS2 method.

PLS can be summarised in two interdependent steps:

- Dimensionality reduction of **X** and **Y** in latent subspaces using scores projections
- Integration of a vector, called weights (**w**), connecting both subspaces and providing the covariance between **X** and **Y** scores as optimisation criterion

Two major algorithms can be used for PLS modelling: an iterative procedure called non-linear iterative partial least squares (NIPALS) algorithm (Wold et al. 1983) or a direct procedure, known as the SIMPLS algorithm, which is faster (De Jong 1993).

2.4.2.5 Model Optimisation and Validation

The number of k latent variables (also called LVs) must be carefully chosen, to reduce the modelling subspace. If too many latent variables are used, noisy or non-relevant part of X is considered in the calibration stage, leading to an overfitted and unstable model, unable to predict accurately unknown data. On the contrary, underfitting a model leads to poor performances in both calibration and prediction stages, as the model misses a relevant part of the signal.

There are different strategies for the calibration and validation of models, in order to avoid under- and overfitting.

The predictive power of a model can be estimated by a set of criteria called figures of merit (FOM) (see Annex: Figures of Merit). As there is no universal criterion, several statistics have to be considered before any final decision. The classical ones are the root-mean-squared error (RMSE; see Eq. 2.33), the bias (systematic error, see Eq. 2.34) and the coefficient of determination R^2 (Eq. 2.35).The criteria are then calculated on calibration and validation sets to compare performances. A classical diagnostic tool is the graph showing the evolution of the RMSEs against the number of latent variables (Fig. 2.11a). The optimal number of latent variables is then given by the minimum of error for validation or cross-validation (RMSEV or RMSECV), checking that the error on the calibration set (RMSEC) remains close to avoid overfitting.

The most used validation technique for optimisation is the cross-validation method which involves an *internal validation set*. The principle is to divide the

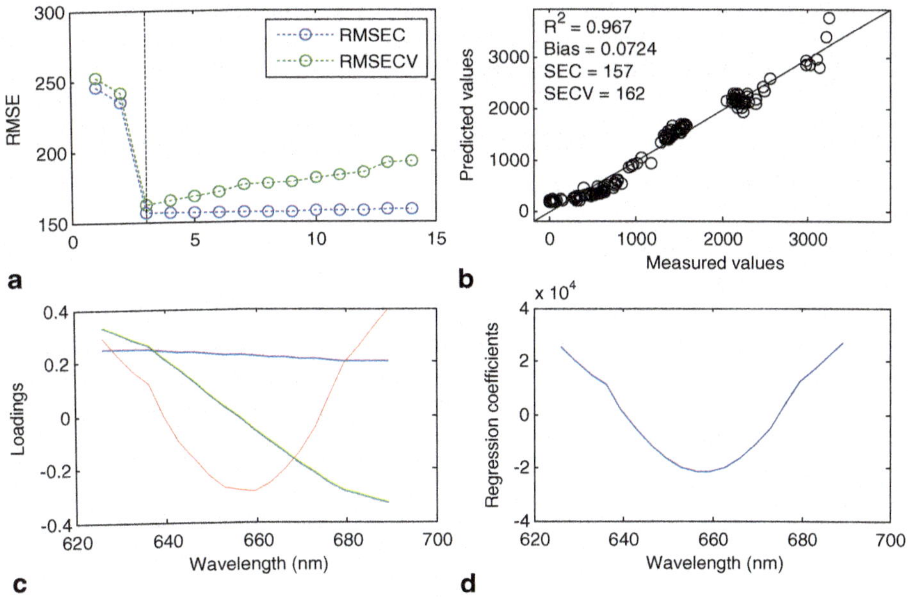

Fig. 2.11 PLS model overview

calibration set into c blocs (contiguous or at random). c models are then built on the calibration set without the cth bloc. For each model, predictions on the removed bloc are recorded. The choice of c depends on the size of the data set. For very small data sets, one can use till n blocs, i.e. remove one sample at a time. This method is known as leave-one-out cross-validation (or Jackknifing). Cross-validation is commonly used since it does not require extra samples to evaluate the model. If only a small number of measurements are available, cross-validation will be the method of choice. However, the higher the number of blocs, the more optimistic (and thus less realistic) are the performance results for the future prediction of new samples.

As a result, it is better to use when possible an *external validation set*. This can be done by splitting the original calibration data set in two, providing an independent validation set. Several methods have been developed to perform the splitting (Kennard and Stone 1969; Snee 1977; Clark 1997; Dantas-Filho et al. 2004; Galvao et al. 2005). However, the use of a brand new set of samples, completely independent from the calibration set (e.g. samples measured later, samples of a new batch…) is preferable to be more representative of the prediction error on future unknown samples.

Thus, to resume, when optimising a prediction model, the following features are monitored (see Fig. 2.11):

1. The evolution of the error in calibration and (cross-) validation against the number of LVs.
2. The prediction plot showing the real **y** values versus predicted **y** values, with the corresponding FOMs for the calibration and validation sets.

3. The regression vector must not be noisy, otherwise *k* must be reduced.
4. The loading vectors must not be noisy. **X** and **y** scores, residuals, leverages and weights can also be checked (not shown).

2.4.2.6 Science-Based Calibration

A spectrum **x** of a sample can be decomposed by a combination of pure spectra of each component present in the sample (**K**), multiplied by their concentrations (**c**), plus a noise spectrum (**e**), as shown in Eq. 2.16.

If all *m* pure spectra are known, the corresponding concentrations can be estimated using the regression coefficient $\mathbf{B}_{(p,\ m)}$ calculated in Eq. 2.17. This method is called classical calibration (or direct calibration). As this assumption is not often encountered due to the complexity of studied samples, inverse calibration is much more widespread in chemometrics (see Sect. 2.4.2.2, Eq. 2.13):

$$\mathbf{x}_{(1,p)} = c_1\mathbf{k}_{1(1,p)} + c_2\mathbf{k}_{2(1,p)} + \ldots + c_m\mathbf{k}_{m(1,p)} + \mathbf{e}_{(1,p)} = \mathbf{c}_{(1,m)}\mathbf{K}_{(m,p)} + \mathbf{e}_{(1,p)} \quad (2.16)$$

$$\mathbf{B} = \mathbf{x}\mathbf{K}^T(\mathbf{K}\mathbf{K}^T)^{-1}. \quad (2.17)$$

However, new methods are emerging. Among them, we can cite the science-based calibration (SBC) which can be described as a combination of both calibration methods, i.e. classical and inverse calibrations (Marbach 2005).

As the pure spectrum of the component of interest ($\mathbf{k}_1$) can often been measured or estimated, the principle of SBC is to replace the unknown components (i.e. replacing $c_2\mathbf{k}_2$ to $c_m\mathbf{k}_m$ and **e**, in Eq. 2.16, by only one signal (**e***). This non-relevant signal can be described by a set of spectra **X**, measured on samples containing none or little variability of the component of interest. The mean spectrum $\overline{x}$, the mean concentration of interest $\overline{y}$ (if any) and the inverse of the covariance matrix Σ^- are extracted from **X**. The regression coefficient **b** is estimated as in Eq. 2.19 and a new sample is then predicted as shown in Eq. 2.20:

$$\mathbf{x}_{(1,p)} = c_1\mathbf{k}_{1(1,p)} + \mathbf{e}^*_{(1,p)} \quad (2.18)$$

$$\mathbf{b}_{(1,p)} = \frac{\mathbf{\Sigma}^-\mathbf{k}_1}{\mathbf{k}_1{}^T\mathbf{\Sigma}^-\mathbf{k}_1} \quad (2.19)$$

$$y_i = \overline{y} + (\mathbf{x}_i - \overline{\mathbf{x}})\mathbf{b}^T. \quad (2.20)$$

Assuming the knowledge of $\mathbf{k}_1$, SBC is a method allowing using spectroscopy as a primary method, i.e. with no need of measuring the reference values *y* and establishing a calibration model. This method is quite appropriate for PAT applications and is already used in this field (Marbach 2007a, b).

2.4.3 *Non-Linear Modelling*

2.4.3.1 Non-Linear PLS

Data often suffer from non-linear behaviour with respect to the **y** reference. PLS generally deals well with weak or moderate non-linearity.

A classic methodology to circumvent the problem is to preprocess the X data (see Sect. 2.3.1 Data preprocessing section), making the relationship linear. PLS modelling can also be applied to non-linear X data by analysing the square or cube of the x values. The original **y** data in calibration can also be transformed with a non-linear function such as logarithm or exponential functions before modelling.

Non-linear PLS models have also been developed, such as quadratic PLS (QPLS; Wold et al. 1989), which are PLS models with polynomial inner relations, or spline-PLS (Wold 1992) to cope with non-linearity in the data (X or y).

2.4.3.2 Local Modelling

The principle of local modelling is to select from the calibration database a subset of similar samples (in terms of X and/or y values) to the unknown sample and use only them for prediction. To predict each new sample, a new model is built. Several such methods have been implemented: locally weighted regression (Cleveland and Devlin 1988), CARNAC-D (Davies et al. 1988), or LOCAL (Shenk et al. 1998).

2.4.3.3 Least-Squares Support Vector Machines

Support vector machines (SVM) were originally established for classification purposes (see Sect. 2.5.2.3). More recently, SVM have been adapted for regression purposes, in particular with least-squares support vector machines (LS-SVM) regression models (Cogdill and Dardenne 2003).

SVM is a local modelling method, focusing on similarity between samples. Instead of basing the model on variables, like the most common multivariate modelling methods (PCA, PLS, ANN), the SVM replace the matrix $\mathbf{X}_{(n,\ p)}$ with a "kernel matrix" $\mathbf{K}_{(n,\ n)}$ made up of similarity measurements between the n calibration samples. The computation of the kernel **K** is often based on a non-linear Gaussian distance based on a radial basis function (RBF) as shown in Eq. 2.21. The parameter σ tunes the kernel width and so the degree of non-linearity (a larger σ decreases the non-linearity degree):

$$k_{i,j} = \exp\left(\frac{-\left\|\mathbf{x}_i^{T} - \mathbf{x}_j^{T}\right\|^2}{\sigma^2}\right). \tag{2.21}$$

In LS-SVM, **K** is then used as explicative input for the prediction of the **y** response (Eq. 2.22). The objective is then to find the regression coefficient vector **b** of dimensionality *n*. For this, the minimisation criterion is quite different from classical least-squares regression (as with MLR) by the addition of an extra term (last part of the Eq. 2.23). The parameter γ regulates the importance of the minimisation of this second term:

$$\hat{\mathbf{y}} = \mathbf{K}\mathbf{b}^T \tag{2.22}$$

$$min(e) = min\left(\frac{\sum_{i=1}^{n} (y_i - \hat{y}_i)^2}{2} + \frac{1}{\gamma}\frac{\sum (\mathbf{b}^T\mathbf{b})}{2} \right). \tag{2.23}$$

The predicted value of a new sample $\mathbf{x}_{new}$ is assessed according to its similarity with the calibration samples.

An application for the prediction of acidity in grapes is discussed in Chauchard et al. (2004a).

2.4.3.4 Artificial Neural Networks

There are many types of Artificial Neural Networks (ANN), but they are all non-parametric models, i.e. they do not require to assume any (Gaussian) distribution assumption of the data, in opposition to most of the classical models (such as MLR, PCR or PLS). ANN are stochastic non-linear modelling tools, i.e. each ANN modelling process will lead to a different result. They can provide either non-linear quantitative prediction or classification models.

They can be separated into two groups:

1. The unsupervised networks, i.e. that are not using the *y* values in their prediction (see Sect. 2.5.1.3), such as Kohonen Networks (self-organising map; SOM)
2. The supervised networks, the most widespread, using the *y* values to build the model

In the latter category, the most employed ANN is the multilayer perceptron (MLP). The MLP is composed of basic elements, called neurons or nodes, interconnected in layers. They are generally composed of three layers as shown in Fig. 2.12:

1. The input layer is composed of n_1 nodes, corresponding to the n_1 input variables (raw or compressed variables to reduce the number of nodes and risk of overfitting).
2. The hidden layer is made of n_2 nodes; n_2 is not determined by the database and must be tuned either (1) to increase the complexity of the NN if it is enable

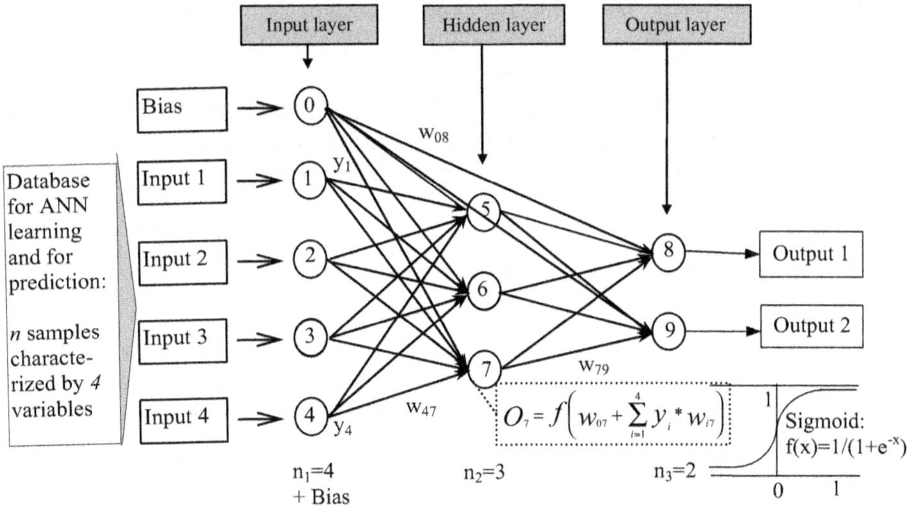

Fig. 2.12 Multilayer perceptron scheme

to model the non-linear relationship or (2) to reduce its complexity in order to decrease the overdetermination and overfitting.

3. The output layer contains as many nodes (n_3) as the number of *y* values to be predicted. Neural networks can also be used as classifiers, in encoding the classes in a complete disjunctive coding.

The parameters of the ANN model are tuned iteratively during the ANN learning process.

The ANN calibration procedure consists in tuning the weight (w_{ij}) linking all the node pairs (i, j) by an iterative learning procedure in order to provide the least prediction error rate. Generally, the initial weight levels are set at random.

At each iteration, in a feed-forward propagation procedure, the output estimation O_k of each node k is computed as the weighted linear combination (weigths: w_{ik}) of the node outputs y_i ($i = 1{:}n_l$) in the previous layer l and the adding of a bias (w_{0k}). This combination is then transformed by an activation function f. This function must have a binary threshold and is generally non-linear with a sigmoid shape (see example of O_7 in Fig. 2.12).

At each iteration, the weights are tuned to reduce the prediction error given by the difference of the output values and the real ones. The most widespread NN learning technique is the back-propagation of the error gradient (Rumelhart et al. 1986).

ANN are very powerful modelling tools; however, one of the main issues during neural network training is to prevent overfitting, i.e. make sure that the resulting model will be able to generalise its prediction to unknown samples (Ghosh and Turner 1994).

To minimise this risk, the number of calibration samples must be rather large (at least around 30 per weight). Furthermore, the ANN size (number of nodes) must be as low as possible, by using compression methods (e.g. PCA scores) or node-pruning techniques (Thimm and Fiesler 1997). The learning phase can also be tuned to avoid overfitting, for instance, by stopped-learning techniques (Weigend et al. 1990) and dynamic-learning procedure or noise introduction.

2.4.4 Robustness Issue and Calibration Transfer

"Robustness" is a term used here to define the ability of a model to be stable or insensitive when facing operational condition changes. This meaning is different to the one used in the chapter dealing with the outliers (see Sect. 2.3.3.3), where "robust chemometrics" refers to the mathematical algorithm stability regarding sample variability and outliers.

First, the stability of the model will depend on some good modelling practices which have to be performed systematically.

The first step is to ensure a fair representation of the calibration data set, using, for example, a DoE (see Sect. 2.2), and to check for outliers (see Sect. 2.3.3). It is also highly recommended to have an external validation set, measured later than the calibration set, for the model performance assessment.

The data must then be preprocessed correctly in order to remove noise or scatter effects. Variable selection can also be carried out to focus on the important features.

The final step is to choose the dimensionality of the model, i.e. the number of latent variables as explained in Sect. 2.4.2.5.

In addition to these good practices, we can consider robustness problem as a *calibration transfer* issue. Two methods are commonly used in many cases. On the one hand, the exhaustive calibration (or global modelling) consists in concatenating the samples containing the new variability or condition to be modelled to the initial calibration database. A new model is thus built. This method requires a large number of new samples to be efficient. On the other hand, the bias and slope correction method adjusts directly the prediction estimations by matching the predictions in the different conditions (Bouveresse et al. 1996). However, this method does not perform well when the perturbation disappears.

Some other specific methods have been developed for calibration transfer or robustness problems, such as model transfer between instruments or drifts due to changes in the process environment (e.g. temperature). These robustness issues can be considered depending on the information availability.

2.4.4.1 Models Using a Standardisation Set

If the same samples are measured in different conditions, knowing the perturbation, a standardisation set is available (Fig. 2.13). In that case, prior correction methods, such as the direct standardisation (DS), alter the perturbed spectra to be as close

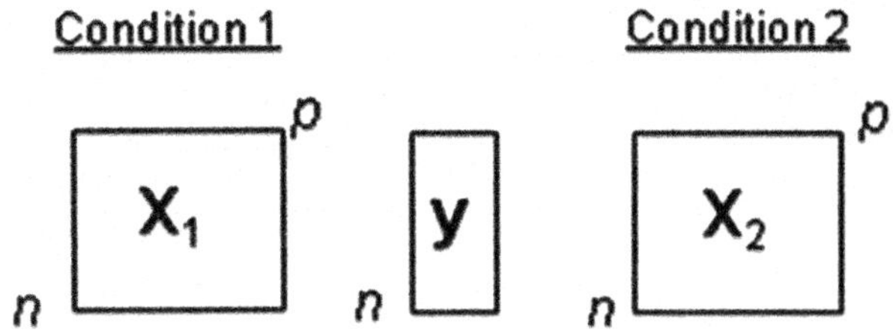

Fig. 2.13 Example of the construction of a standardisation set on two conditions

as possible to those of the calibration set. A more popular version of DS is the piecewise direct standardisation (PDS), which uses a moving window for the fitting (Wang et al. 1991; Bouveresse and Massart 1996). The inverse fitting (i.e. fitting old spectra to the new ones), known as (Piecewise) Reverse Standardisation ((P)RS), allows a definitive correction, i.e. new spectra do not have to be pretreated (Lima and Borges 2002). As these methods act on discrete perturbations, they are often used in model transfer between instruments. A PDS-like method, continuous PDS (CPDS), for correcting quantitative perturbations, such as temperature variation, has also been developed (Wülfert et al. 2000a).

Other methods try to take the variation into account in the modelling stage. The Repeatability file (Rep file) method uses spectral differences between the standardisation samples measured in several conditions and adds them to the calibration database with a response value of zero (Westerhaus 1991; Tillmann and Paul 1998).

2.4.4.2 Models Using a Small Experimental Design

It is possible to build a small experimental design X_{DoE} containing variations due to the perturbation (Fig. 2.14). For these methods, neither the perturbation level nor the y response needs to be measured. The aim of this small DoE is to model the perturbation subspace. The external parameter orthogonalisation (EPO) orthogonalises the calibration database $\mathbf{X}_0$ to this perturbation subspace (Roger et al. 2003). The new model is thus insensitive to the perturbation. If the perturbation disappears, the model performance is not affected. Similar methods have been developed, such as transfer by orthogonal projections (TOP), which has been used for calibration transfer between instruments (Andrew and Fearn 2004), or error removal by orthogonal subtraction (EROS), which uses replicates to model perturbations (Zhu et al. 2008). Some other variants using small DoEs have been developed, such as the OSC on DoE (Preys et al. 2008). All these methods have the additional advantage to interpret and diagnose the perturbed (removed) subspace.

2.4.4.3 Models When Only a Few Reference Control Points are Available

When neither a standardisation set nor a small experimental design is available, the direct orthogonal projection (DOP) method can be used as an alternative to exhaustive

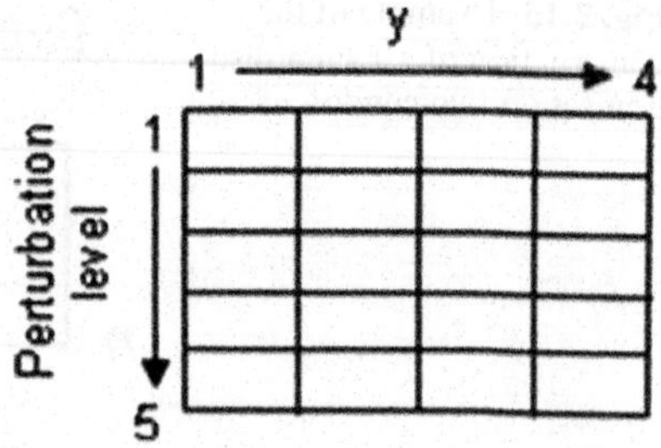

Fig. 2.14 Example of a small experimental design crossing four y levels with five perturbation levels

Fig. 2.15 Representation of few reference control points

calibration and bias and slope correction methods. DOP has been developed to exploit the information of very few real-time reference control points (Fig. 2.15). Virtual standards of $\mathbf{X}_{control}$ are built using samples of the initial calibration database by means of a kernel (Zeaiter et al. 2006). The difference between the measured spectra ($\mathbf{X}_{control}$) and their corresponding standard allows the modelling of the perturbation similarly to EPO. The calibration database is then corrected and a new model is rebuilt, now insensitive to the perturbation.

Reviews and comparisons of methods for robustness improvement or calibration transfer can be found in Zeaiter et al. (2005), Fearn (2001), Feudale et al. (2002b), Wülfert et al. (2000b), Chauchard et al. (2004b) and Igne et al. (2009).

2.5 Classification

In the PAT approach, the issue of sample classification is often encountered. For classification purposes, there are two ensembles of methods. The first one is the "unsupervised classification" methods (or clustering) which aim at classifying similar samples without the use of prior knowledge (see Sect. 2.5.1). The second one is the "supervised classification" methods (or discrimination), where class memberships are used to build a model (see Sect. 2.5.2). The latter ensemble contains also quantitative prediction methods adapted to qualitative issues.

2.5.1 Clustering Techniques

2.5.1.1 Introduction

The clustering techniques are also called "unsupervised classification" techniques in the chemometrics field or "data-mining" techniques in the machine learning field. They are exploratory tools which aim at finding "natural" clustering trends out of the inner data structure (**X**) without any other prior knowledge on the class assignment of the samples. Thus, all the methods described in this section measure the similarities between samples only according to their X values.

As mentioned in Sect. 2.3.2, PCA can be seen as a basis for cluster analysis since PCA scatter plots show the spread of samples in a low dimensional space. The groups can then be visualised by the operator to determine the sample clusters. The only parameter to set is the number of PC to retain, containing meaningful information and not residual noise. The interpretation of these classification results is based on the coordinates of the samples on the scatter plot relatively to the relevant loadings.

In genuine clustering methods, the crucial task is to tune the initial parameters, set a stopping criterion and assess the validity of the obtained clusters.

Some tools have been developed to help the user make final decisions. As an example, the cluster validity can be assessed by calculating indices or by statistical testing; it then helps to choose the best partition (Halkidi et al. 2002a, b).

Jain et al. have reviewed the different clustering techniques in 1999 (Jain et al. 1999).

2.5.1.2 Hierarchical Clustering Analysis

Hierarchical Clustering Analysis (HCA) is a method which assembles or dissociates successively the ensembles of samples (Johnson 1966). In agglomerative hierarchical classification, n classes are considered at the beginning, i.e. one class per sample, and are regrouped successively until it constitutes a unique class. The result is given in a form of a classification tree, called dendrogram (Fig. 2.16), where the length of the branches represents the distance between groups. The choice of the final groups is decided by cutting at a threshold; thus, the number of clusters is not a parameter to be set beforehand, contrarily to the k-means techniques (see Sect. 2.5.1.3).

The inverse procedure, known as divisive hierarchical classification, considers samples as one unique group which is then split successively into two groups as different as possible until all samples are separated. The two approaches are not equivalent, and the latter is used much less than the former because it is computationally intensive.

In hierarchical classifications, two parameters must be set:

- The distance between samples: Euclidean distance (see Eq. 2.24) or Mahalanobis distance (see Eq. 2.25). It is noted that the distance can be computed on scores of a PCA if **X** is collinear.
- The clustering or "linkage" criterion. Classical clustering criteria are based on the minimisation of one of the following terms: the minimum distance (single linkage), the maximum distance (complete linkage), the average distance (average linkage), the distance between gravity centres or the increase of within-class inertia (Ward criterion), favouring tight groups.

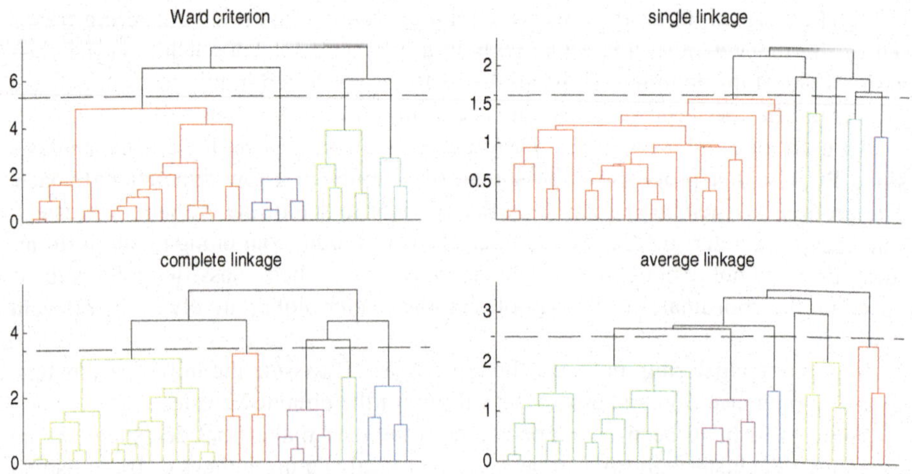

Fig. 2.16 Dendrograms built with Mahalanobis distance. Ward criterion (*upper left*), single linkage (*upper right*), complete linkage (*lower left*) and average linkage (*lower right*)

$$d_e(\mathbf{x}_a, \mathbf{x}_b) = \sqrt{(\mathbf{x}_a - \mathbf{x}_b)(\mathbf{x}_a - \mathbf{x}_b)^T} \tag{2.24}$$

$$d_m(\mathbf{x}_a, \mathbf{x}_b) = \sqrt{(\mathbf{x}_a - \mathbf{x}_b)\mathbf{C}^{-1}(\mathbf{x}_a - \mathbf{x}_b)^T}, \tag{2.25}$$

where x_a and x_b are the responses of samples ***a*** and ***b***, and C^{-1} is the inverse of the variance–covariance matrix of all the data X.

Depending on the chosen parameters, the classification results can be different, as shown in Fig. 2.16.

As for an application example, agglomerative hierarchical clustering analysis has been applied to identify consumer tomato preferences (Serrano-Megías and López-Nicolás 2006).

2.5.1.3 Non-hierarchical Clustering Methods

Non-hierarchical clustering methods aim at building one final partition of the data. Contrary to the hierarchical approach, the user must set a fixed number of groups out of his prior knowledge, which can be a strong limitation to these techniques.

Many techniques of non-hierarchical clustering methods have been developed, especially in the data-mining field:

- Methods based on distance measurement to assess the similarities between samples, such as *K*-means
- Methods based on sample density, such as DB-scan (Ester et al. 1996; Daszykowski et al. 2001)

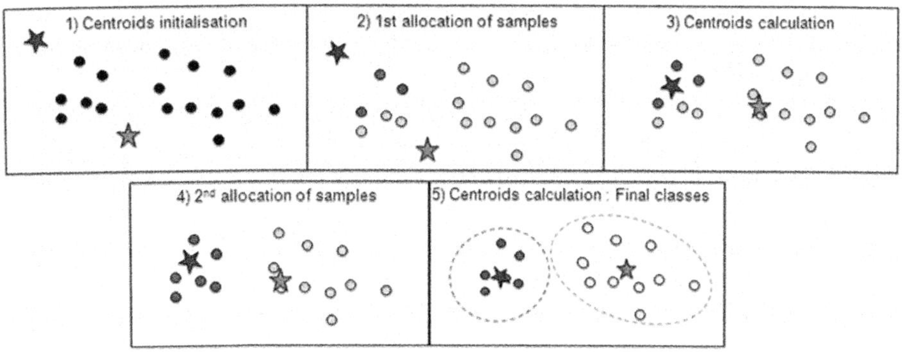

Fig. 2.17 *k*-means algorithm principle

- Methods based on the relationship between variables and samples, such as OPTICS (Ankerst et al. 1999) or Kohonen Artificial Neural Networks (e.g. SOM)

This section will describe the most commonly used techniques in chemometrics, i.e. the *K*-means and Kohonen SOM.

K-means

K-means (MacQueen 1967) is an iterative procedure (see Fig. 2.17). As in HCA, the method can be used on scores of PCA depending on the kind of data:

- Initial conditions: The initial partition of *k* groups is generally generated at random. *K*-means results are highly dependent on the initial partition and on the choice of the number of classes *k*.
- For each iteration, the barycentre of each class (i.e. the mean vector of the class) is recalculated and the samples are assigned to the nearest centre.
- Stopping criterion: This procedure is carried out until the termination criterion is reached (for instance: no assignment changes or maximum number of iterations reached).

Jain has reviewed the *K*-means algorithms (Jain 2009). In this article, criteria to find the optimal number of clusters are mentioned.

Kohonen SOM

The SOM method is a non-linear classification method based on ANN (Kohonen 1990, 1998). As opposed as classical ANN techniques (see Sect. 2.5.2.3), this one is an unsupervised technique where *p* input neurons correspond to the *p* variables of **X** and *k* output neurons represent the *k* classes. It results in a 2D map, where similar patterns are found close to each other and the most dissimilar ones are far away from each others. In 2001, an extension of SOM has been proposed to improve the application of the method in clustering analysis by using the agglomerative approach (Kiang 2001).

2.5.2 *Supervised Discrimination*

2.5.2.1 Introduction

In supervised discrimination methods, the sample classes of the calibration set are known and used. The goal is to build a discrimination model able to classify new samples. The model performance is assessed by a confusion matrix, showing the number of misclassified samples (see Annex: Figures Of Merit).

In the food industry, the main applications are product authentication, conformity to a standard quality or product sorting. A review of methods and their use in the food industry can be found in (Berrueta et al. 2007).

2.5.2.2 Linear Supervised Discrimination

Factorial Discriminant Analysis

The Factorial Discriminant Analysis (FDA; Fisher 1936) is a projection on the latent variables method which maximises the between-group variance (B) and minimises the global within-group variance (W) instead of focusing on the global variance (variance–covariance matrix) as in PCA.

A new sample is attributed to the group for which the Mahalanobis distance is minimal. Linear discriminant analysis (LDA) is very similar to FDA, and quadratic discriminant analysis (QDA) is performed when the heterogeneity of the groups has to be taken into account by computing a within-group variance per group.

FDA is very commonly used for discrimination applications. This method requires however more samples than variables and as a result it is not directly suitable for spectroscopic data. To overcome this issue, FDA can be applied to selected wavelengths or to PCA or PLS scores. With the latter solutions, discriminant factors can be reconstructed into discriminant spectra for class interpretation (Devaux et al. 1988). Figure 2.18 shows the application of FDA for the discrimination of milling products. Extreme particle sizes (E35 and E50) were well discriminated by the first discriminant factor, whereas medium particle sizes (E40 and E45) overlapped (Fig. 2.18).

For example, FDA has been applied to PCA scores computed out of near-infrared spectra to discriminate between wheat flour types and thus authenticate products in the flour milling industry (Sirieix and Downey 1993).

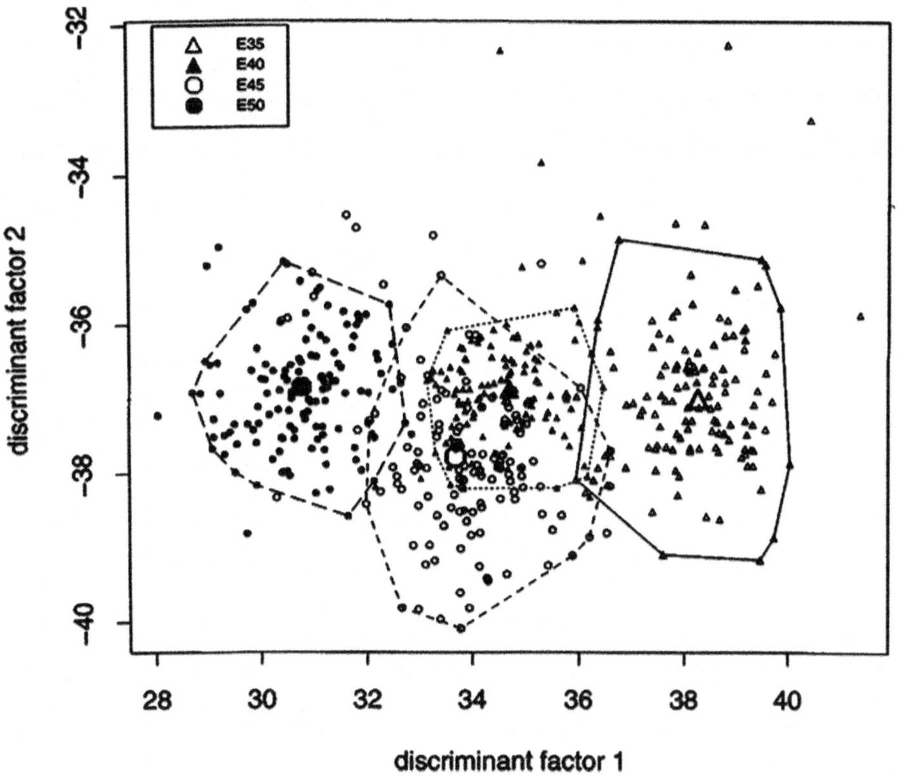

Fig. 2.18 Example of FDA for the discrimination of milling products with different particle sizes. (Source: Novales et al. 1998)

Soft Independent Modelling of Class Analogy

Soft independent modelling of class analogy (SIMCA) (Wold and Sjostrom 1977) is based on PCA and is thus suitable for high dimensional data.

Each class k is modelled by a specific PCA. Then, for each model a confidence interval is built to define the membership limit of the class. This membership criterion limit can be based on the Euclidean distance of the X-residuals (Q), or, more often, on the combination of Q and Hotelling's T^2 criteria. An unknown sample is then classified in class k if it falls within the class limits. A sample can be assigned to a "rejection class", i.e. to any existing class, which is a specificity of SIMCA. However, a sample can be attributed to several classes if they overlap or are very close to each other.

The Cooman's plot shows the residuals (or the T^2) of two models (i.e. two classes) against each other with their respective confidence limits, determining the class overlap. The plot is split into four regions limited by the confidence interval of the X-residuals (Fig. 2.19):

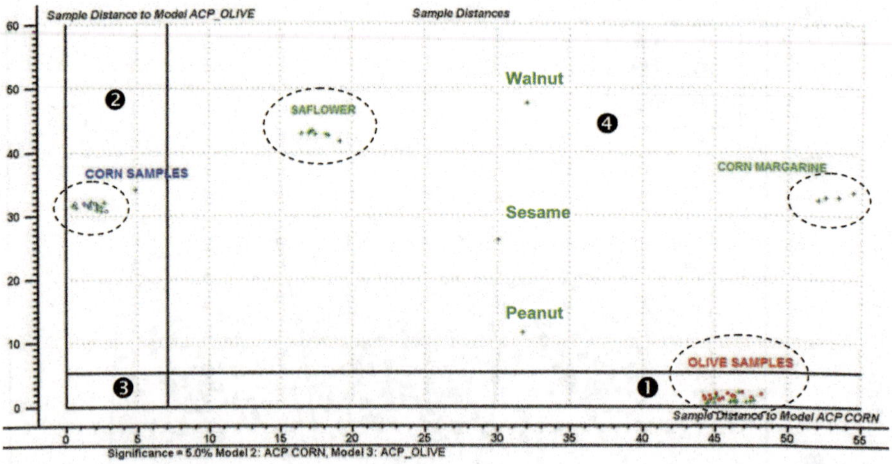

Fig. 2.19 Cooman's plot between classes A (olive oil) and B (corn oil). (On the courtesy of CAMO, Oslo, Norway)

1. The lower-right part corresponds to samples assigned to class A (olive oil).
2. The upper-left part corresponds to samples assigned to class B (corn oil).
3. The lower-left part corresponds to an overlap between classes A and B, containing samples matching both models.
4. The upper-right part corresponds to rejected samples from classes A and B, i.e. outliers or samples belonging to another class if SIMCA includes more than two PCA class models.

Kohler et al. (2002) show an application for sorting different qualities of fish using SIMCA based on sample imaging.

Partial Least-Square Discriminant Analysis

Partial least-square discriminant analysis (PLS-DA) corresponds to a PLS regression on a set of $\mathbf{Y}(n, k)$ dummy binary variables (Fig. 2.20, right) describing k classes (Fig. 2.20, left; Barker and Rayens 2003). PLS models predict continuous values; the sample membership is then determined by the highest predicted value among the m values.

Roussel et al. have applied PLS-DA for the classification of white grape musts, testing the relevance of different analytical methods such as electronic nose, ultraviolet (UV) and FT infrared spectroscopy and various classification methods (Roussel et al. 2003).

Classification and Regression Trees

Classification and Regression Trees (CART; Breiman et al. 1984) are similar to divisive HCA (see Sect. 2.5.1.2) in the fact that they successively split the data

Fig. 2.20 Conversion of classical encoding of three classes A, B and C, into a binary Y matrix

$$G = \begin{bmatrix} A \\ A \\ B \\ C \\ C \\ C \end{bmatrix} \qquad \mathbf{Y} = \begin{matrix} A & B & C \\ \end{matrix} \begin{bmatrix} 1 & 0 & 0 \\ 1 & 0 & 0 \\ 0 & 1 & 0 \\ 0 & 0 & 1 \\ 0 & 0 & 1 \\ 0 & 0 & 1 \end{bmatrix}$$

into two groups, forming a binary tree. At each node of the tree, the data are split according to one variable of the X values chosen by a forward selection. The predicted class of a new sample is then determined by the class majority of the samples belonging to the terminal node in which it falls.

2.5.2.3 Non-linear Supervised Discrimination

Support Vector Machines

SVM are non-linear discrimination methods for splitting two classes (encoded $[-1;+1]$) of samples (Cortes and Vapnik 1995; Burges 1998). When a linear separation is not possible, SVM allow the user to switch from $\mathbf{X}_{(n,\ p)}$ matrix to the Kernel matrix of similarity $\mathbf{K}_{(\mathbf{n},\ \mathbf{n})}$, as defined in Sect. 2.4.3.3. The Kernel matrix takes into account the non-linearity and is able to draw a linear separation into this new space. SVM is a local method, as samples which are too far away from the separation do not participate in the model elaboration (weights equal to zero). Samples with non-zero weights are called support vectors. The choice of the linear separation (between all combinations giving the same error) is made by maximising the margin between classes, i.e. the chosen separation is the farthest away from all points in the space.

SVM is valid for the separation of two classes, but some tools have been established to extend the application to multiple class data sets. An example on the determination of feed composition with hyperspectral NIR imaging is given by Fernández Pierna et al. (2006).

ANN

The ANN principle for classification is exactly the same as for quantitative prediction (see Sect. 2.4.3.4). Instead of having only one output node, the ANN is composed of as many output nodes as classes, with a disjunctive encoding.

MLP has been compared to other classification techniques by Roussel and Hardy et al. (2001) for genetically modified organism (GMO) soybean discrimination with NIR spectroscopy, or applied to olive oil adulteration detection with mass spectroscopy (Goodacre et al. 1993).

2.5.2.4 A Particular Case: k-Nearest Neighbours (k-NN)

k-Nearest Neighbours (k-NN; Hart 1967) is more a classification rule than a modelling method. Starting from a calibration set, where all samples belong to a known class, a new sample is classified according to the *k*-nearest neighbours present in the calibration data set. The neighbourhood is generally assessed by the Euclidean distance. The unknown sample is assigned to the class majority among the *k* samples. If no majority is established, then the new sample is attributed to the class of its first nearest neighbours. The value of *k* is generally small; 3 or 5 is often encountered.

k-NN can be useful when updating the calibration data set, i.e. for adding new calibration samples, since no new model building is needed. Furthermore, the method can be useful after a hierarchical clustering for the class assignment of new samples. The lack of information on the characterisation of the classes and the potentially long computation time constitute the main drawbacks of the method.

We can mention an application in the development and validation of spectral libraries for the characterisation of ingredients in animal feed (Fernández-Ibáñez et al. 2010).

2.6 Multivariate Process Monitoring

2.6.1 Multivariate Statistical Process Control

2.6.1.1 Introduction

Since the quality of an end product is multivariate, generally not enough properties are measured to assess it correctly. Targeting the end-product specifications is thus not sufficient. Timely process measurements are hence important to be additionally monitored and kept in control (Kourti 2006). Real-time release can also be performed in the frame of the PAT, instead of waiting the end of the process and the measurement of the end-product quality.

Process databases are usually composed of initial conditions (raw material characterisation, information available before processing, etc.; **Z**), process variables (**X**) and quality attributes for the end product (**Y**) (Fig. 2.21). The variables are timely measurements, including real-time basic sensors, such as thermometers, and more complex analysers, such as NIR spectrometers. These process databases have usually a low statistical rank, since the variables are correlated, auto-correlated (i.e. with themselves over time) and cross-correlated (i.e. correlated with other variables at different time lags). These data usually contain a large amount of missing data as well, and present a low signal-to-noise ratio. These are all the reasons why the multivariate data-based latent variable methods, such as PCA and PLS, are powerful tools to handle and monitor process data.

Different objectives at different levels are considered when dealing with process data. The process is usually first analysed, then monitored and finally controlled if

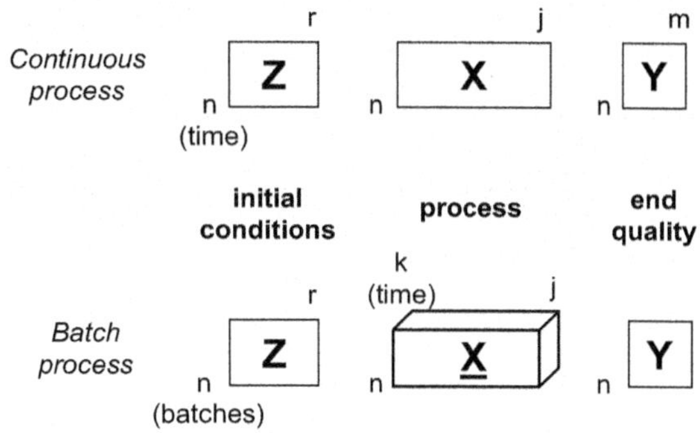

Fig. 2.21 Matrix structures for process data

the process is no more in its state of statistical control, i.e. the process variables and product properties are no more close to their target values (MacGregor et al. 2005).

2.6.1.2 Process Analysis

In a process understanding and monitoring framework, the first step consists in analysing historical databases. The objective is to explore the data, differentiating "good" and "bad" processes or batches, and identifying the reasons of eventual clusters, performing troubleshooting in hindsight. It allows improving the process and determining the reference database containing the "good" or in-control processes or batches for the next step. A PLS is usually performed using the **X** and **Y** matrices (see Sect. 2.4.2.4).

When working with batches, the $\underline{\mathrm{X}}$ cube can be rowwise unfolded, so that each row corresponds to a batch. This method also allows handling of the cases when the different process variables are not measured with the same frequency or duration. The resulting score plot discriminates the clusters of processes or batches and the loading plot gives an interpretation of the clusters. When the batches do not always run the same duration, some algorithms attempt to align the batches, using for example dynamic time warping (Kassidas et al. 1998) or the indicator variable approach (Garcia-Munoz et al. 2003).

2.6.1.3 Process Monitoring and Fault Diagnosis

Monitoring the process aims at identifying if the process is in-control in real time. The nature of the deviation is then diagnosed; this corresponds to early fault diagnosing or fault detection and isolation (FDI). This helps the operator to perform corrective actions and avoids reliance on the end-product quality measurement.

The proposed modelling methodology is an extension of the univariate statistical process control (SPC) and the Shewhart control charts to the multivariate data (Montgomery 1997). Figure 2.22 shows why it is more accurate to build multi-

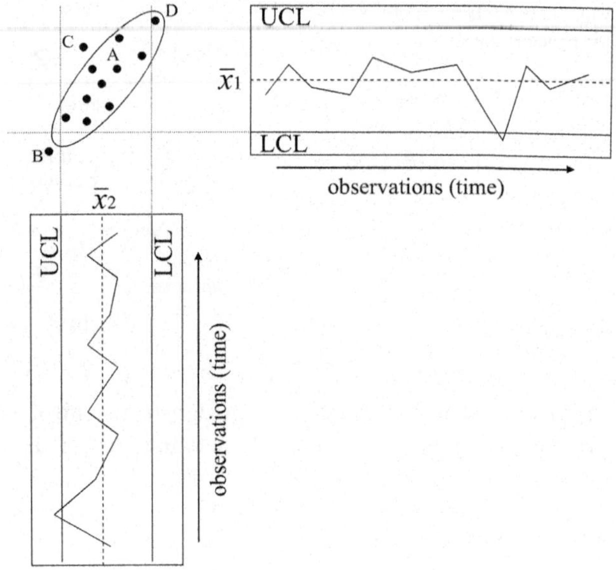

Fig. 2.22 Univariate versus multivariate control charts. *A* is in-control and *B* out-of-control in both univariate and multivariate cases. *C* is in-control in the univariate charts but not in the multivariate case, and conversely for *D*. *UCL* upper control limit, *LCL* lower control limit

variate control charts, using the covariance between the variables, instead of many univariate charts as variables.

A first multivariate control chart, Hotelling's T^2, corresponds to the multivariate distance of one sample to the centre of the model considering the *A* first components using the reference (in-control) database (Eq. 2.26). It should be emphasized that selecting the reference processes (see Sect. 2.6.1.2) is of a great importance, since the monitoring model is built upon them:

$$T_A^2 = \sum_{i=1}^{A} \frac{\mathbf{t}_i^2}{\mathbf{s}_{\mathbf{t}_i}^2}, \tag{2.26}$$

where $\mathbf{s}_{\mathbf{t}_i}^2$ is the variance of the **t** score for the ith component.

A second multivariate control chart, built for each sample, represents the residual variance, i.e. the squared prediction error (SPE), detecting abnormal situations, which are not modelled with the reference processes.

Upper control limits (UCL) are calculated for each of the two multivariate control charts with a certain risk α (Tracy et al. 1992). Contribution plots, stemming from the loadings, are available for each sample and each chart in order to immediately diagnose the fault when the limits are exceeded (Fig. 2.23).

Monitoring and fault diagnosis using MSPC for a milk pasteurisation process can be found in Tokatli et al. (2005).

Batch statistical process control (BSPC) is an extension of MSPC to the batch processes, where batch trajectories are monitored and compared to a target. Several algorithms are compared in Dahl et al. (1999) and van Sprang and Ramaker et al. (2002). Batch cube data are either unfolded or analysed as is with multi-way tech-

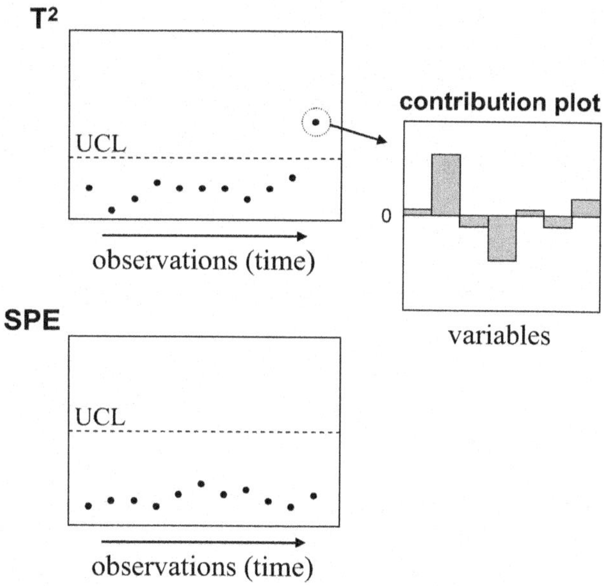

Fig. 2.23 Hotelling's T^2 and SPE multivariate control charts and the corresponding contribution plot for the deviating point

niques (see Sect. 2.7.2). When using the unfolding methods, Wold et al. propose that a first observation level models the **X** matrix, i.e. the batch evolution, according to columnwise unfolding, where each row represents an observation (Wold et al. 1998b). The local time of each observation can be used as a dummy variable as well. The second level, called the batch level, corresponds to rowwise unfolding, where each row represents a batch, and models **Z**, **X** and **Y** simultaneously, i.e. the final batch results. Nomikos *et al.* proposed earlier a rowwise unfolding method using multi-way PCA (MPCA) and multi-way PLS (MPLS; Nomikos and MacGregor 1995a, b). These methodologies can be used in real time for batch monitoring or afterwards for real-time release of the batches (Kourti 2006). In this case, the model is run at the end of the batch to see if the product can be released or not.

Multi-block PLS (MB-PLS; see Sect. 2.7.1) is used when sequential process operations can be separated and analysed together. Several algorithms are compared in Westerhuis et al. (1998).

2.6.1.4 Process Control

The processes can also be controlled, meaning that manipulated variables such as temperature or a valve opening can be set from a multivariate modelling. A feedback control consists in deciding a corrective action from a model output, such as a multivariate distance which deviates at a certain time of the process. A feed-forward control is based on a model input such as the raw material characterisation. Finally, end point determination is the estimation of the process settings to get the desired

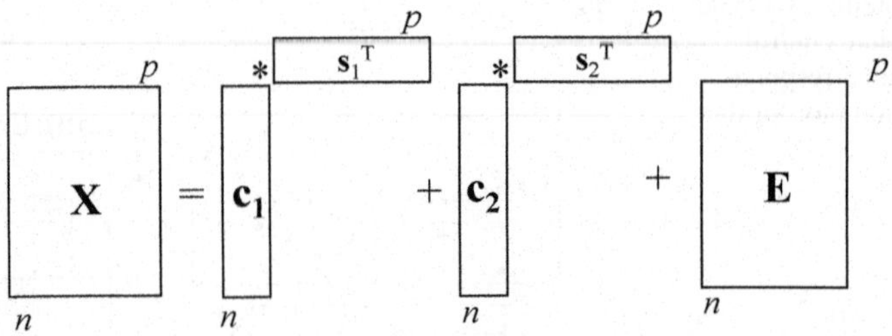

Fig. 2.24 Decomposition in spectra and concentration profiles

product; manipulation of process variables and other applications are in the scope of process control based on latent variable methods.

2.6.2 *Multivariate Curve Resolution*

Multivariate curve resolution (MCR) methods aim at recovering or deconvolving "pure spectra" $\mathbf{S}^T$ and their corresponding "concentration profiles" $\mathbf{C}$ in mixtures of several chemical compounds (see Eq. 2.27 and Fig. 2.24). Unlike PCA (see Sect. 2.3.2), MCR methods do not search for orthogonal latent variables, since two pure spectra may present overlapped area. To reduce rotational ambiguities, leading to the non-uniqueness of the solution, constraints on both spectra and concentration profiles can be applied. This helps the algorithm to find an optimal solution with increasing physicochemical meanings.

The classical constraints are the following:

- The non-negativity for spectra and/or concentration profiles
- The unimodality for concentration profiles, allowing only one maximum or minimum
- The closure, forcing the sum of the concentrations to be always constant
- Hard modelling can also be applied to the concentration profiles if a physicochemical model or kinetics are known
- Equality constraint is used if a pure spectrum is known.

$$\mathbf{X} = \mathbf{C}\mathbf{S}^T. \tag{2.27}$$

Different kinds of MCR methods have been developed since the first one (Lawton and Sylvestre 1971), and most of them are iterative (Jiang et al. 2004). The most commonly used iterative MCR method is the MCR-alternative least square (MCR-ALS; Tauler 1995; de Juan and Tauler 2003) and includes four steps:

- Determination of the number of components (using, e.g. PCA)
- Initialisation with a set of $\mathbf{C}$ or $\mathbf{S}^T$ initial estimates, using, e.g. simple-to-use interactive self-modelling analysis (SIMPLISMA), evolving factor analysis (EFA) or prior knowledge

- Iterative calculation of **C** and $\mathbf{S}^T$ using ALS with the chosen constraints (Eqs. 2.28 and 2.29):

$$\mathbf{C} = \mathbf{XS}(\mathbf{S}^T\mathbf{S})^{-1} \tag{2.28}$$

$$\mathbf{S} = (\mathbf{C}^T\mathbf{C})^{-1}\mathbf{C}^T\mathbf{X} \tag{2.29}$$

- Checking convergence

MCR-ALS was more recently used with trilinear data as an alternative to multi-way methods (see Sect. 2.7.2), by applying the strong trilinear constraint, hence reducing totally the rotational ambiguity (de Juan and Tauler 2006).

Another extension of the MCR-ALS use is the application to multi-block matrices. Rowwise augmented matrices (e.g. describing a multi-sensor analysis of a set of samples) or columnwise augmented matrices (e.g. describing a multi-condition experiment) can be handled with MCR-ALS, leading to additional decrease to rotational ambiguity. Applications in biochemistry and environment science can be found in Felipe-Sotelo et al. (2006) and Navea et al. (2006).

2.7 Multi-block and Multi-way Analyses

Multivariate data are often structured in different tables (rowwise or columnwise) or in cubes, i.e. with a third way, such as time or wavelength. Some chemometrics tools were developed to deal with this kind of data.

2.7.1 Multi-block Analysis

2.7.1.1 Definition of Multi-block Data Sets

Multi-block data sets include data sets where (1) the same samples are characterised with different blocks of variables, and the nature and the number of the variables of these different blocks can vary, or (2) several blocks of samples are characterised with the same variables, and the number of these samples in each block can be different (Fig. 2.25).

2.7.1.2 Exploratory Multi-block Analyses

Different chemometrics methods have been developed to handle multi-block data sets for exploratory analysis, i.e. to identify common and specific information within the different blocks of data. Correlation canonical analysis (CCA) (Hotelling

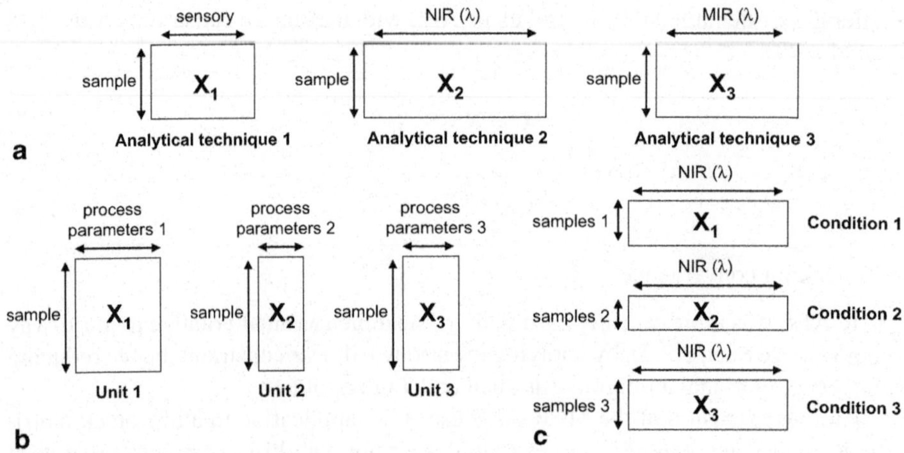

Fig. 2.25 Examples of multi-block data sets: **a** characterisation with different analytical methods, **b** in different process units and **c** with different experimental conditions

1936) and procuste analysis (PA), which attempts to find the subspace where a block is close to a reference one, have been historically used. PCA on concatenated blocks, sometimes called SUM-PCA, or consensus-PCA (C-PCA) and multi-block PCA (MB-PCA), depending on the way the blocks are weighted, are direct extensions of PCA. These methods however do not explicitly take into account the relationships between blocks. Hierarchical PCA (H-PCA; Wold et al. 1996), multiple co-inertia analysis (MCoA; Chessel and Hanafi 1996) and common component and specific weights analysis (CCSWA; Qannari et al. 2000) build common subspaces with "super-scores" where the different blocks show similar or specific pieces of information.

Different applications have been published. CCA has been applied to oil spectra after PCA preprocessing (Devaux et al. 1993). C-PCA and MB-PCA have been used for process monitoring (Qin et al. 2001). Dairy products have been characterised by different analytical techniques using CCSWA (Mazerolles et al. 2006).

2.7.1.3 Predictive Multi-block Analyses

For predictive purposes, MB-PCA has been extended to MB-PLS (Wold et al. 1984; Wangen and Kowalski 1988) and H-PCA to H-PLS (Wold et al. 1996). Different ways to deal with predictive multi-block applications were also developed, e.g. the PLS-path modelling (PLS-PM) approach (Wold 1982) and more recently the structural equation exploratory regression (SEER; Bry et al. 2009). L-PLS has been recently developed to handle data blocks linked through rows and columns simultaneously (Martens et al. 2005; Westad et al. 2008).

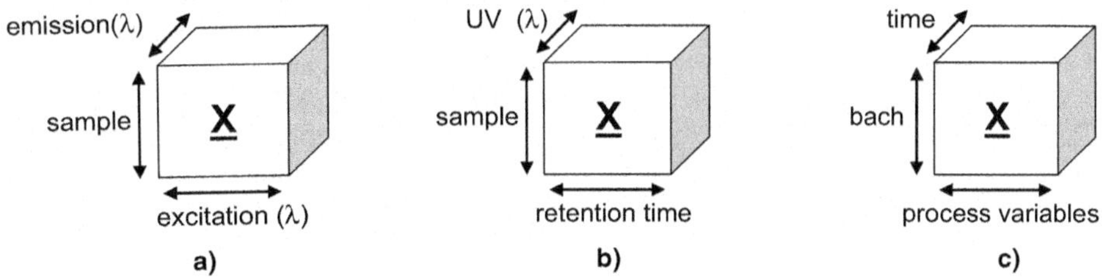

Fig. 2.26 Examples of multi-way data sets: **a** fluorescence spectroscopy, **b** HPLC-DAD and **c** batch process data

2.7.2 *Multi-way Analysis*

2.7.2.1 Definition of Trilinear Data Sets

When data have a (hyper-) cube structure, i.e. a trilinear (or more) relationship between the sample way and two (or more) other variable ways (or modes), multi-way tools are relevant to handle these multi-way data sets, also called N-way data sets (Fig. 2.26).

2.7.2.2 Exploratory Multi-way Analyses

A straightforward extension of PCA to three-way data is the PARAFAC (parallel factor analysis) trilinear model, developed by Harshman (Harshman 1970) and Carroll et al. (Carroll and Chang 1970; Harshman 1970):

$$x_{ijk} = \sum_{f=1}^{F} a_{if} b_{jf} c_{kf} + e_{ijk}, \tag{2.30}$$

where F is the number of components chosen for the model.

PARAFAC is a way to perform curve resolution, i.e. to estimate the profiles and concentrations of mixtures during a process, for example. Many related topics for PARAFAC have been developed by Bro, such as the uniqueness of the solution, the constraints (non-negativity, unimodality, etc.), the missing values and validation (Bro 1997). Due to the different chosen constraints, PARAFAC solutions often lead to meaningful physicochemical meanings, such as "pure" spectra.

PARAFAC has been applied to several issues, including process monitoring. For instance, the production of sugar beets has been monitored using fluorescence spectroscopy and PARAFAC (Bro 1999).

To handle shift problems in multivariate signals or batches with different durations, PARAFAC2 has been developed, assigning one set of profiles for each sample (Bro et al. 1999; Kiers et al. 1999).

The Tucker3 model is another way to handle multi-way data sets (Tucker 1966). The solutions are not unique, and more difficult to interpret, since the optimal number of components is not the same for each variable mode. All kinds of multi-way data sets, and not only the trilinear (or more) data, can be modelled by the Tucker3 model.

2.7.2.3 Predictive Multi-way Analyses

When the aim is to build predictive models, N-PLS (or N-way PLS) is commonly used as an extension to PLS for multi-way data sets (Bro 1996). For three-way data, two loading weight vectors for each component are calculated, one for each variable mode.

De Belie et al. studied the chewing sound of different dry-crisp snacks using PARAFAC, Tucker3 and N-PLS multi-way methods (De Belie et al. 2003). A review of the applications for chemical compound quantification in complex matrices using chromatography and multi-way analyses can be found in Ortiz and Sarabia (2007).

2.8 Conclusion

Within the framework of the process analytical technology (PAT) approach, multivariate data processing (chemometrics) techniques are very useful. They enable the chemometrics user to effectively plan experiments, model the very numerous sensor data and extract useful information for process monitoring or product characterisation.

Acknowledgments The authors want to thank Dr. Mazerolles from INRA for his multi-block section review, Dr. Williams from the Canadian Grain Commission, CAMO (Oslo, Norway) and Dr. Guillaume from Cemagref for authorising the usage of their data.

Annex: Figures of Merit

Some statistical criteria (Figures of Merit) encountered during this chapter are mathematically defined here.

Root-mean-squared error (RMSE) is defined as the root square of the ratio of the prediction error sum of squares (PRESS) and the estimated degrees of freedom of the set. This square root value is generally preferred because the unit is the same as the original data. RMSE is calculated for a calibration set (RMSEC) and a prediction (or validation) set (RMSEP):

Table 2.4 Confusion matrix

		Real classes		
		Class A	Class B	Second-order error
	Number of samples in each class	N1	N2	
Predicted classes	Class A	n_{11}	n_{12}	Impurities in class A
	Class B	n_{21}	n_{22}	Impurities in class B
	First-order error	Error in class A	Error in class B	Total error

$$\text{PRESS} = \sum_{i=1}^{n} (y_i - \hat{y}_i)^2 \tag{2.31}$$

$$\text{RMSEC} = \sqrt{\frac{\text{PRESS}}{n-p-1}} \tag{2.32}$$

$$\text{RMSEP} = \sqrt{\frac{\text{PRESS}}{n}}. \tag{2.33}$$

The bias is the mean error, i.e. the systematic part of the error:

$$bs = \frac{1}{n}\sum_{i=1}^{n} (y_i - \hat{y}_i). \tag{2.34}$$

The coefficient of determination (R^2) represents the spread of the predictions. It is important not to consider it alone. For example, the R^2 value could be almost 1 whereas the bias and/or the PRESS could be high:

$$R^2 = 1 - \frac{\text{PRESS}}{\sum_{i=1}^{n} (y_i - \bar{y})^2}. \tag{2.35}$$

In classification, standard errors are defined as the proportion of misclassified samples. A confusion matrix is generally built to summarise the results. The number of correctly classified objects corresponds to n_{11} and n_{22}, and the misclassified ones to n_{21} and n_{12}. It can also be seen in terms of first-order errors, which are similar to the lack of sensitivity (e.g. the proportion of samples A not classified in A), or second order, which represents the lack of specificity (e.g. the proportion of B classified in A) (Table 2.4).

It is important to note that some other figures of merit are widely used in certain applications and can be found in Olivieri et al. (2006).

References

Abrahamsson C, Johansson J et al (2003) Comparison of different variable selection methods conducted on NIR transmission measurements on intact tablets. Chemometr Intell Lab Syst 69:3–12

Alsberg BK, Woodward AM et al (1997) An introduction to wavelet transforms for chemometricians: a time-frequency approach. Chemometr Intell Lab Syst 37(2):215–239

Andersson CA (1999) Direct orthogonalization. Chemometr Intell Lab Syst 47(1):51–63

Andrew A, Fearn T (2004) Transfer by orthogonal projection: making near-infrared calibrations robust to between-instrument variation. Chemometr Intell Lab Syst 72(1):51–56

Ankerst M, Breunig MM et al (1999) OPTICS: ordering points to identify the clustering structure. ACM SIGMOD international conference on management of data (SIGMOD'99), Philadelphia

Araujo PW, Brereton RG (1996a) Experimental design I. Screening. TrAC Trends Anal Chem 15(1):26–31

Araujo PW, Brereton RG (1996b) Experimental design II. Optimization. TrAC Trends Anal Chem 15(2):63–70

Azzouz T, Puigdoménech A et al (2003) Comparison between different data pretreatment methods in the analysis of forage samples using near-infrared diffuse reflectance spectroscopy and partial least-squares multivariate calibration method. Anal Chim Acta 484:121–134

Barker M, Rayens W (2003) Partial least squares for discrimination. J Chemometr 17:166–173

Barnes RJ, Dhanoa MS et al (1989) Standard normal variate transformation and de-trending of near-infrared diffuse reflectance spectra. Appl Spectrosc 43(5):772–777

Berrueta LA, Alonso-Salces RM et al (2007) Supervised pattern recognition in food analysis. J Chromatogr A 1158(1–2):196–214

Bouveresse E, Massart DL (1996) Improvement of the piecewise direct standardisation procedure for the transfer of NIR spectra for multivariate calibration. Chemometr Intell Lab Syst 32:201–213

Bouveresse E, Hartmann C et al (1996) Standardization of near-infrared spectrometric instruments. Anal Chem 68(6):982–990

Box GEP, Draper NR (1987) Empirical model-building and response surfaces. Wiley, New York

Breiman L, Friedman JH et al (1984) Classification and regression trees. Wadsworth International Group, Belmont

Bro R (1996) Multiway calibration, multilinear PLS. J Chemometr 10:47–61

Bro R (1997) PARAFAC. Tutorial and applications. Chemometr Intell Lab Syst 38(2):149–171

Bro R (1999) Exploratory study of sugar production using fluorescence spectroscopy and multiway analysis. Chemometr Intell Lab Syst 46(2):133–147

Bro R, Andersson CA et al (1999) PARAFAC2—Part II. Modeling chromatographic data with retention time shifts. J Chemometr 13(3-4):295–309

Bry X, Verron T et al (2009) Exploring a physico-chemical multi-array explanatory model with a new multiple covariance-based technique: structural equation exploratory regression. Anal Chim Acta 642(1–2):45–58

Burges CJC (1998) A tutorial on support vector machines for pattern recognition. Data Min Knowl Disc 2(2):121–167

Carroll JD, Chang J (1970) Analysis of individual differences in multidimensional scaling via an N-way generalization of 'Eckart-Young' decomposition. Psychometrika 35:283–319

Centner V, Massart D-L et al (1996) Elimination of uninformative variables for multivariate calibration. Anal Chem 68(21):3851–3858

Chauchard F, Cogdill R et al (2004a) Application of LS-SVM to non-linear phenomena in NIR spectroscopy: development of a robust and portable sensor for acidity prediction in grapes. Chemometr Intell Lab Syst 71(2):141–150

Chauchard F, Roger JM et al (2004b) Correction of the temperature effect on near infrared calibration—application to soluble solid content prediction. J Near Infrared Spectrosc 12:199–205

Chessel D, Hanafi M (1996) Analyses de la co-inertie de K nuages de points. Revue de Statistique Appliquée XLIV(2):35–60
Christiansen KF, Vegarud G et al (2004) Hydrolyzed whey proteins as emulsifiers and stabilizers in high-pressure processed dressings. Food Hydrocoll 18(5):757–767
Clark RD (1997) OptiSim: an extended dissimilarity selection method for finding diverse representative subsets. J Chem Inf Comput Sci 37:1181–1188
Cleveland WS, Devlin SJ (1988) Locally weighted regression: an approach to regression analysis by local fitting. J Am Statistical Assoc 83(403):596–610
Cogdill RP, Dardenne P (2003) Least-squares support vector machines for chemometrics: an introduction and evaluation. J Near Infrared Spectrosc 12(2):93–100
Cornell JA (1990) Experiments with mixtures. Wiley, New York
Cortes C, Vapnik V (1995) Support vector networks. Mach Learning 20:273–297
Dahl KS, Piovoso MJ et al (1999) Translating third-order data analysis methods to chemical batch processes. Chemometr Intell Lab Syst 46(2):161–180
Dantas-Filho HA, Galvao RKH et al (2004) A strategy for selecting calibration samples for multivariate modelling. Chemometr Intell Lab Syst 72:83–91
Daszykowski M, Walczak B et al (2001) Looking for natural patterns in data: part 1. Density-based approach. Chemometr Intell Lab Syst 56(2):83–92
Daszykowski M, Kaczmarek K et al (2007) Robust statistics in data analysis—a review. Basic concepts. Chemometr Intell Lab Syst 85:203–219
Daubechies I (1990) The wavelet transform, time-frequency localization and signal analysis. IEEE Trans Inform Theory 36(5):961–1005
Davies AMC, Britcher HV et al (1988) The application of fourier-transformed near-infrared spectra to quantitative analysis by comparison of similarity indices (CARNAC). Microchim Acta 94(1–6):61–64
De Belie N, Sivertsvik M et al (2003) Differences in chewing sounds of dry-crisp snacks by multivariate data analysis. J Sound Vib 266(3):625–643
De Jong S (1993) SIMPLS: an alternative approach to partial least squares regression. Chemometr Intell Lab Syst 18(3):251–263
de Juan A, Tauler R (2003) Chemometrics applied to unravel multicomponent processes and mixtures: revisiting latest trends in multivariate resolution. Anal Chim Acta 500(1-2):195–210
de Juan A, Tauler R (2006) Multivariate curve resolution (MCR) from 2000: progress in concepts and applications. Crit Rev Anal Chem 36:163–176
De Maesschalck R, Jouan-Rimbaud D et al (2000) The Mahalanobis distance. Chemometr Intell Lab Syst 50(1):1–18
Devaux MF, Bertrand D et al (1988) Application of multidimensional analyses to the extraction of discriminant spectral patterns from NIR spectra. Appl Spectrosc 42(6):941–1132
Devaux MF, Robert P et al (1993) Canonical correlation analysis of mid and near infrared oil spectra. Appl Spectrosc 47:1024–1028
Ellekjær MR, Ilseng MA et al (1996) A case study of the use of experimental design and multivariate analysis in product improvement. Food Qual Prefer 7(1):29–36
Eriksson L, Johansson E et al (1998) Mixture design-design generation, PLS analysis, and model usage. Chemometr Intell Lab Syst 43(1–2):1–24
Ester M, Kriegel H-P et al (1996) A density-based algorithm for discovering clusters in large spatial databases with noise. 2nd international conference on knowledge discovery and data mining
Fearn T (2001) Standardisation and calibration transfer for near infrared instrument: a review. J Near Infrared Spectrosc 9:229–244
Felipe-Sotelo M, Gustems L et al (2006) Investigation of geographical and temporal distribution of tropospheric ozone in Catalonia (North-East Spain) during the period 2000–2004 using multivariate data analysis methods. Atmos Environ 40(38):7421–7436
Fernández Pierna JA, Wahl F et al (2002) Methods for outlier detection in prediction. Chemometr Intell Lab Syst 63(1):27–39
Fernández Pierna JA, Baeten V et al (2006) Screening of compound feeds using NIR hyperspectral data. Chemometr Intell Lab Syst 84(1–2):114–118

Fernández-Ibáñez V, Fearn T et al (2010) Development and validation of near infrared microscopy spectral libraries of ingredients in animal feed as a first step to adopting traceability and authenticity as guarantors of food safety. Food Chemistry. (In press, corrected proof).
Ferreira SLC, Santos WNL dos et al (2004) Doehlert matrix: a chemometric tool for analytical chemistry-review. Talanta 63(4):1061–1067
Ferreira SLC, Bruns RE et al (2007) Box-Behnken design: an alternative for the optimization of analytical methods. Anal Chim Acta 597(2):179–186
Feudale RN, Tan H et al (2002a) Piecewise orthogonal signal correction. Chemometr Intell Lab Syst 63:129–138
Feudale RN, Woody NA et al (2002b) Transfer of multivariate calibration models: a review. Chemometr Intell Lab Syst 64(2):181–192
Fisher R (1936) The use of multiple measurements in taxonomic problems. Annals Eugenics 7:179–188
Gacula MC, Jagbir Singh JR (1984) Statistical methods in food and consumer research. Academic, New York
Galvao RKH, Araujo MCU et al (2005) A method for calibration and validation subset partitioning. Talanta 67:736–740
Garcia-Munoz S, Kourti T et al (2003) Troubleshooting of an industrial batch process using multivariate methods. Ind Eng Chem Res 42:3592–3601
Geladi P, MacDougall D et al (1985) Linearization and scatter-correction for near-infrared reflectance spectra of meat. Appl Spectrosc 39(3):491–500
Ghosh J, Turner K (1994) Structural adaptation and generalization in supervised feed-forward networks. J Artificial Neural Netw 1(4):431–458
Gnanadesikan R, Kettenring JR (1972) Robust estimates, residuals, and outlier detection with multiresponse data. Biometrics 28:81–124
Goodacre R, Kell DB et al (1993) Rapid assessment of the adulteration of virgin olive oils by other seed oils using pyrolysis mass spectrometry and artificial neural networks. J Sci Food Agric 63:297–307
Halkidi M, Batistakis Y et al (2002a) Cluster validity methods: part I. ACM SIGMOD Rec 31(2):40–45
Halkidi M, Batistakis Y et al (2002b) Clustering validity checking methods: part II. ACM SIGMOD Rec 31(3):19–27
Harshman RA (1970) Foundations of the PARAFAC procedure. UCLA Working Papers in Phonetics 16:1–84.
Hart P (1967). Nearest neighbour pattern classifcation. IEEE Trans Inform Theory 13(1):21–27
Hotelling H (1936). Relations between two sets of variants. Biometrika 28:321–377
Igne B, Roger J-M et al (2009) Improving the transfer of near infrared prediction models by orthogonal methods. Chemometr Intell Lab Syst 99(1):57–65
Jain AK (2009). Data clustering: 50 years beyond K-means. Pattern Recognit Lett. (In press, corrected proof)
Jain AK, Murty MN et al (1999) Data clustering: a review. ACM Comput Surv 31(3):264–323
Jiang J-H, Liang Y et al (2004) Principles and methodologies in self-modeling curve resolution. Chemometr Intell Lab Syst 71(1):1–12
Johnson SC (1966) Hierarchical clustering schemes. Psychometrika 32(3):241–254
Kassidas A, MacGregor JF et al (1998) Synchronization of batch trajectories using dynamic time warping. AIChE J 44:864–875
Kennard RW, Stone LA (1969) Computer aided design of experiments. Technometrics 11(1):137–148
Kiang MY (2001) Extending the Kohonen self-organizing map networks for clustering analysis. Comput Stat Data Anal 38(2):161–180
Kiers HAL, Berge JMFt et al (1999) PARAFAC2—part I. A direct fitting algorithm for the PARAFAC2 model. J Chemometr 13(3–4):275–294
Kohler A, Skaga A et al (2002) Sorting salted cod fillets by computer vision: a pilot study. Comput Electr Agric 36(1):3–16

Kohonen T (1990) The self organizing map. Proceedings of the IEEE 78(9):1464–1480
Kohonen T (1998) The self-organizing map. Neurocomputing 21(1–3):1–6
Kourti T (2006) Process analytical technology beyond real-time analyzers: the role of multivariate analysis. Crit Rev Anal Chem 36:257–278
Lawton WH, Sylvestre EA (1971) Self modeling curve resolution. Technometrics 13:617–633
Leardi R (2009) Experimental design in chemistry: a tutorial. Anal Chim Acta 652(1–2):161–172
Leardi R, González AL (1998) Genetic algorithms applied to feature selection in PLS regression: how and when to use them. Chemometr Intell Lab Syst 41(2):195–207
Liang Y-Z, Kvalheim OM (1996) Robust methods for multivariate analysis—a tutorial review. Chemometr Intell Lab Syst 32(1):1–10
Lima FSG, Borges LEP (2002) Evaluation of standardisation methods of near infrared calibration models. J Near Infrared Spectrosc 10(4):269–278
Lundstedt T, Seifert E et al (1998) Experimental design and optimization. Chemometr Intell Lab Syst 42(1–2):3–40
MacGregor J, Yu H et al (2005) Data-based latent variable methods for process analysis, monitoring and control. Comput Chem Eng 29(6):1217–1223
MacQueen J (1967) Some methods for classification and analysis of multivariate observations. Fifth Berkeley symposium on mathematical statistics and probability
Marbach R (2005) A new method for multiariate calibration. J Near Infrared Spectrosc 13:241–254
Marbach R (2007a) Multivariate calibration: a science-based method—part 1. Pharmaceutical Manufacturing 6(1):42–47
Marbach R (2007b) Multivariate calibration: a science-based method—part 2. Pharmaceutical Manufacturing 6(2):44–47
Martens H, Stark E (2001) Extended multiplicative signal correction and spectral interference subtraction: new preprocessing methods for near infrared spectroscopy. J Pharm Biomed Anal 9(8):625–635
Martens H, Anderssen E et al (2005) Regression of a data matrix on descriptors of both its rows and of its columns via latent variables: L-PLSR. Comput Stat Data Anal 48(1):103–123
Mazerolles G, Hanafi M et al (2006) Common components and specific weights analysis: a chemometric method for dealing with complexity of food products. Chemometr Intell Lab Syst 81(1):41–49
McClure WF, Norris KH et al (1977) Rapid spectrophotometric analysis of the chemical composition of tobacco. Part 1. Total reducing sugars. Beitr Tabakforsch 9(1):13–18
Montgomery DC (1997) Multivariate quality control. Introduction to statistical quality control. John Wiley & Sons Inc, New York, pp 360–373
Navea S, Tauler R et al (2006) Monitoring and modelling of protein processes using mass spectrometry, circular dichroism and multivariate curve resolution methods. Anal Chem 78:4768–4778
Nomikos P, MacGregor JF (1995a) Multi-way partial least squares in monitoring batch processes. Chemometrics Intelligent Laboratory Systems 30(1):97–108
Nomikos P, MacGregor JF (1995b) Multivariate SPC charts for monitoring batch processes. Technometrics 37:41–59
Nørgaard L, Saudland A et al (2000) Interval partial least-squares regression (iPLS): a comparative chemometric study with an example from near-infrared spectroscopy. Appl Spectrosc 54(3):413–419
Norris KH, Williams PC (1984) Optimization of mathematical treatments of raw near-infrared signal in the measurement of protein in hard red spring wheat: I. Influence of particle size. Cereal Chem 61:158–165
Novales B, Guillaume S et al (1998) Particle size characterisation of in-flow milling products by video image analysis using global features. J Sci Food Agric 78(2):187–195
Olivieri A, Faber NM et al (2006) Guidelines for calibration in analytical chemistry. Part 3: uncertainty estimation and figures of merit for multivariate calibration. Pure Appl Chem 78(3):633–661
Ortiz MC, Sarabia L (2007) Quantitative determination in chromatographic analysis based on n-way calibration strategies. J Chromatogr A 1158(1–2):94–110

Plackett RL, Burman JP (1946) The design of optimal multifactorial experiments. Biometrika 33:305–325

Preys S, Roger JM et al (2008) Robust calibration using orthogonal projection and experimental design. Application to the correction of the light scattering effect on turbid NIR spectra. Chemometr Intell Lab Syst 91:28–33

Qannari EM, Wakeling I et al (2000) Defining the underlying sensory dimensions. Food Qual Prefer 11(1–2):151–154

Qin SJ, Valle S et al (2001) On unifying multiblock analysis with application to decentralized process monitoring. J Chemometr 15:715–742

Rechtschaffner RL (1967) Saturated fractions of 2n and 3n fractional designs. Technometrics 9:569–575

Rinnan A, Berg Fvd et al (2009) Review of the most common pre-processing techniques for near-infrared spectra. TrAC Trends Analytical Chem 28(10):1201–1222

Roger J-M, Chauchard F et al (2003) EPO-PLS external parameter orthogonalisation of PLS application to temperature-independent measurement of sugar content of intact fruits. Chemometr Intell Lab Syst 66(2):191–204

Rouillé J, Le Bail A et al (2000) Influence of formulation and mixing conditions on breadmaking qualities of French frozen dough. J Food Eng 43(4):197–203

Rousseeuw PJ (1984) Least median of squares regression. J Am Stat Assoc 79:871–880

Roussel SA, Hardy CL et al (2001) Detection of Roundup Ready™ soybeans by near-infrared spectroscopy. Appl Spectrosc 55(10):1425–1430

Roussel S, Bellon-Maurel V et al (2003) Authenticating white grape must variety with classification models based on aroma sensors, FT-IR and UV spectrometry. J Food Eng 60(4):407–419

Rumelhart DE, Hinton GE et al (1986) Learning internal representations by error propagation. Parallel Distributed Processing, Cambridge

Savitzky A, Golay MJE (1964) Smoothing and differentiation of data by simplified least squares procedures. Anal Chem 36(8):1627–1639

Serrano-Megías M, López-Nicolás JM (2006) Application of agglomerative hierarchical clustering to identify consumer tomato preferences: influence of physicochemical and sensory characteristics on consumer response. J Sci Food Agric 86(4):493–499

Shenk JS, Westerhaus MO et al (1998) Investigation of a LOCAL calibration procedure for near infrared instruments. J Near Infrared Spectrosc 5(4):223–232

Sirieix A, Downey G (1993) Commercial wheatflour authentication by discriminant analysis of near infrared reflectance spectra. J Near Infrared Spectrosc 1:187–197

Snee RD (1977) Validation of regression models: methods and examples. Technometrics 19(4):415–428

Svensson O, Kourti T et al (2002) An investigation of orthogonal signal correction algorithms and their characteristics. J Chemometr 16:176–188

Tauler R (1995) Multivariate curve resolution applied to second order data. Chemometr Intell Lab Syst 30(1):133–146

Thimm G, Fiesler E (1997) Pruning of neural networks. I.-R. R. 97-03. Valais, Switzerland, Dalle Molle Institute for perceptive artificial intelligence

Tillmann P, Paul C (1998) The repeatability file-a tool for reducing the sensitivity of near infrared spectroscopy calibrations to moisture variation. J Near Infrared Spectrosc 6(1):61–68

Tokatli F, Cinar A et al (2005) HACCP with multivariate process monitoring and fault diagnosis techniques: application to a food pasteurization process. Food Control 16(5):411–422

Tracy ND, Young JC et al (1992) Multivariate control charts for individual observations. J Qual Technol 24:88–95

Trygg J, Wold S (2002) Orthogonal projections to latent structures (O-PLS). J Chemometr 16:119–128

Tucker LR (1966) Some mathematical notes on three-mode factor analysis. Psychometrika 31:279–311

van Sprang ENM, Ramaker H-J et al (2002) Critical evaluation of approaches for on-line batch process monitoring. Chem Eng Sci 57(18):3979–3991

Wang Y, Veltkamp DJ et al (1991) Multivariate instrument standardization. Anal Chem 63(23):2750–2756
Wangen LE, Kowalski BR (1988) A multiblock partial least squares algorithm for investigating complex chemical systems. J Chemometr 3:3–20
Weigend AS, Huberman BA et al (1990) Predicting the future: a connectionist approach. Int J Neural Syst 1(3):193–209
Westad F, Schmidt A et al (2008) Incorporating chemical band-assignment in near infrared spectroscopy regression models. J Near Infrared Spectrosc 16:265–273
Westerhaus MO (1991) Improving repeatability of calibrations across instruments. 3rd International conference on near infrared spectroscopy, Gembloux, Belgium
Westerhuis JA, Kourti T et al (1998) Analysis of multiblock and hierarchical PCA and PLS models. J Chemometr 12:301–321
Westerhuis JA, De Jong S et al (2001) Direct orthogonal signal correction. Chemometr Intell Lab Syst 56:13–25
Wold H (1982) Soft modelling: the basic design and some extensions. System under indirect observation, vol 2. (H Wold, KG Jöreskog (eds)). Amsterdam, North Holland, pp 1–54
Wold S (1992) Nonlinear partial least squares modelling II. Spline inner relation. Chemometr Intell Lab Syst 14(1–3):71–84
Wold S, Sjostrom M (1977) SIMCA: a method for analyzing chemical data in terms of similarity and analogy—(Chapter Book). Chemometr Theory Appl 52:243–282
Wold S, Martens H et al (1983) The multivariate calibration problem in chemistry solved by the PLS method. Matrix Pencils. Springer, Heidenberg
Wold S, Martens H, & Wold H (1984) In S. Wold (Ed.), Muldast Proceedings, Technical Report, Research Group for Chemometrics, Umeå University, Sweden
Wold S, Kettaneh-Wold N et al (1989) Nonlinear PLS modeling. Chemometr Intell Lab Syst 7(1–2):53–65
Wold S, Kettaneh N et al (1996) Hierarchical multiblock PLS and PC models for easier model interpretation and as an alternative to variable selection. J Chemometr 10(5-6):463–482
Wold S, Antti H et al (1998a) Orthogonal signal correction of near-infrared spectra. Chemometr Intell Lab Syst 44(1–2):175–185
Wold S, Kettaneh N et al (1998b) Modelling and diagnostics of batch processes and analogous kinetic experiments. Chemometr Intell Lab Syst 44(1–2):331–340
Wold S, Cheney J et al (2006) The chernometric analysis of point and dynamic data in pharmaceutical and biotech production (PAT)—some objectives and approaches. Chemometr Intell Lab Syst 84(1–2):159–163
Wu W, Walczak B et al (1996) Feature reduction by Fourier transform in pattern recognition of NIR data. Anal Chim Acta 331:75–83
Wülfert F, Kok WT et al (2000a) Correction of Temperature-Induced Spectral Variation by Continuous Piecewise Direct Standardization. Anal Chem 72:1639–1644
Wülfert F, Kok WT et al (2000b) Linear techniques to correct for temperature-induced spectral variation in multivariate calibration. Chemometr Intell Lab Syst 51:189–200
Zeaiter M, Roger JM et al (2005) Robustness of models developed by multivariate calibration. Part II: The influence of pre-processing methods. TrAC Trends Anal Chem 24(5):437–445
Zeaiter M, Roger JM et al (2006) Dynamic orthogonal projection. A new method to maintain the on-line robustness of multivariate calibrations. Application to NIR-based monitoring of wine fermentations. Chemom Intell Lab Syst 80(2):227–235
Zhu Y, Fearn T et al (2008) Error removal by orthogonal subtraction (EROS): a customised pretreatment for spectroscopic data. J Chemometr 22:130–134

Chapter 3
Data Management Systems

Jarka Glassey

List of Abbreviations

BSE	Bovine spongiform encephalopathy
CMM	Corrective maintenance management
CRM	Customer relations management
DCS	Distributed control systems
ERP	Enterprise resource planning
FDA	Food and drug administration
HACCP	Hazard analysis critical control point
HMI	Human machine interfaces
IEC	International electrotechnical society
ISA	Instrumentation, systems and automation society
LIMS	Laboratory information management systems
MES	Manufacturing execution system
OMAC	Open modular architecture controls
OPC	Object linking and embedding for process control
PAT	Process analytical technologies
PLC	Programmable logic controllers
PLM	Product lifecycle management
QbD	Quality by design
QM	Quality management
SCADA	Supervisory control and data acquisition
SCM	Supply chain management
WCS	Warehouse control systems
WSN	Wireless sensor networks

J. Glassey (✉)
School of Chemical Engineering and Advanced Materials, Newcastle University, Newcastle upon Tyne NE1 7RU, UK
e-mail: jarka.glassey@ncl.ac.uk

C. P. O'Donnell et al. (eds.), *Process Analytical Technology for the Food Industry*, Food Engineering Series, DOI 10.1007/978-1-4939-0311-5_3,

3.1 Introduction

Monitoring process performance is essential in any manufacturing process as producing quality product within specification reproducibly is a prerequisite of an economically viable and safe process. Effective monitoring and control strategies lead to large amounts of diverse data regardless of the type of a manufacturing process. Monitoring is required in all stages of processing—the quality of raw materials is usually tested on intake, process equipment has to be rigorously qualified, environment is controlled by implementing manufacturing area classification where relevant, waste is treated prior to release and the quality of the final product is tested before release. Initiatives, such as Quality by Design (QbD) and a supporting enabling technology of process analytical technology (PAT) championed by the US Food and Drug Administration (FDA), aim to shift the focus for manufacturing from end-product-quality testing to building the quality in the process. Such a shift in emphasis would not be possible without reliable and effective monitoring. Indeed, PAT has been defined as 'a system for designing, analyzing, and controlling manufacturing through timely measurements (that is, during processing) of critical quality and performance attributes of raw and in-process materials and processes, with the goal of ensuring final product quality' (FDA 2004). Traditional process control strategies based upon information from laboratory assays and supervisory computer systems (supervisory control and data acquisition, SCADA) are routinely used to regulate process operation and correct disturbances resulting from raw material variations through to production plant variations. If PAT can provide additional information on disturbances and deviations, giving greater plant insight, then the effects of disturbances can be reduced and quality control tightened. However, greater benefits are to be gained by the systematic use of PAT tools in process development to increase fundamental understanding and more robust definition of the design and control space of the process operation as outlined in Chap. 1.

A particular challenge in attaining these benefits is the handling of highly heterogeneous processes and product quality data. Such data are characterised by varying frequency of various measurements typically with significant delays in the laboratory measurements. They are also often highly correlated, non-linear in nature and with high levels of redundancy and noise. Whilst software products and data management systems described briefly in Sect. 3.2.1 can aid in data pipelining and preprocessing necessary for appropriate analysis, significant challenges in this area still remain to be addressed.

The Hazard Analysis Critical Control Point (HACCP) food safety standard and the ISO 22000:2005 food safety management systems—requirements for any organization in the food chain standard—are placing additional requirements upon the food industry in terms of process monitoring, control and data management that will be dealt with in Sect. 3.3.

3.2 Information Management Systems

Rapidly developing computing, automation and measurement technologies resulted in increasingly large amounts of data being collected and stored by companies. It soon became apparent that competitive advantage can be gained by exploiting the wealth of this information for more effective decision making and as a result an area of *business intelligence* (BI) has emerged (Yeoh and Koronios 2010). The main tasks of a BI system include 'intelligent exploration, integration, aggregation and a multidimensional analysis of data originating from various information resources' (Olszak and Ziemba 2007). Accordingly, data should be treated as a highly valuable corporate resource which can be transformed from data mountain to quality information (Wang and Wang 2008). BI applications include the activities of data warehousing, decision support systems, query and reporting, online analytical processing (OLAP), statistical analysis, forecasting and data mining. Technical aspects of data warehousing, OLAP and data mining are detailed in a range of computing science textbooks (e.g. Connolly and Begg 2010), although some data analysis methods used in data mining are also described in Chap. 2.

The number of process monitoring, data acquisition and processing systems available on the market, either targeted to particular business requirements or to general manufacturing industries, is increasing rapidly. The concerns over the communication between systems from different vendors at different levels even within the same company lead to the introduction of International Society of Automation (ISA) 95 standard. This standard defines activities in the manufacturing environment as a multilevel model with relevant standards at each level or level interface. This model is often utilized to associate various software applications and networks with individual levels as indicated in Fig. 3.1.

In this multilevel model, level 0 refers to the actual production process with direct sensing; control of the process is defined as level 1; and monitoring, supervisory and automated control as level 2. Batch, continuous and discrete control is classed as overlapping levels 1 and 2. Manufacturing operations management, including workflow or recipe control, record maintenance and optimization of the production process is defined as level 3 and business planning and logistics covering plant schedule, resource use, production, delivery and shipping is defined as level 4.

As Fig. 3.1 indicates, there is an overlap between levels in terms of applications with, for example, distributed control system (DCS) capabilities often supporting functions within level 3 or enterprise resource planning (ERP) systems (level 4) integrating with certain functions of the Laboratory Information Management System (LIMS), as illustrated in Sect. 3.2.1.

3.2.1 Information Management Systems

Food and beverage industry is becoming more and more tightly regulated as the public is increasingly demanding rigorous safety standards, at the same time as

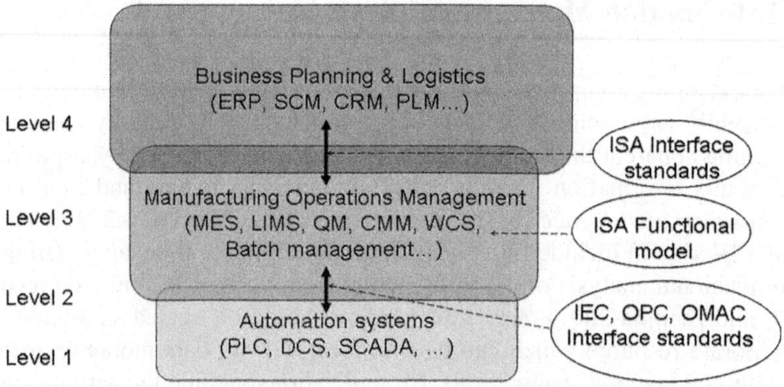

Fig. 3.1 ISA 95 applications and interfaces at individual levels

becoming increasingly conscious of 'processed' foods. Regulatory agencies, such as FDA, are now placing requirements on the food products to be tracked throughout the manufacturing process and distribution thus driving the development of LIMSs and electronic laboratory notebooks (ELNs). Some of the basic functions that a LIMS should possess are outlined in Table 3.1. These functions enable seamless, secure and meaningful flow of analytical data to plant systems and decision makers.

Whilst laboratories traditionally operated relatively autonomously within organizations, it is clear that the increasing sophistication of manufacturing processes and increasing regulation demand that the number of laboratory analyses required increases dramatically. To achieve operational excellence under such conditions, the laboratory data must be managed effectively. This will become even more critical with the introduction of PAT methodologies to improve process monitoring. Techniques such as infrared spectroscopy (described in detail in Chap. 4), Raman spectroscopy (Chap. 5), magnetic resonance imaging and nuclear magnetic resonance (Chap. 6), computer imaging (Chap. 7), thermal (Chap. 8) and hyperspectral imaging (Chap. 9), and ultrasound (Chap. 10) as well the emerging technologies described in Chap. 11 tend to produce large sets of multidimensional data, which have to be effectively integrated with other on-line, off-line, continuous or discrete data monitored during the process. Figure 3.2 illustrates some of the issues with data dimensionality that are encountered even currently with raw material quality data relating to a range of processing batches, from which continuous operating data (e.g. temperature) are collected together with laboratory-based, discrete measurements of various quality attributes of the product. Packaging and distribution data must also be linked to this array of process-related data to ensure traceability (see Sect. 3.4).

Traditional PAT methodologies tend to involve fewer independent samples (e.g. batches of raw materials or process batches) than measured variables and often require multivariate data analysis and preprocessing to present sensible data to the end user at any level in the company. Whilst in some areas of food manufacturing the volume of continuous on-line data collected during the process may be limited

Table 3.1 Laboratory information management system (LIMS) functions. (Reprinted from Cagindi and Otles (2004) with permission from Elsevier)

Function	Capability
Instrument management	Centralized storage of maintenance and calibration records
Data management	Information about personnel, instruments, analytical methods, work procedures and costs mapped onto the database, organised
Validation	Ensuring the integrity of data
Sample management	Techniques for registration, processing, authorisation and archiving of routine and non-routine samples, standards and reference materials and commonly used test sequences
Resource management	Instrument backlog reporting, as well as personnel time management, costs associated with analyses may be calculated and invoiced to client accounts
Communication management	Ensuring that important information reaches decision makers with minimal delay
Quality management	Achieved through audit trail and validation facilities; Quality Control is enhanced through specification libraries and action triggers with graphical data interpretation
Security	Setting up passwords, authority levels and menus for each user, etc. to create a secure system

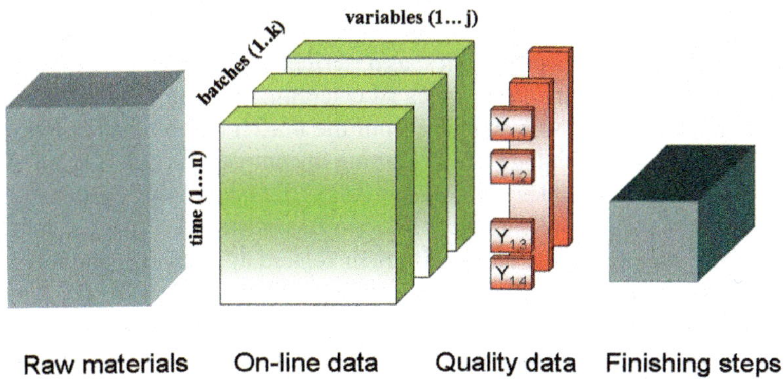

Fig. 3.2 Diagrammatic representation of data dimensionality in process monitoring

relative to the discrete off-line laboratory measurements, data analysis methods applied within PAT may be equally applicable. However, the implementation of PAT concept and real-time release of product, resulting from real-time product quality measurements during production, will require the development of novel data analysis and modelling methodologies. These approaches will provide deeper understanding of the relationships between raw material characteristics, process parameters and the final product quality (Stenlund et al 2009). In terms of data management systems, such as LIMS, this means that any software environment with which PAT methods would be integrated will have to be capable of streaming required data and/or preprocessing it as required and then presenting the resulting model estimations/predictions in appropriate format (e.g. score plots in the case of principal component analysis; Mohan et al. 2006). The model structures would also have to be stored for future use and validation purposes.

A wide range of LIMS products are already available and used by food industries. Sansom (2008) and Blackman (2009) describe applications of LIMS systems from a range of providers by various food companies. For example, Labware's LIMS product is being used by Kraft Foods since 1996, brewer Foster's as well as by British Sugar, which uses it to carry out a wide range of roles from product quality analysis through to complex analytical testing, such as heavy metal analysis and process support investigations. Thermo Fisher's food and beverage clients include Coca-Cola, the South African Sugar Association, ALcontrol Laboratories (a large European analytical company) or dairy product producer Muller using Nautilus LIMS in its UK laboratories to manage quality control data for raw materials, in-process samples and finished dairy desserts. PerkinElmer's Labworks are used by, for example, Ajinomoto (a Japanese-owned multination company producing food products, amino acids and pharmaceuticals) and Australian Wine Research Institute.

Siemens combines LIMS and ELN functionality in its Simatic IT R&D suite, which is marketed as a complete, integrated solution that can control the complete R&D process from design through commercialisation to manufacturing (Sansom 2008). Siemen's food and beverage customers include the brewers Inbev or Birra Peroni, the Chinese dairy company Mengniu or Loders Croklaan, producer of fats and oils.

There are already examples of LIMS systems integrating with various advanced analytical methods, such as proteomics, as described in Ganjei et al., 2003, detailing the application of Sapphire Proteomics Accelerator. Proteomics samples are frequently annotated, split and fractioned from the master sample and subjected to a series of extractions from image files, resulting in complex multidimensional arrays of data from singular samples—not dissimilar to the data arrays produced by PAT methodologies. In this application, 'LIMS unifies vast and disparate volumes of biological and chemical data, along with their related applications and tools, into a single, browser-based, scientific interface' (Ganjei et al. 2003).

A natural extension for systems like LIMS is towards ERP systems. LIMSs typically operate at level 3 defined by ISA 95 standard, providing traceability in the micro supply chain processes and managing the data relating to individual analytical test results carried out on raw materials, finished products and environmental samples; the macro supply chain management processes are typically handled by level 4 systems, such as ERP. Such successful integration has already been reported by Chin (2003) in case of StarLIMS and SAP as implemented by Novozymes—a leading enzyme manufacturer—and BASF Agricultural Products.

3.3 HACCP, Food Safety and ISO 22000 Issues

Ropkins and Beck (2000) argue that traditional end-point food testing could not effectively ensure food safety due to a variety of reasons. These include substantial subsampling of food for analysis to ensure representative sample; limited assurance of safety (only tested hazards can be assured); issues with current testing procedures, e.g. time and resource demand, destructive nature and the difficulty of interpretation; reactive nature of control; limiting the number of employees responsible for quality assurance and control personnel and the obvious fact of product safety being assured only at end point. Thus, HACCP system was proposed by the Codex Alimentarius Commission in 1993 as a systematic approach to the identification, assessment and control of hazards. The seven basic principles of HACCP implementation consist of:

1. Conduct hazard analysis, identifying all potential hazards—microbiological, chemical and physical—in terms of food safety considering all ingredients, processing steps, handling procedures and other activities involved in foodstuff production.
2. Identify critical control points (CCPs). These are procedures or operational steps that can be controlled to minimize the likelihood, eliminate the identified food safety hazards or reduce them to an acceptable level.
3. Define critical limits for ensuring the control of each CCP. These are the maximum or minimum values to which individual hazards must be controlled at a CCP to prevent, eliminate or reduce it to an acceptable level.
4. Establish monitoring procedures and their frequencies to ensure that critical limits are not exceeded at any of the CCPs and define procedure(s) for maintaining control.
5. Define corrective actions to be taken if control is lost (i.e. monitoring indicates that critical limits have been exceeded). Corrective actions are intended to ensure that the product does not become unsafe for human consumption given the observed deviations.
6. Establish effective documentation and record-keeping procedures for developed HACCP procedure. All organisations have to maintain documentation on their hazard analysis, written HACCP plan, records documenting the monitoring of CCPs, critical limits, verification activities and the handling of processing deviations.
7. Establish verification procedures for routinely assessing the effectiveness of the HACCP procedure, once implemented. This is intended to ensure that the organisations carry out the procedures that were designed and that they are successful in ensuring the production of safe products. The validation procedures include test and programmes to ensure the HACCP system is working effectively.

The application of these principles is detailed in the Codex Alimentarius Commission (1997) guidelines as a sequence of 12 implementation steps which are illustrated in Fig. 3.3.

In 2005, a new international standard, ISO 22000:2005, was introduced to ensure safe food supply chains worldwide and to extend the recommendations of HACCP system (Frost 2005). This standard places more emphasis on interactive communication with suppliers and customers along the food chain and introduces prerequisite programmes (PRPs) as basic conditions and activities necessary to maintain a hygienic environment throughout the food chain suitable for the production, handling and provision of safe end products and safe food for human consumption.

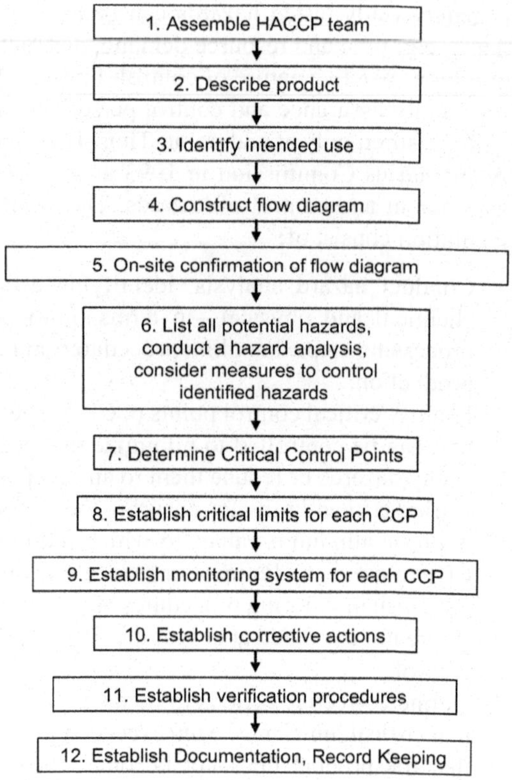

Fig. 3.3 Schematic of HACCP implementation steps

Clearly, effective monitoring is critical to ensuring product quality from the point of view of food safety. Essential components of effective monitoring include representative measurement and a robust representation of the obtained information, allowing appropriate action to be taken. This provides another critical area of PAT methodology application in food and beverage industry. Techniques described in subsequent chapters are capable of monitoring process behaviour and product characteristics throughout the manufacturing chain and capable of indicating non-compliance with CCPs or PRPs.

In analogy to the application of PAT framework within the QbD strategy in the biopharmaceutical industry, PAT methodology provides a critical capability within the food industry to aid in product design and testing as well as in ensuring full compliance with the HACCP and ISO 22000:2005 requirements during processing. PAT methodologies have the potential to aid the identification of CCPs and their critical limits, their effective monitoring and control, but also effective communication with suppliers and customers. Multivariate data analysis techniques described in Chap. 2 have long been used successfully by chemometricians for interpretation of multidimensional data sets in various subject areas including food industry, for example Perez-Martinez et al. (2008). Their ability to reduce dimensionality by removing the redundancy and noise leads to the identification of salient features in the data. These features can be used in process monitoring, fault detection and process optimization and also information exchange with stakeholders in the supply chain.

3.4 Traceability

Despite the introduction of food safety legislation, incidents of food-borne diseases and contaminations, such as bovine spongiform encephalopathy (BSE) and dioxin poisoning in chicken feed, are widely publicised and contribute to the public's wariness of industrially produced food. Equally ethical considerations and fears regarding the presence of genetically modified components of food products lead to increasing demands on complete traceability of food products. In addition, in cases where a non-compliant batch is inadvertently released for distribution, traceability facilitates fast and efficient recall of the product.

European Commission (regulation No 178/2002) defines traceability as the *ability to trace and follow a food, feed, food-producing animal or substance intended to be, or expected to be incorporated into a food or feed, through all stages of production, processing and distribution.* This means that traceability has to be implemented both vertically, reliably tracing and documenting the origin of all components of food (referred to in agricultural supply system as 'from farm to fork' or in fisheries as 'from fish to dish'), and horizontally, carrying out the same stringent analysis and documentation steps throughout the processing steps.

There exists a range of legislation requiring various levels of stringency in traceability. For example, European regulations relating to general food and food-producing animals (European regulation No 178/2002, article 18) or to specific industries (fishery products regulation 104/2000, article 4), US Bioterrorism and Response Act 2002 or Farm Security and Rural Investment Act 2002. Some of the legislation and standards (EU regulation 178/2002 and ISO 22005:2007) require a so-called one step up/one step down approach to information flow, although an aggregated information flow is also championed for particular foodstuffs (Folinas et al. 2006). In the one step up/one step down flow, only some traceability information follows the product to the next stage of the chain, whilst the rest is kept at each stage. The information is not accessible directly to the final consumer, but can be recovered, should this be required. Such 'filtering' of information makes the traceability system potentially much more flexible and easy to use. On the other hand, for organic products, fresh fish and meat or foods free of genetic modification, the aggregate information flow provides the customer with immediate access to information about all stages of production and treatment.

Traceability, in particular for the latter model of information flow, places much more stringent requirements on the data management systems employed by the food industry. Thompson et al. (2005) argue that comprehensive planning during the initial stages of development is critical if the traceability system is to be successful. They identify three critical issues: (1) compatibility, (2) data standardisation and (3) the definition of a traceable resource unit (TRU; Moe 1998), i.e. a unit of trade that has unique characteristics from a traceability point of view that no other unit can have.

The issues of compatibility and data standardisation place certain requirements upon data management systems employed by the companies within the supply chain. There are reports of advanced scanning technology, such as radio frequency identification (RFID) tags (Regattieri et al. 2007) for 'Parmigiano Reggiano' cheese traceability system. They argue that RFID can be particularly helpful in the management of perishable items with continuous monitoring of item routing reducing waste and improving customer satisfaction amongst other benefits. Another example is the use of wireless sensor networks (WSN) for monitoring fruit storage and transport conditions (Ruiz-Garcia et al 2008) or the use of georeferenced data from geographical information systems (GIS) in improving traceability of high-quality honey using an open-source code-based GIS web site (Serrano et al. 2008).

Integration of current traceability activities with PAT technologies has already been reported in a number of applications. For example, Bollen et al. (2007) describe the use of optical image analysis (see Chap. 7 for more details on this method) in tracing in feed and packing lane mixing in fruit packing. However, the success of any application of a traceability system, whether integrated with PAT technologies or not, relies heavily on effective communication and data exchange between individual components of the system.

The issues of data compatibility, security and confidentiality have to be carefully considered in the development stage of a traceability system. Incorporating traceability into the ERP system within individual companies addresses these issues at the level of individual entity in the supply chain. However, the diversity of ERP systems employed by different companies within a supply chain complicates integration across various entities. Electronic data interchange (EDI; Bechini et al. 2008) enables data exchange in standardised format, as do programmes capable of Open Database Connectivity (ODBC) or standard Structure Query Language (SQL) enabling information access from various database formats (Thompson et al. 2005). Web services and electronic business using eXtensible Markup Language (ebXML) are presented by Bechini et al. (2008) as the most promising technologies for effective inter-enterprise business collaboration that would be required for successful traceability in food supply chain.

References

Bechini A, Cimino MGCA, Marcelloni F, Tomasi A (2008) Patterns and technologies for enabling supply chain traceability through collaborative e-business. Inf Software Tech 50:342–359

Blackman G (2009) A taste for LIMS, Scientific Computing World, August/September

Bollen AF, Riden CP, Cox NR (2007) Agricultural supply system traceability, part I: role of packing procedures and effects of fruit mixing. Biosyst Eng 98:391–400

Cagindi O, Otles S (2004) Importance of laboratory information management systems (LIMS) software for food processing factories. J Food Eng 65(4):565–568

Chin W (2003) Leverage LIMS intelligence to improve OpX, Hydrocarbon Processing, July

Codex Alimentarius Commission (1997) Hazard Analysis Critical Control Points (HACCP) system guide-lines for its application. Roma, 1997. http://www.fao.org/docrep/005/y1579e/y1579e03.html Accessed 2 Oct 2011

Connolly TM, Begg CE (2010) Database systems: a practical approach to design, implementation, and management. Addison-Wesley, Reading. ISBN 13:978-0-321-52306-8

European regulation No 178/2002 (2002) European Parliament 28/01/2002 for food safety issues, Official Journal of the European Communities, 1.2.2002 L 31/1, http://eur-lex.europa.eu/LexUriServ/LexUriServ.do?uri=OJ:L:2002:031:0001:0024:EN:PDF. Accessed 21 Oct 2011

Folinas D, Manikas I, Manos B (2006) Traceability data management for food chains. Br Food J 108(8):622–633

Frost R (2005) ISO 22000 is first in family of food safety management system standards, ISO Management Systems, Nov/Dec., 16–19.

Ganjei JK, Smallmon T, Lee C (2003) Food firm benefits from information system. Scientific Computing World, Nov/Dec

Le Dréau L, Dupuy N, Artaud J, Ollivier D, Kister J (2009) Infrared study of aging of edible oils by oxidative spectroscopic index and MCR-ALS chemometric method. Talanta 77:1748–1756

Moe T (1998) Perspective on traceability in food manufacture. Trends Food Sci Technol 9:211–214

Mohan P, Glassey J, Montague GA (2006) Pharmaceutical operations management. McGraw-Hill, New York. ISBN 0071472495/97:80071472494

Olszak C, Ziemba E (2007) Approach to Building and Implementing Business Intelligence Systems. Interdiscip J Inf Knowl Manag 2:135–148

Perez-Martinez M, Sopelana P, Paz dePM, Cid C (2008) Application of multivariate analysis to the effects of additives on chemical and sensory quality of stored coffee brew. J Agric Food Chem 56:11845–11853

Regattieri A, Gamberi M, Manzini R (2007) Traceability of food products: general framework and experimental evidence. J Food Eng 81:347–356

Ropkins K, Beck AJ (2000) Evaluation of worldwide approaches to the use of HACCP to control food safety. Trends Food Sci Technol 11:10–21

Ruiz-Garcia L, Barreiro P, Robla JL (2008) Performance of ZigBee-based wireless sensor nodes for real-time monitoring of fruit logistics. J Food Eng 87:405–415

Sansom C (2008) Quality control is food and drink for LIMS, Scientific Computing World, June/July

Serrano S, Jimenez-Hornero FJ, Gutierrez de Rave E, Jodral ML (2008) GIS desing application for "Sierra Morena Honey" designation of origin. Comput Electron Agric 64:307–317

Stenlund H, Johansson E, Gottfries J, Trygg J (2009) Unlocking Interpretation in near infrared multivariate calibrations by orthogonal partial least squares. Anal Chem 81:203–209

Thompson M, Sylvia G, Morrissey MT (2005) Seafood traceability in the United States: current trends, system design, and potential applications. Comp Rev Food Sci Food Saf 1:1–7

US Food and Drug Administration Guidance for Industry. PAT–A Framework for Innovative Pharmaceutical Development, Manufacturing, and Quality Assurance (FDA, Rockville, MD, Sep 2004). http://www.fda.gov/downloads/Drugs/GuidanceComplianceRegulatoryInformation/Guidances/UCM070305.pdf. Accessed 21 Oct 2011

Wang H, Wang S (2008) A knowledge management approach to data mining process for business intelligence. Ind Manage Data Syst 108(5):622–634

Yeoh W, Koronios A (2010) Critical success factors for business intelligence systems. J Comput Inf Syst 50:23–32

Chapter 4
Infrared Spectroscopy

Colette C. Fagan

4.1 Introduction

Infrared spectra of food products can help to reveal information pertaining to molecular bonds present and hence provide details of their molecular structures. This ultimately can be related to various quality indices. Infrared spectroscopy is an ideal process analytical technology (PAT) tool that can rapidly, accurately and usually non-destructively assess the quality and functional properties of raw, in-process and final product materials. In addition to the need for efficiency, there is an emerging need in food processing for all major compositional and quality parameters to be determined, on-line and in real time. In addition to this, there is a need for food manufacturers to be able to demonstrate the authenticity of their products (Woodcock 2008).

Spectroscopic techniques, other than infrared spectroscopy, have been investigated as potential PAT technologies in the food industry. These include Raman spectroscopy (Chap. 5), fluorescence spectroscopy (Chap. 12) and UV–Vis spectroscopy. UV–Vis has been employed to detect adulterated and authentic spirits (Contreras et al. 2010),discriminate between brands (Barbosa-García et al. 2007), classify coffee (Souto et al. 2010) and quantify β-carotene (Biswas et al 2011). However, the focus of this chapter is infrared spectroscopy, and it will provide an overview of its theory, its instrumentation and its applicability as a PAT tool. Finally, it will review applications of infrared spectroscopy to food products.

C. C. Fagan (✉)
Food and Nutritional Science, Department of Food and Nutritional Sciences,
University of Reading, Reading RG6 6AP, P.O. Box 226, Reading, UK
e-mail: c.c.fagan@reading.ac.uk

C. P. O'Donnell et al. (eds.), *Process Analytical Technology for the Food Industry*,
Food Engineering Series, DOI 10.1007/978-1-4939-0311-5_4,

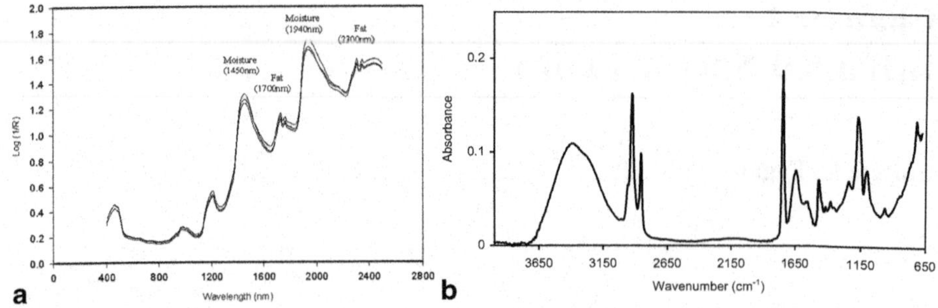

Fig. 4.1 Characteristic **a** NIR spectra and **b** MIR spectra of cheese

4.2 Theory of Near- and Mid-infrared Spectroscopy

Infrared spectroscopy results from the interaction of infrared radiation and matter. The energy provided by the infrared radiations results in transitions between quantized vibrational energy states of molecules, i.e. resulting in molecular vibration. Atoms in a molecule can have a number of vibrational modes. Each mode (i) involves approximately harmonic displacements of the atoms from their equilibrium positions (Griffiths 2010). When atoms vibrate as a simple harmonic oscillator, i.e. according to Hooke's law (Eq. 4.1) where x is the displacement away from equilibrium, k is the proportionality (or force) constant and F is the force in newtons, the vibrational energy states (V_{iv}) can be described according to Eq. 4.2, where h is Planck's constant, v_i is the fundamental frequency of the particular mode and v_i is the vibrational quantum number of the ith mode (0, 1, 2, etc.):

$$F = kx \tag{4.1}$$

$$V_{iv} = hv_i\left(v_i + \frac{1}{2}\right) \tag{4.2}$$

While the energy difference between $v_i = 0$ and $v_i = 1$ of most vibrational modes corresponds to the energy of radiation in the mid-infrared (MIR) range, overtone bands which relate to the transition between $v_i = 0$ and states higher than $v_i = 1$ are located in the near-infrared (NIR) region. Combination bands in the NIR region occur when there is a simultaneous promotion of two modes (Griffiths 2010).

A number of studies have assigned various food constituents (lipids, amides, moisture, sugars) to specific bands in MIR and NIR spectra. A selection of these regions and their associated mode of vibration of some food constituents are given in Tables 4.1 and 4.2. The characteristic broad peaks, resulting from overtone and combination bands, observed in the NIR spectra of a food product are shown in Fig. 4.1a; a corresponding MIR spectrum is shown in Fig. 4.1b. Such infrared spectra (Fig. 4.1) can contain a wealth of information on the molecular make-up of a food product. However, the spectral response of a molecular group can be influenced

Table 4.1 Selected molecular group absorption frequencies in the MIR region

Peak wave number (cm^{-1})	Functional group	Mode of vibration	Constituent
Fingerprint region			
1036, 1088	C–O	Stretch	
1060	C–O	Stretch	Carbohydrates
900–1200	C–O, C–C, O–H	Stretch	Carbohydrates
1115–1170	C–O	Stretch	
1232	C–H	Bend	
1240	C–O	Stretch	
1371	C–H	Bend	
1274, 1372, 1445, 1486	O–C–H, C–C–H, C–O–H	Bend	
1400–1477	C–H	Bend	
Functional group region			
1535–1570	Amide II	Stretch	Protein
1620–1690	Amide I	Stretch	Protein
1640	O–H	Bend	Moisture
1600–1900			Organic acids
1700–1765	C=O	Stretch	Lipids
2869	CH_2	Symmetric stretch	Lipid
2926	CH_3	Anti-symmetric stretch	Lipid
3047–3703	O–H	Stretch	Moisture

Table 4.2 Selected chemical assignments of absorption frequencies in the NIR region

Wavelength (nm)	Functional group	Functional group assignment	Constituent
982	OH	Second overtone; stretch	Water
1458	OH	First overtone; stretch	Water
1940	OH	Combination; asymmetric and scissoring stretch	Water
1210	C–H	Second overtone; stretch	Lipids
1728	C–H	First overtone; stretch	Lipids
1762	C–H	First overtone; stretch	Lipids
2308	C–H, CH_2	Combination; stretch and deformation	Lipids
2348	C–H, $=CH_2$	Combination; stretch and deformation	Lipids
1000–1020	N–H, Amide I	Stretch	Proteins

by neighbouring molecular groups (Reh 2001). The complexity of food substances enhances these difficulties as the presence of various substances can result in peak shifts (Fagan and O'Donnell 2007). Therefore, powerful statistical techniques, for example, principal component analysis (PCA) and partial least squares (PLS) regression, can be used for data compression and model development (Chap. 2).

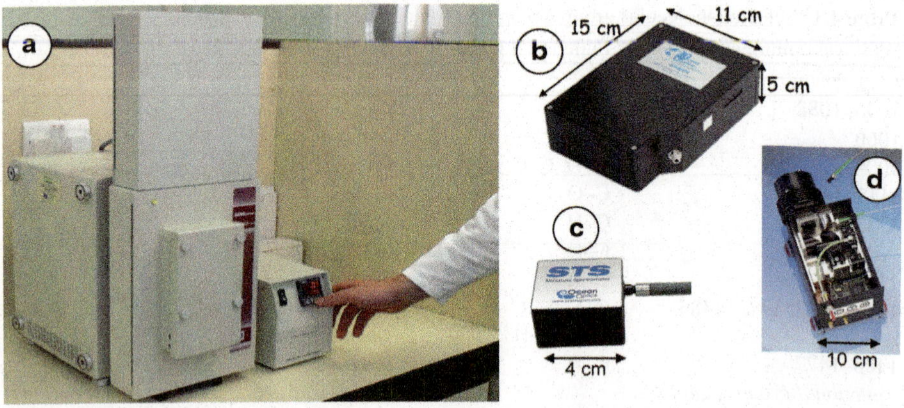

Fig. 4.2 A selection of **a** benchtop and **b–d** miniature, microspectrometer and portable spectrometers

4.3 Instrumentation

There have been significant developments in the field of infrared instrumentation over the past decades. Initially, equipment focused on the use of monochromator or filter (fixed or tunable)-based systems. However, developments in instrumentation, such as Fourier transform infrared (FTIR) spectrometers and polychromators with InGaAs detectors, substantially improved the instrumentation performance, the range of applications and therefore the popularity of such equipment. The principles of such MIR and NIR instrumentation have been reviewed previously and will not be discussed further (Fagan and O'Donnell 2007; Griffiths 2010). However, technical developments in infrared spectroscopy instrumentation will facilitate the transfer of this technology from laboratory to on-line application, thereby enhancing its potential as a PAT tool. Equipment manufacturers have moved from benchtop laboratory instruments (Fig. 4.2a) to the manufacturing of portable miniature-type spectrometers (Fig. 4.2b) to microspectrometers (Fig. 4.2c). These have been driven in part by the requirement of end users who want to have the facility to bring the "spectrometer to the samples" rather than the "samples to the laboratory". This has, for example, opened up opportunities for pre-harvest fruit and vegetable inspection. Such equipment may also include the added functionality of integrated global positioning system (GPS) measurements which are acquired simultaneously with the infrared spectra. Such facilities can allow for the "mapping" of produce quality in situ, thereby allowing the producer to make corrective decisions. In such applications, interference of the environment, such as ambient light and fluctuating temperatures, should be either minimized or accounted for by appropriate data processing (Nicolaï et al. 2007). Another related emerging platform technology is hyperspectral imaging. It has the advantage of acquiring both spectral and spatial information of sample simultaneously. It has shown considerable potential in the pharmaceutical industry in terms of mapping active ingredients in tablets. Its potential as a PAT tool in the food industry is discussed in detail in Chap. 9.

Another significant consideration of the end user must be the development and maintenance of calibration equations. Many portable spectrometers rely on the end user to provide and maintain the calibration equations required. However, this can require a substantial investment in time, labour and cost. Companies are emerging, however, which offer transferable NIR calibration solutions. Such companies have developed calibration models over many years using benchtop spectrometers, and they license them out for transfer to portable spectrometers. A service contract can also be entered into whereby the company maintains, updates and ensures set accuracy levels for the calibration model over time.

Continued research into the development of robust, fast miniature and micro-spectrometers will facilitate the continued adoption of this technology as a PAT tool in the food industry.

4.4 Infrared Spectroscopy as a PAT Technology

Infrared spectroscopy has been widely investigated as a rapid non-destructive assessment tool for food products. Fruit, vegetable, dairy and meat products have been the most widely investigated. However, the majority of these studies have been laboratory based. The greatest advantage in the use of infrared-based technology as PAT tools will be their implementation in the form of on-line/at-line process analysers, which take advantage of rapid analysis times and the minimal sample presentation required. However, it should be noted that the requirements for laboratory-based analysis will differ in comparison with on-line technology. Infrared spectroscopy also has the capacity to predict numerous indices of a material simultaneously. In order to realize the potential of such data-rich tools in the food industry, appropriate data analysis (Chap. 2) and data management strategies (Chap. 3) are required. Food quality, however, cannot be considered as a single, well-defined attribute. In fact, it encompasses a number of properties or characteristics, which are often referred to as quality indices, of the product under test (Abbott 1999). While infrared spectroscopy can offer a solution to this challenge, one must ensure that the basis for the prediction of quality is fully understood, as well as its inherent limitations.

4.5 Applications

4.5.1 Dairy

The dairy industry has seen significant advances towards automation of production processes. For example, the move to closed commercial cheese vats versus the traditional open cheese vat drove the desire for on-line milk coagulation monitoring systems.

4.5.1.1 Raw Material

Milk composition and quality can vary depending on a number of factors, including animal genetics, health, and (in some countries) season. Such variability could significantly impact the quality of the final product. For example, milk fat to protein ratio will significantly affect a number of processing steps (coagulation, syneresis) which ultimately affects cheese quality and quantity. Therefore, it is usual that processors would standardise milk fat and protein content prior to use. Therefore, the use of infrared spectroscopy to facilitate the production of high quality of milk has been investigated in applications ranging from monitoring rumen metabolism through to standardisation of milk in the milk processing plant (Fagan et al. 2009b).

Off-line rapid analysis of milk composition using the FTIR measuring principle has been successfully commercialized with products such as the MilkoScan™ FT 120 (Foss Analytical, Denmark). It utilizes FTIR technology to analyse up to 600 samples/h and can be used for routine analysis, such as fat, protein, lactose, total solids and solids-non-fat, density, freezing point depression, urea and casein analysis, in compliance with International Dairy Federation (IDF) and Association of Analytical Communities (AOAC) standards.

The further development of on-line determination of milk composition and quality would be advantageous as such knowledge is essential for the efficient management of dairy herds. Brandt et al. (2010), however, stated that while a number of sensors are available or in development which can be used for management support in improving mastitis detection, monitoring fertility and reproduction and adapting individual diets, there is still a requirement to adapt these sensors to the particular requirements of on-farm utilization such as robustness, calibration and maintenance, costs, operating cycle duration, and high sensitivity and specificity.

Tsenkova et al. (2001) examined the potential of predicting somatic cell count (SCC) of milk using NIR transflectance spectra obtained using a benchtop spectrophotometer. They stated that the results indicated that NIR spectroscopy would be a suitable screening tool in such an application as the differentiation between healthy and mastitic milk samples was possible. More recently, an NIR spectroscopic sensing system for on-line monitoring of milk quality during milking has been developed (Kawamura et al. 2007). The system was installed between a teatcup cluster and a milk bucket of a milking machine. The authors developed models for the prediction of fat, protein, lactose, SCC and milk urea nitrogen (MUN) during milking with sufficient precision and accuracy (R^2=0.82-0.95), although only four cows were monitored over time. Following this study, the sensing system was installed in an automatic milking system. The system recorded diffusion transmittance spectra (600-1050nm) with a 1-nm interval every 10 s during milking. Seventeen cows were used in this study. The models developed for fat, protein, lactose, SCC and MUN had R^2 values of 0.95, 0.83, 0.72, 0.68 and 0.53, respectively. The authors used the SCC calibration model to discriminate between healthy cow samples and other cow samples. The resulting classification gave a probability, for classifying correctly, of 82%. In both studies, the samples were divided into calibration (2/3) and validation (1/3) sets. Further validation of the models is therefore recommended in conjunction with testing on a wider range of animals.

MIR spectroscopy has also been explored for the offline determination of milk traits (Cecchinato et al. 2009; Dal Zotto et al. 2008; De Marchi et al. 2009). Milk coagulation properties (MCP) will vary depending on a number of factors, including heritable parameters (Cassandro et al. 2008). However, if this information is to be fully exploited, there would be a requirement for a rapid method of determining MCP in milk-recording systems. Dal Zotto et al. (2008) found that MIR spectroscopy could predict the rennet coagulation time (RCT) of milk samples albeit with an R^2 of 0.73, which suggested approximate quantitative predictions were possible. De Marchi et al. (2009) carried out a further examination of this approach. Using a dataset of over a thousand samples, RCT was predicted with an R of 0.79. In both studies, the range error ratio (RER) was similar: 9.2 and 10.6. Cecchinato et al. (2009) investigated the variation of MCP predictions obtained by MIR spectroscopy, as well as estimating the expected response from a breeding program focusing on the enhancement of MCP using MIR predictions as indicator traits. They found that estimated genetic correlations between measure and predictions of RCT were very high.

4.5.1.2 Process Monitoring

NIR technology has been successfully applied at laboratory and commercial scales for monitoring processes during cheese manufacture. In particular, the milk coagulation process during cheese production has received a great deal of attention, and cutting the coagulum either before or after the optimum point results in losses of curd and fat. An increase in cheese moisture also occurs if the gel is too firm when cut. Originally, the determination of the cutting time was established by the cheese maker. Although accurate this method is not feasible in closed commercial vats and, together with an increased desire for automation in the cheese industry, has led to the need for an on-line objective method for the monitoring of milk coagulation. Instruments have been developed based on several technologies to this end. Ideally, a sensor to monitor milk coagulation could be installed on-line to allow for automation of the production process, without causing damage to the forming curd, and NIR sensors meet these requirements. Early methods, which utilized the changes in the optical properties of the milk, were reflection photometry (Hardy and Fanni 1981) and absorbance (McMahon et al. 1984). Although the reflection photometry and absorbance methods were found to monitor coagulation, they found little usage. However, developments in fibre optics have overcome many of the problems associated with these techniques. Light in the NIR spectral region can be transmitted through a fibre optic bundle and diffuse reflectance or transmission monitor. As the gel is formed, reflectance will increase while transmission will decrease. Payne et al. (1993) developed a method based on changes in diffuse reflectance during milk coagulation. Reflectance was measured using a fibre optic probe, utilizing a photodiode light source at a wavelength of 940 nm. The time to the inflection point (t_{max}) was determined from the first derivative and was found to correlate well with Formograph cutting times. Linear prediction equations, which were considered to be of the form required for predicting cutting time, were also developed

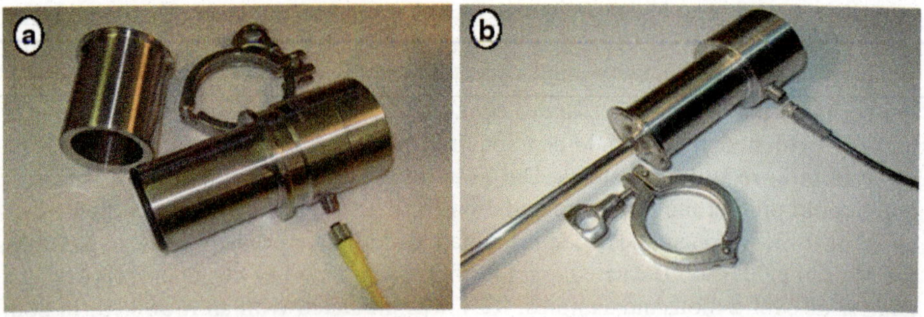

Fig. 4.3 **a** The CoAguLite sensor for predicting the optimal cutting time. **b** The FiberView Dairy Waste Sensor System (Reflectronics Inc, KY)

using t_{max}. This technology has been commercialized as the CoAguLite sensor (Reflectronics Inc, Lexington, KY) (Fig. 4.3a). This technology could also be used in conjunction with other sensors; for example, the FiberView dairy waste sensor system (Fig. 4.3b) could be used to monitor waste streams in dairy facilities. This enables the location, occurrence or concentration of the discharge to be determined. It monitors solids concentration in dairy plant effluents in the range of 0–1 % solids (or higher), and due to its quick response to loss events, it allows operators to take corrective actions.

Syneresis is a critical phase in cheese manufacture, with the rate and extent of syneresis playing a fundamental role in determining the moisture, mineral and lactose content of drained curd and hence that of the final cheese (Lawrence and Gilles 1980; Pearse and Mackinlay 1989). Therefore, research is ongoing into the development of a syneresis control technology. A number of potentially non-invasive technologies have been investigated for such an application, including ultrasound and computer vision (Everard et al. 2007; Fagan et al. 2008a; Taifi et al. 2006; Tellier et al. 1993) and NIR sensing (Castillo et al. 2005a; Fagan et al. 2009a; Fagan et al. 2007a). Initial studies focused on offline optical sensing of whey samples (Castillo et al. 2005b). An adaption of this technology led to the development of a sensor which could be installed in the wall of a cheese vat for on-line continuous monitoring of both coagulation and syneresis (Fagan et al. 2007a). The sensor operated at 980 nm and was sensitive to casein micelle aggregation and curd firming during coagulation and to changes in curd moisture and whey fat contents during syneresis. This sensor was also used to predict whey fat content (i.e. fat losses), curd yield and curd moisture content with standard error predictions (SEPs) of 2.37 g, 0.91 and 1.28 %, respectively (Fagan et al. 2008b). Further work used a wider spectral range (300–1100 nm) in conjunction with PLS regression to predict whey fat and curd moisture with root mean square error of cross-validation (RMSECV) values of 0.094 and 4.066 %, respectively (Fagan et al. 2009a). Mateo et al. (2009) developed another set of models which predicted the yield of whey (R^2=0.83, error=6.13 g/100 g) using three terms, namely light backscatter, milk fat content and cutting intensity. These studies were carried out in laboratory-scale cheese vats (7–

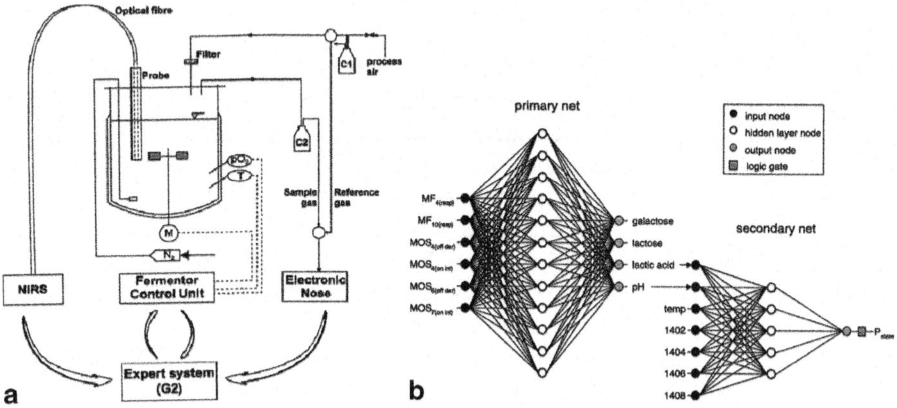

Fig. 4.4 The multi-analyzer setup **a** applied during yogurt fermentations: *C1*, compensator bottle 1, to trap condensating vapour and to compensate for minor flow rate variations; *C2*, compensator bottle 2 and **b** the neural network topology used for sensor fusion. The primary network received six input signals from the electronic nose and was cascaded by the secondary network, which received seven input signals: the output signals from the primary network for pH and lactic acid, four second-derivative NIRS signals (1402–1408 nm) and the first derivative of the reactor temperature signal. A logic gate made the final decision for the state variable. (Cimander et al. 2002)

11 L). Therefore, further scaling up and development under commercial conditions of the technology would be required if it is to become viable at a commercial scale.

NIR spectroscopy has also been investigated as a process control tool in yogurt production. Cimander et al. (2002) studied the potential of NIR spectroscopy to monitor yogurt fermentation in a 4.2-L laboratory-scale vat. A sensor signal fusion approach was adopted with NIR (400–2500 nm), electronic nose, and standard bioreactor sensors installed as part of a multi-analyzer setup (Fig. 4.4a). While the electronic nose followed changes in galactose, lactic acid, lactose and pH, the NIR sensor signal correlated well with the changes in the physical properties during fermentation. Therefore, the signals from the sensors were fused using a cascade artificial neural network (ANN) as detailed in Fig. 4.4b. Results suggested that the accuracy of the neural network prediction was acceptable. This approach was further investigated by Navrátil et al. (2004) under industrial conditions in a 1000-L vat. Signal responses from NIR and electronic nose sensors were subjected to PCA separately. The scores of the first principal component from each PCA were then used to make a trajectory plot for each fermentation batch. PLS regression of the NIR spectra was also used to predict pH and titratable acidity (expressed as Thorner degrees, °Th) during fermentation with reasonable success (SEPs of 0.17 and 6.6 °Th, respectively). MIR spectroscopy has also been employed to monitor the sorghum fermentation process (Correia et al. 2005). They used FTIR spectroscopy to detect differences due to the effect of lactic bacteria on sorghum fermentation. They found it was possible to differentiate between samples which used natural yogurt and *Lactobacillus fermentum* as inocula due to variations in protein and starch structure.

4.5.1.3 Final Product Quality and Authenticity

It is stated in Chap. 12 that "Quality attributes for dairy products can be both the chemical composition of a given product like protein, moisture and fat content, and the sensory quality attributes like taste, smell and consistency." Therefore, the integration of sensing technologies which provide information on such attributes is critical. Table 4.3 summarizes a number of studies which have examined the potential of NIR and MIR spectroscopies to predict the composition of dairy products. These have primarily been laboratory based. Offline laboratory-based infrared sensing systems which provide rapid compositional analysis of dairy products are available. These systems should conform to relevant standards such as ISO Standard 21543:2006 (ISO 2006). A number of studies have also investigated the prediction of sensory quality attributes. Downey et al. (2005) predicted the maturity and sensory attributes of Cheddar cheese using NIR spectroscopy. Generally, second derivative spectra in the region of 750–1098 nm produced the most accurate models with age predicted with an RMSECV of 0.61 months, while the most successfully predicted sensory texture attributes were rubbery, chewy, mouthcoating and massforming with RER values of 8.8, 6.3, 7.6 and 8.5, respectively. NIR spectroscopy was also employed to predict both the sensory and instrumental attributes of processed cheese using NIR spectroscopy (Blazquez et al. 2006). In general, they found that the models developed for predicting sensory texture in processed cheese were stronger than those for Cheddar cheese, with rubbery, chewy, mouthcoating and massforming predicted with RER values of 9.1, 12.0, 8.1 and 8.1, respectively. Fagan et al. (2007b) compared the NIR models developed by Blazquez et al. (2006) to models developed using MIR spectroscopy, which also predict sensory texture parameters of processed cheese. NIR spectroscopy was better at predicting creamy, chewy, and melting, with the R^2 values of the NIR models indicating excellent predictions as opposed to the good predictions of the MIR models. The RER values for the NIR reflectance models indicated a high utility value, whereas the RER values obtained by Fagan et al. (2007b) had a good practical utility. However, the MIR-derived fragmentable model had better accuracy than the NIR model, with excellent and good predictions, respectively.

The requirement to demonstrate the authenticity and safety of dairy products has also led to research into the use of infrared technology for such applications. Determination of the geographic origin and manufacturing conditions of cheese has received a great deal of attention (Boubellouta et al. 2010; Cattaneo et al. 2008; Karoui et al. 2004, 2005a, b, 2007a, 2008; Kocaoglu-Vurma et al. 2009; Pillonel et al. 2003). For example, Pillonel et al. (2003) studied the potential of MIR and NIR spectroscopies to discriminate between Emmental cheeses ($n=20$) based on geographic origin. Samples were obtained from six regions, and they found that MIR transmission spectra could be used to discriminate (i.e. 100% correct classification) Swiss cheese from the other regions, while NIR spectra classified the samples by the six regions of origin. Karoui et al. (2007a) also examined MIR spectroscopy to determine the authentication of 25 Gruyère "protected designation of origin" (PDO) and L'Etivaz PDO cheeses. They found that the spectral regions

Table 4.3 Application of near- and mid-infrared spectroscopy in cheese and yogurt composition analysis. (Modified from Woodcock (2008)

Composition Parameter	Product	Spectral Region	Mode	Wavelength/wave number Range	Prediction Error	Ref.
Moisture	Cheese	NIR	R	400–2498 nm	SECV=0.5	Blazquez et al. (2004)
Content	Cheese	NIR	R	900–2500 nm	SEP=0.429	Čurda and Kukačková (2004)
	Cheese	NIR	R	515–1700 nm	RMSEP=1.72–2.21	da Costa Filho and Volery (2005)
	Cheese	NIR	R	1900–2320 nm	SEP=0.889	Lee et al. (1997)
	Cheese	NIR	R	1000–4000 nm	SEP=0.12–0.35	McKenna (2001)
	Cheese	NIR	T	1000–4000 nm	SEP=0.12–0.35	McKenna (2001)
	Cheese	MIR	ATR	5000–400 cm^{-1}	SEP=0.04–0.09	McQueen et al. (1995)
	Cheese	NIR	R	1740–2280 nm	SEP=0.02–0.05	McQueen et al. (1995)
	Cheese	NIR	R	400–2500 nm	SECV=0.05–0.92	Pérez-Marín et al. (2001)
	Cheese	NIR	R	400–2498 nm	SEC=0.412	Rodriguez Otero et al. (1994)
	Cheese	NIR	R	400–2500 nm	RMSEP=0.58	Wittrup and Nørgaard (1998)
Fat	Cheese	NIR	R	1100–1498 nm	SECV=0.45	Blazquez et al. (2004)
Content	Cheese	FT-NIR	R	900–2500 nm	SEP=0.997	Čurda and Kukačková (2004)
	Cheese	NIR	R	1000–2500 nm	RMSEP=3.61	Karoui et al. (2007b)
	Cheese	NIR	R	1900–2320 nm	SPE=0.855	Lee et al. (1997)
	Cheese	MIR	ATR	5000–400 cm^{-1}	SEP=0.12–0.35	McQueen et al. (1995)
	Cheese	NIR	R	1740–2280 nm	SEP=0.12–0.35	McQueen et al. (1995)
	Cheese	NIR	R	400–2500 nm	SECV=0.05–0.92	Pérez-Marín et al. (2001)
	Cheese	NIR	R	400–2498 nm	SEC=0.388	Rodriguez Otero et al. (1994)
	Cheese	NIR	R	400–2500 nm	RMSEP=0.52	Wittrup and Nørgaard (1998)
Protein	Cheese	FT-NIR	R	900–2500 nm	SEP=0.303	Čurda and Kukačková (2004)
Content	Cheese	NIR	R	1000–2500 nm	RMSEP=2.34	Karoui et al. (2006)

Table 4.3 (continued)

Composition Parameter	Product	Spectral Region	Mode	Wavelength/wave number Range	Prediction Error	Ref.
	Cheese	NIR	R	1900–2320 nm	SEP=0.608	Lee et al. (1997)
	Cheese	MIR	ATR	5000–400 cm^{-1}	SEP=0.04–0.09	McQueen et al. (1995)
	Cheese	NIR	R	1740–2280 nm	SEP=0.04–0.09	McQueen et al. (1995)
	Cheese	NIR	R	400–2500 nm	SECV=0.05–0.92	Pérez-Marín et al. (2001)
	Cheese	NIR	R	400–2498 nm	SEC=0.397	Rodriguez Otero et al. (1994)
	Yogurt	MIR	ATR	1800–1500 cm^{-1}	REP=7.25	Khanmohammadi et al. (2009)
	Yogurt	MIR	ATR	1800–1500 cm^{-1}	REP=3.7	Khanmohammadi et al. (2009)
	Yogurt	MIR	ATR	1515–1800 cm^{-1}	RMSEP=0.2	Moros et al. (2006)
Sugar	Yogurt	NIR	R	400–1000 nm	RMSEP=0.2621	Shao and He (2009)
Content	Yogurt	MIR	ATR	1500–900 cm^{-1}	SEP=0.105–0.05	Khurana et al. (2008)
	Yogurt	NIR	R	400–1000 nm	SEP=0.389	He et al. (2007)
	Yogurt	NIR	R	400–1000 nm	RMSEP=0.33–0.36	Shao et al. (2007)
Carbohydrate	Yogurt	MIR	ATR	2850–1083 cm^{-1}	RMSEP=36	Moros et al. (2006)
Calcium Content	Yogurt	MIR	ATR	1461–1636 cm^{-1}	RMSEP=9	Moros et al. (2006)

R reflection, *T* transmission, *ATR* attenuated total reflection

3000–2800 cm^{-1} and 1500–900 cm^{-1} were most useful with 90.5 and 90.9 % correct classification results achieved, respectively. MIR spectroscopy (supplemented by partial 16S rDNA sequencing) has also been employed to monitor the population dynamics of microorganisms during cheese ripening (Oberreuter et al. 2003).

4.5.2 Cereal Grains and Seeds

NIR spectroscopy has been widely used in routine quality control analysis in the grain industry since the 1960s (Scotter 1990). This has included the assessment of moisture and protein content (Downey and Byrne 1987; Norris and Williams 1979;

Williams 1979; Williams and Cordeiro 1979, 1981). More recently, developments in this area have focused on assessment of grain quality at harvest, grain quality classification and sorting and grain blending.

Kawamura et al. (2003) developed an automated rice quality inspection system which utilized both visible and NIR technology. The objective was to develop a system which measured not only moisture content but also other rice quality indices in order to grade rough rice according to quality when it arrives at the drying facility. The system they developed consisted of a rice huller, a rice cleaner, an NIR instrument and a Vis segregator. This system enabled rough rice transported to a rice-drying facility to be classified into six qualitative grades.

Grain quality at harvesting is also a critical parameter as there can be significant within-field variability of grain quality parameters, for example, protein and moisture content. Maertens et al. (2004) described some of the requirement for online grain quality assessment at harvest. They included the use of a robust NIR spectrometer, design of a measurement configuration that guarantees a constant grain sample presentation while also avoiding dirt and blockages, that the sensor should be calibrated on the harvester and not under simulated conditions in the laboratory and finally that appropriate signal processing techniques should be employed to filter the spectral data, both in the time and wavelength domain They also studied the potential of an NIR sensor mounted on the bypass of the grain elevator of a combine harvester for online prediction of wheat moisture and protein content. They found that the average prediction errors were 0.56 and 0.31 % for protein and moisture content, respectively, where moisture content was below 18 %.

Detection and removal of internal insects and fungal contamination from seeds (grains, beans and nuts) are important control measures for ensuring storage longevity, seed quality and food safety (Pasikatan and Dowell 2001). NIR spectroscopy has been applied to the detection of infestation of such products. NIR spectroscopy has been used to differentiate among individual wheat kernels that are uninfested, those infested with weevil larvae or pupae, or those that contain a parasitoid pupa (Baker et al. 1999). Wang et al. (2002) recorded single-seed NIR spectra of a total of 1600 soya bean seeds, i.e. 700 sound seeds and 900 seeds damaged by weather, frost, sprout, heat or mould. The regions 750–1690 nm and 450–1690 nm gave the best classification of seeds into "sound" and "damaged" categories. They also found that an optimally developed neural network (parameters: momentum = 0.6, learning rate = 0.7, learning cycles = 150,000, wavelength region = 490–1690 nm) could classify seed according to six categories, i.e. "sound" (100 %) and five damage categories, "weather" (98 %), "frost" (97 %), "sprout" (64 %), "heat" (79 %) and "mold" (83 %), with reasonable success.

Aflatoxin B1 is recognized by the International Agency of Research on Cancer as a group 1 carcinogen for animals and humans, and Fernández-Ibañez et al. (2009) investigated the potential of Fourier transform NIR spectroscopy to detect aflatoxin B1 in cereal grains. They analysed maize and barley samples ($n = 152$) and developed models ($R^2 = 0.82$–0.85) for prediction of the presence of aflatoxin B1, which suggested that NIR spectroscopy could be a suitable alternative for fast detection of aflatoxin B1 in cereals.

4.5.3 *Fruit and Vegetables*

The application of NIR and MIR to quality assessment of fruit and vegetables has been widely studied (Table 4.4). In terms of infrared spectroscopy's role as a PAT tool in this industry, it could be employed for the optimization of harvesting, defect identification, disease control, process control applications and overall quality classification.

4.5.3.1 Harvest Optimization

Prediction of the optimal harvest time of apples will minimize the occurrence of quality losses. Peirs et al. (2001) predicted the optimal harvest date of apples harvested no more than 8 weeks before the commercial picking date using Vis–NIR spectra collected post harvest in the laboratory (measurements were carried out on the same day or the day after picking). They stated that it was possible to measure apple maturity for harvest of individual cultivars within an orchard and that the number of days before the optimum harvest date was well predicted (R=0.90–0.93). Further work examined the potential of Vis–NIR spectroscopy to estimated apple pre- and post-storage quality indices at harvest (McGlone et al. 2002). The apples were harvested 1–3 weeks before and up to 1 week after the commercial harvest period. Spectral analysis in this case took place between 16 and 24 h after harvest. The authors found that although models were developed to predict quality indices of the apples they were still very poor in terms of prediction accuracies. Therefore, they were unlikely to be useful for sorting or grading due to the high rate of prediction errors that would result. They also stated that the prediction models, with the exception of soluble solids content, may be almost solely dependent on changes in the apple chlorophyll level and not have any direct sensitivity to the constituents or properties of interest.

Clark et al. (2004) examined the potential of Vis–NIR spectroscopy to predict the storage potential of kiwifruit. They employed canonical discriminant analysis (CDA) to optimize the separation between the two categories, i.e. "sound fruit" and "fruit developing a disorder during storage". They estimated that the overall incidence of disorders could have been reduced from 33.9 to 17.9% and 14.7 to 8.5% depending on the harvest or when using all harvests from 13.7 to 6.8%.

A similar approach has also been investigated for mango (Saranwong et al. 2004). Vis–NIR spectra of mango were collected on the day of harvest and models were developed to predict harvest and eating quality using multiple linear regression and PLS regression. They stated that the calibration equations developed were sufficiently accurate to determine the harvest quality, dry matter and starch content of hard green mango fruit non-destructively. Using this information, the soluble solids content of the ripe fruit, which is an eating quality index, could be precisely predicted at the time of harvest.

Table 4.4 Examples of applications of near- and mid-infrared spectroscopy to quality assessment of fruit

Product	Parameter	Technology	Mode	Range	Result	Equipment	Ref.
Avocado	Moisture content	NIR	R		$R^2=0.92$, SEFV = 2 %, RER = 16	NIRS6500 spectrophotometer	Blakey et al. (2009)
Apple, grape, pear; apple-chewy and apple-banana juices	Soluble solids and total solids/total moisture	NIR	R		$R^2=0.90$	Katrina Inc designed NIR sensor	Singh et al. (1996)
Zucchini squash	Chilling injury	FTIR	R		Spectral shift of the maxima to 3400 cm^{-1}	Nicolet 60 SX FTIR spectrometer	Buta et al. (1997)
Fruit concentrates	Total sugar, glucose, fructose and sucrose	NIR	T		Relative standard deviation values obtained vary from 0.4 to 2.3 %	Perkin-Elmer Lambda 9 double beam spectrophotometer	Rambla et al. (1997)
Ginseng	Moisture content	NIR	R		$R^2=0.998$, SEP = 0.12 %	Model 6500, Perstorp Analytical Inc	Ren and Chen (1997)
Kiwifruit	Soluble solids distribution	NIR imaging	R		Prediction error of 1.2 Brixt	Author design	Martinsen and Schaare (1998)
Apple	Soluble solids	NIR	R		$R^2=0.56$, SEP 1.14 Brix	Ocean Optics SD-1000	Ventura et al. (1998)
Fig	Sorting defect and acceptable categories	NIR	R	400–1700 nm	Classification 83 to 100 %	Diode-array Perten Instruments	Burks et al. (2000)
Kiwifruit	Discrimination of pre-harvest fruit management treatment	NIR	R	516–998 nm	Best classifications based on fast Fourier transform features	PS1000 Ocean Optics	Kim et al. (2000)

Table 4.4 (continued)

Product	Parameter	Technology	Mode	Range	Result	Equipment	Ref.
Potato	Elimination of interference from peel	Vis–NIR	R/T	600–1100 nm	RMSEC = 3–4.1 %	Modular spectrophotometer, Bentham Instruments Ltd	Krivoshiev et al. (2000)
Apple	Comparison of two optical configurations for measuring internal apple quality attributes, describe interaction of skin and the incident radiation, determine penetration depth values in apple tissue	Vis–NIR	R	500–1900 nm	Can provide information about the state of the fruit flesh; 0.55 Brix	OSA 6602, Rees Instruments Ltd	Lammertyn et al. (2000)
Mango	Predict physiological properties and quality indices	NIR	R	1200–2400 nm	MLR: SEP = 1.223, 0.161, 17.14, 37.03, R^2 = 0.93, 0.61, 0.82, 0.94 for TSS, acidity, firmness and storage period	Quantum 1200, LTI	Schmilovitch et al. (2000)
Peach	Assess woolliness	NIR (in conjunction with non-destructive impact response (NDIR))	R	900–1400 nm	NIR classified into juicy categories. NIR + NDIR classified woolly peaches at 80 %	OSA 6602, Monolight	Ortiz et al. (2001)

Table 4.4 (continued)

Product	Parameter	Technology	Mode	Range	Result	Equipment	Ref.
Apple	Predict optimal harvest date	Vis/NIR	R	380–2000 nm	Maturity was orchard-dependent; $R = 0.80–0.90$ (soluble solids, acidity)	OSA 6602, Rees Instruments Ltd	Peirs et al. (2001)
Fruit juice	Predict sugar levels	FT-NIR	R/TR	1000–2500 nm	SEP < 0.10 %; $R^2 = 0.999$	Perkin–Elmer Spectrum Identicheck	Rodriguez-Saona et al. (2001)
Papaya	Methylation level of pectin fractions	FTIR	A	4000–500 cm^{-1}	No significant difference	Bomem MB-100	Manrique and Lajolo (2002)
Apple	Estimation of pre- and post-storage quality indices	Vis/NIR	T	300–1140 nm	Poor predictions (chlorophyll level dependent)	Specially developed laboratory system	McGlone et al. (2002)
Red paprika	Mycotoxin detection	NIR	R	1100–2000 nm	RPD = 5.2, 2.8, 4.4	Foss NIRS system 5000	Hernández-Hierro et al. (2008)

It has also been demonstrated that infrared technology can be used for fruit assessment prior to harvesting. Pérez-Marín et al. (2009) used a handheld micro-electro-mechanical system (MEMS) spectrometer and a diode-array Vis–NIR spectrophotometer to collect the spectra of nectarine during on-tree ripening ($n=144$). They developed models to quantify changes in soluble solids content, flesh firmness, fruit weight and diameter. Both instruments provided good precision for soluble solids content ($R^2=0.89$; SEP=0.75–0.81 %) and for firmness ($R^2=0.84$–0.86; SEP=11.6–12.7 N). The diode-array instrument predicted the two other physical parameters well ($R^2=0.98$ and SEP=5.40 g for fruit weight and $R^2=0.75$ and SEP=0.46 cm for diameter), while the handheld MEMS instrument proved less accurate in this respect (Pérez-Marín et al. 2009).

A portable non-invasive instrument based on NIR spectroscopy has also been developed to measure the ripeness of wine grapes (Larrain et al. 2008). It was used to predict three ripeness variables with excellent success for Brix and pH ($R^2=0.87$–0.93) and with less accuracy ($R^2=0.56$–0.80) for pH.

4.5.3.2 Defect Identification

Burks et al. (2000) applied NIR spectroscopy to the sorting and classification of figs. They classified the figs according to the number of categories ("passable", "infested", "rotten", "sour", "dirty") with correct classifications ranging from 83 to 100 %. However, 20 PLS factors were required which might limit the robustness to the models. Vis–NIR spectroscopy in both transmission and reflectance modes has been employed to detect brown heart of pears (Fu et al. 2007). They found that, using discriminant analysis, they could discriminate between brown heart pears and non-brown heart pears. Transmission spectra were more successful than reflectance spectra in this classification: a classification rate of 91.2 % using transmission spectra.

A conceptual view of an NIR transmission-based system for apple assessment (Fig. 4.5) has been proposed by McGlone and Martinsen (2004). They employed two prototype on-line NIR transmission systems to determine the percentage of internal tissue browning in apples. One prototype used time-delayed integration spectroscopy (TDIS) in which light transmitted through a moving object was electronically tracked as it moved through the spectrometer's field of view. The other used a large aperture spectrometer (LAS) in which the light from the object is accumulated in a series of one-shot measurements as the fruit progresses through the field of view (McGlone and Martinsen 2004). The systems operated 500 mm s^{-1}. The LAS system gave the best results ($R^2=0.9$) for fast on-line assessment of apples.

Further developments in defect identification have focused on the use of multispectral or hyperspectral imaging (Ariana et al. 2006; Blasco et al. 2007). This emerging platform technology is discussed in Chap. 9.

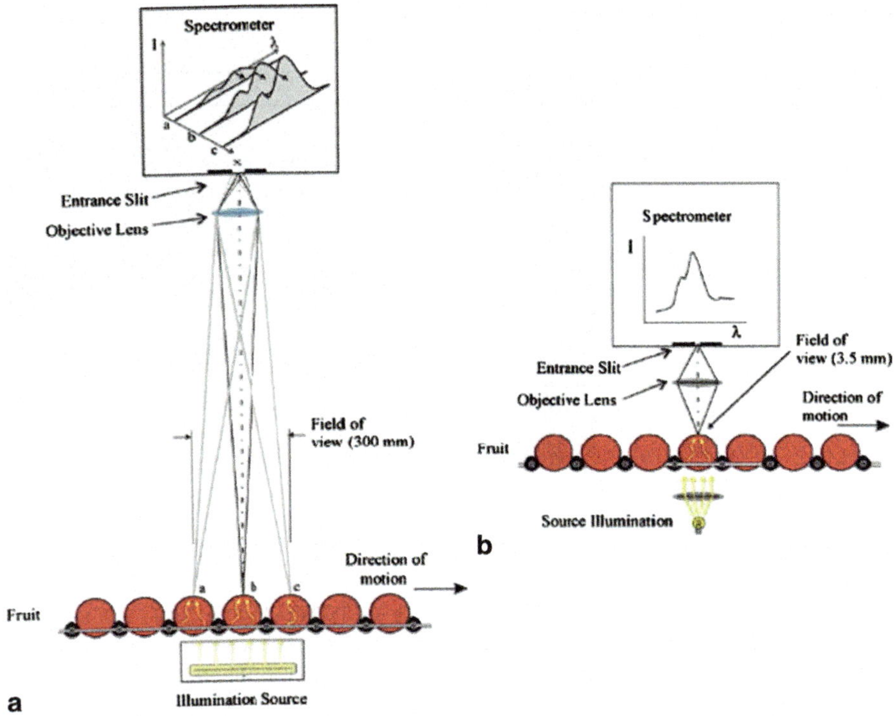

Fig. 4.5 A conceptual view of NIR transmission system. As the fruit passes through a relatively large field-of-view in the TDIS system (**a**), a detector simultaneously accumulates many sequential points over three apples. In contrast, the LAS system (**b**) takes a simple snapshot, like a camera, over a much shorter time for a small portion of one fruit (McGlone and Martinsen 2004). (Reprinted with permission from *Journal of Near Infrared Spectroscopy* 12(1), 37–43 (2004). Copyright: IM Publications LLP 2004)

4.5.3.3 Quality Classification

A key quality characteristic of fruit is SSC. As fruit ripen, there is conversion of insoluble starch into soluble solids, to which the simple sugars (glucose, fructose and sucrose) make the largest contribution (Martinsen and Schaare 1998). Numerous studies have investigated infrared spectroscopy to predict this parameter non-destructively and have been summarized in Table 4.4. The majority of such studies have utilized NIR spectroscopy. A study by Lammertyn (2000) compared two optical configurations, i.e. a bifurcated and a 0°/45° optical configuration. They found that while the former configuration gave slightly better performance for the prediction of SSC, they recommended 0°/45° configuration for commercial applications as it had a lower cost and could be used for non-contact measurements. However, bifurcated reflectance-based instruments have found an array of applications (Fig. 4.6). It should be noted that numerous variables (e.g. cultivar, geographic

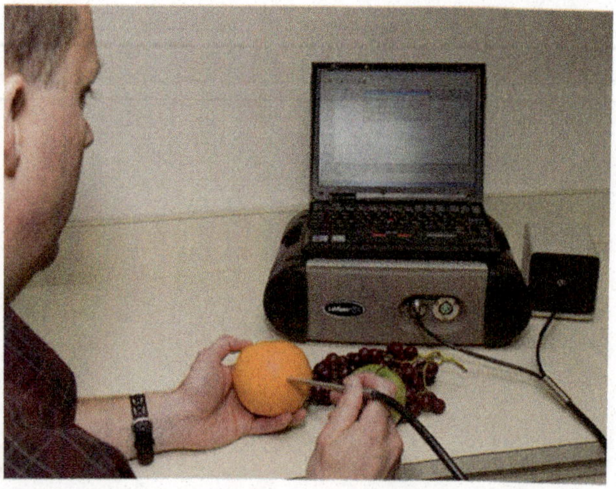

Fig. 4.6 A NIR (LabSpec) with bifurcated fibre optic probe for contact reflectance measurement

origin, etc.) can affect the performance of such predictive models, and therefore studies which have independently validated models, for example, over and within seasons, are crucial to an assessment of model robustness (Golic and Walsh 2006). Golic and Walsh (2006) collected NIR spectrum of peaches, nectarines and plums and found that model performance for SSC was acceptable when peaches and nectarines were combined, but it was best if a separate plum model was employed. They also stated that model performance was stable over several seasons in terms of R^2 (typical $R^2 > 0.8$).

4.5.4 *Meat and Poultry*

4.5.4.1 Fresh Meat

A number of studies examined the application of infrared spectroscopy to fat extracts to predict meat quality as fatty acid composition of meat can determine its processing quality. Villé et al. (1995) developed a method for the determination of total fat and phospholipid content in intramuscular pig meat using FTIR spectroscopy. They employed an extraction using chloroform and methanol. FTIR spectra were subsequently recorded in transmission mode, and utilizing selected regions of the FTIR spectra related to the C=O bond (1785–1697 cm^{-1}) developed linear regression equation to predict total fat ($R^2 = 0.99$). A study has also examined the use of FTIR spectroscopy in the NIR and MIR regions of fat extracts and non-processed pork to determine the fatty acid content in fat slices and fat extracts (Ripoche and Guillard 2001). They found that MIR spectra using an attenuated total reflectance samples accessory ($R^2 \sim 0.91$–0.98) and NIR transmission spectra ($R^2 \sim 0.85$–0.96) of fat extracts could be used to predict saturated fatty acids (SFA), monounsaturated fatty acids (MUFA), polyunsaturated fatty acids (PUFA), palmitic acid (C16:0), oleic acid (C18:1) and linoleic acid (C18:2). However, with 9–15 latent variables

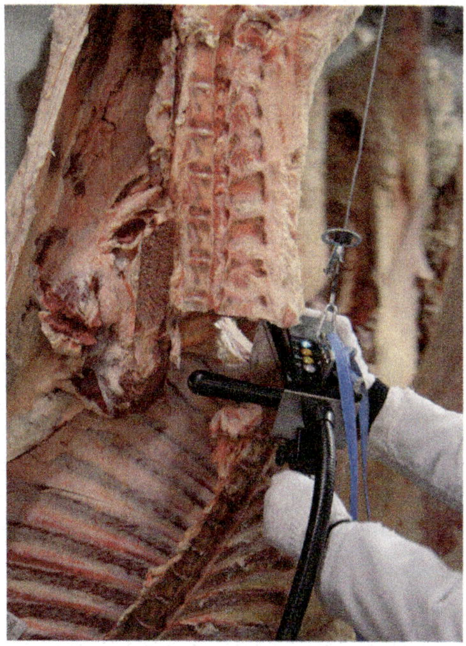

Fig. 4.7 The QualitySpec BT Spectrometer from Analytical Spectral Devices Inc. for measuring meat quality

included in the models, they may not be very robust. While NIR reflectance spectroscopy successfully predicted SFA, PUFA, C18:1 and C18:2 from spectral measurements of the back and breast fat, MUFA and C16:0 could not be predicted. Mitsumoto et al. (1991) used NIR spectroscopy in reflectance and transmittance mode to predict the quality of beef cuts Warner–Bratzler shear value (tenderness) (R=0.798–0.826), protein (R=0.822–0.904), moisture (R=0.895–0.941), fat (R=0.890–0.965) and energy content (R=0.899–0.961) were successfully predicted using both modes. Park et al. (2001) also developed models for predicting the tenderness, i.e. Warner–Bratzler shear value of beef using NIR reflectance spectra and principal component regression (PCR). The coefficient of determination of the developed models were of a similar order (R^2=0.612–0.692). This technology has also been commercially investigated with instruments such as the QualitySpec BT Spectrometer from Analytical Spectral Devices (Fig. 4.7).

NIR spectroscopy has also been investigated at laboratory scale for determination of the maximum temperature to which beef had been subjected to during a heat treatment (Ellekjaer and Isaksson 1992), species identification (Ding and Xu 1999) and authenticity assessment (Fumiere et al. 2000). Other applications of NIR such as the detection of faecal contamination on poultry have been studied. Windham et al. (2003) applied Vis–NIR spectroscopy to discriminate between uncontaminated poultry breast skin and faeces. They found that the developed model could successfully classify faecal-contaminated material due to spectral differences between faecal colour and myoglobin and/or hemoglobin content of the uncontaminated breast skin. However, hyperspectral imaging (Chap. 9) has also been utilized for such an application (Heitschmidt et al. 2007; Liu et al. 2007; Park et al. 2006a, b, 2007).

4.5.4.2 Ground Meat Quality

The quality of ground meat used as a raw material in products such as burgers and sausages is critical as processors must comply with product-type-dependent restrictions, i.e. chemical composition and origin of raw materials (Togersen et al. 1999). Togersen et al. (1999) utilized an on-line NIR sensor to determine the fat, water and protein contents in industrial-scale meat batches (beef and pork) in an industrial environment. The NIR sensor was installed at the outlet of a large meat grinder. The models developed had RMSECV of 0.82–1.49 %, 0.94–1.33 % and 0.35–0.70 % for fat, water and protein, respectively. Togersen et al. (2003) went on to predict the chemical composition of industrial-scale batches of frozen beef using a similar system. The resulting RMSECVs were 0.48–1.11 % (fat), 0.43–0.97 % (moisture) and 0.41–0.47 % (protein).

NIR spectroscopy has also been investigated as a tool for detecting adulteration of hamburgers (Ding and Xu 2000). They found it was possible to predict the level of adulterants in hamburgers with errors of 3.33, 2.99, 0.92 and 0.57 % for the adulterants mutton, pork, skim milk powder and wheat flour, respectively.

4.5.4.3 Meat Emulsion

Optical sensors have also been developed to monitor meat emulsion stability (Alvarez et al. 2007, 2009, 2010a, b). Initial work focused on prediction of meat emulsion stability using reflection photometry (Alvarez et al. 2007). They found that L* values increased at the beginning of chopping associated with reduced cooking losses, following 8 min of chopping there was a reduction in L* and b* values and an associated increase in cooking losses, which suggested the feasibility of an online optical sensor technology to predict the optimum end point of emulsification in the manufacture of finely comminuted meat products. These authors then recorded light backscatter intensity from beef emulsions manufactured with different fat/lean ratio and chopping duration using a dedicated fibre optic prototype (Alvarez et al. 2009). They found several optically derived parameters to be significantly correlated with fat loss during cooking. In subsequent work, they found normalized intensity decreased with increased chopping time as a result of emulsion homogenization, and with increased distance, chopping time had a positive correlation with fat losses during cooking, which in turn had a negative correlation with normalized light intensity and loss of intensity. Therefore, they suggest that light extinction spectroscopy could provide information about emulsion stability (Alvarez et al. 2010).

4.6 Future

Infrared spectroscopy has been demonstrated to be an excellent PAT tool for monitoring critical processes and prediction of quality indices during food processing. Advances in equipment design will assist in the deployment of infrared spectroscopy-based technologies as PAT tools in the food industry. This will include improvements in robustness, cost and advances in microspectrometers. However, where studies have primarily been at laboratory scale, further research is required to ensure appropriate scaling up and transfer of the technology to industry. The combined acquisition of spectral and spatial information through the use of hyperspectral imaging has a number of potential applications. However, further developments are required to reduce the cost and increase the acquisition and processing speed for it to be fully exploited in food quality and safety applications.

References

Abbott JA (1999) Quality measurement of fruits and vegetables. Postharvest Biol Technol 15:207–225

Alvarez D, Castillo M, Payne FA, Garrido MD, Banon S, Xiong YL (2007) Prediction of meat emulsion stability using reflection photometry. J Food Eng 82:310–315

Alvarez D, Castillo M, Payne FA, Xiong YL (2009) A novel fiber optic sensor to monitor beef meat emulsion stability using visible light scattering. Meat Sci 81:456–466

Alvarez D, Castillo M, Xiong YL, Payne FA (2010a) Prediction of beef meat emulsion quality with apparent light backscatter extinction. Food Res Int 43:1260–1266

Alvarez D, Castillo M, Payne FA, Cox RB, Xiong YL (2010b) Application of light extinction to determine stability of beef emulsions. J Food Eng 96:309–315

Ariana DP, Lu R, Guyer DE (2006) Near-infrared hyperspectral reflectance imaging for detection of bruises on pickling cucumbers. Comput Electron Agric 53:60–70

Baker JE, Dowell FE, Throne JE (1999) Detection of parasitized rice weevils in wheat kernels with near-infrared spectroscopy. Biol Control 16:88–90

Barbosa-García O, Ramos-Ortíz G, Maldonado JL, Pichardo-Molina JL, Meneses-Nava MA, Landgrave JEA, Cervantes-Martínez J (2007) UV-vis absorption spectroscopy and multivariate analysis as a method to discriminate tequila. Spectrochim Acta Part A Mol Biomol Spectrosc 66:129–134

Biswas AK, Sahoo J, Chatli MK (2011) A simple UV-Vis spectrophotometric method for determination of [beta]-carotene content in raw carrot, sweet potato and supplemented chicken meat nuggets. LWT—Food Science and Technology. 44:1809–1813

Blakey RJ, Bower JP, Bertling I (2009) Influence of water and ABA supply on the ripening pattern of avocado (Persea americana Mill.) fruit and the prediction of water content using near infrared spectroscopy. Postharvest Biol Technol 53:72–76

Blasco J, Aleixos N, Gómez J, Moltó E (2007) Citrus sorting by identification of the most common defects using multispectral computer vision. J Food Eng 83:384–393

Blazquez C, Downey G, O'Donnell C, O'Callaghan D, Howard V (2004) Prediction of moisture, fat and inorganic salts in processed cheese by near infrared reflectance spectroscopy and multivariate data analysis. J Near Infrared Spectrosc 12:149–157

Blazquez C, Downey G, O'Callaghan D, Howard V, Delahunty C, Sheehan E, Everard C, O'Donnell CP (2006) Modelling of sensory and instrumental texture parameters in processed cheese by near infrared reflectance spectroscopy. J Dairy Res 73:58–69

Boubellouta T, Karoui R, Lebecque A, Dufour E (2010) Utilisation of attenuated total reflectance MIR and front-face fluorescence spectroscopies for the identification of Saint-Nectaire cheeses varying by manufacturing conditions. Eur Food Res Technol 231:873–882

Brandt M, Haeussermann A, Hartung E (2010) Invited review: technical solutions for analysis of milk constituents and abnormal milk. J Dairy Sci 93:427–436

Burks CS, Dowell FE, Xie F (2000) Measuring fig quality using near-infrared spectroscopy. J Stored Prod Res 36:289–296

Buta JG, Qi L, Wang CY (1997) Fourier transform infrared spectra of zucchini squash stored at chilling or non-chilling temperatures. Environ Exp Bot 38:1–6

Cassandro M, Comin A, Ojala M, Zotto RD, De Marchi M, Gallo L, Carnier P, Bittante G (2008) Genetic parameters of milk coagulation properties and their relationships with milk yield and quality traits in Italian Holstein cows. J Dairy Sci 91:371–376

Castillo M, Payne FA, Hicks CL, Lopez MB (2000) Predicting cutting and clotting time of coagulating goat's milk using diffuse reflectance: effect of pH, temperature and enzyme concentration. Int Dairy J 10:551–562

Castillo M, Payne F, Shea A (2005a) Development of a combined sensor technology for monitoring coagulation and syneresis operations in cheese making. J Dairy Sci 88:142–142

Castillo M, Payne FA, Lopez MB, Ferrandini E, Laencina J (2005b) Optical sensor technology for measuring whey fat concentration in cheese making. J Food Eng 71:354–360

Cattaneo TMP, Tornelli C, Erini S, Panarelli EV (2008) Relationship between sensory scores and near infrared absorptions in characterising Bitto, an Italian protected denomination of origin cheese. J Near Infrared Spectrosc 16:173–178

Cecchinato A, De Marchi M, Gallo L, Bittante G, Carnier P (2009) Mid-infrared spectroscopy predictions as indicator traits in breeding programs for enhanced coagulation properties of milk. J Dairy Sci 92:5304–5313

Cimander C, Carlsson M, Mandenius C-F (2002) Sensor fusion for on-line monitoring of yoghurt fermentation. J Biotechnol 99:237–248

Clark CJ, McGlone VA, De Silva HN, Manning MA, Burdon J, Mowat AD (2004) Prediction of storage disorders of kiwifruit (Actinidia chinensis) based on visible-NIR spectral characteristics at harvest. Postharvest Biol Technol 32:147–158

Contreras U, Barbosa-García O, Pichardo-Molina JL, Ramos-Ortíz G, Maldonado JL, Meneses-Nava MA, Ornelas-Soto NE, López-de-Alba PL (2010) Screening method for identification of adulterate and fake tequilas by using UV-VIS spectroscopy and chemometrics. Food Res Int 43:2356–2362

Correia I, Nunes A, Duarte IF, Barros A, Delgadillo I (2005) Sorghum fermentation followed by spectroscopic techniques. Food Chem 90:853–859

Čurda L, Kukačková O (2004) NIR spectroscopy: a useful tool for rapid monitoring of processed cheese manufacture. J Food Eng 61:557–560

da Costa Filho PA, Volery P (2005) Broad-based versus specific NIRS calibration: determination of total solids in fresh cheese. Anal Chim Acta 554:82–88

Dal Zotto R, De Marchi M, Cecchinato A, Penasa M, Cassandro M, Carnier P, Gallo L, Bittante G (2008) Reproducibility and repeatability of measures of milk coagulation properties and predictive ability of mid-infrared reflectance spectroscopy. J Dairy Sci 91:4103–4112

De Marchi M, Fagan CC, O'Donnell CP, Cecchinato A, Dal Zotto R, Cassandro M, Penasa M, Bittante G (2009) Prediction of coagulation properties, titratable acidity, and pH of bovine milk using mid-infrared spectroscopy. J Dairy Sci 92:423–432

Ding HB, Xu RJ (1999) Differentiation of beef and kangaroo meat by visible/near-infrared reflectance spectroscopy. J Food Sci 64:814–817

Ding HB, Xu RJ (2000) Near-infrared spectroscopic technique for detection of beef hamburger adulteration. J Agric Food Chem 48:2193–2198

Downey G, Byrne S (1987) Protein determination of wheat in trade by near-infrared reflectance spectroscopy—calibration and instrument performance over a 4 year period. Sci Des Aliment 7:325–336

Downey G, Sheehan E, Delahunty C, O'Callaghan D, Guinee T, Howard V (2005) Prediction of maturity and sensory attributes of Cheddar cheese using near-infrared spectroscopy. Int Dairy J 15:701–709

Ellekjaer MR, Isaksson T (1992) Assessment of maximum cooking temperatures in previously heat-treated beef.1. Near-infrared spectroscopy. J Sci Food Agric 59:335–343

Everard CD, O'Callaghan DJ, Fagan CC, O'Donnell CP, Castillo M, Payne FA (2007) Computer vision and color measurement techniques for inline monitoring of cheese curd syneresis. J Dairy Sci 90:3162–3170

Fagan CC, O'Donnell CP (2007) Application of mid-infrared spectroscopy to food processing systems. In: Irudayaraj J, Reh C (eds) Nondestructive testing of food quality. Blackwell Publishing, Oxford, pp 119–142

Fagan CC, Castillo M, Payne FA, O'Donnell CP, Leedy M, O'Callaghan DJ (2007a) Novel online sensor technology for continuous monitoring of milk coagulation and whey separation in cheesemaking. J Agric Food Chem 55:8836–8844

Fagan CC, Everard C, O'Donnell CP, Downey G, Sheehan EM, Delahunty CM, O'Callaghan DJ (2007b) Evaluating mid-infrared spectroscopy as a new technique for predicting sensory texture attributes of processed cheese. J Dairy Sci 90:1122–1132

Fagan CC, Du CJ, O'Donnell CP, Castillo M, Everard CD, O'Callagran DJ, Payne FA (2008a) Application of image texture analysis for online determination of curd moisture and whey solids in a laboratory-scale stirred cheese vat. J Food Sci 73:E250–E258

Fagan CC, Castillo M, O'Donnell CP, O'Callaghan DJ, Payne FA (2008b) On-line prediction of cheese making indices using backscatter of near infrared light. Int Dairy J 18:120–128

Fagan C, Castillo M, O'Callaghan D, Payne F, O'Donnell C (2009a) Visible-near infrared spectroscopy sensor for predicting curd and whey composition during cheese processing. Sens Instrum Food Qual Saf 3:62–69

Fagan CC, O'Donnell CP, Rudzik L, Wust E (2009b) Milk and dairy products. In: Sun D-W (ed) Infrared spectroscopy for food quality analysis and control. Elsevier, San Diego, pp 241–273

Fernández-Ibañez V, Soldado A, Martínez-Fernández A, de la Roza-Delgado B (2009) Application of near infrared spectroscopy for rapid detection of aflatoxin B1 in maize and barley as analytical quality assessment. Food Chem 113:629–634

Fu X, Ying Y, Lu H, Xu H (2007) Comparison of diffuse reflectance and transmission mode of visible-near infrared spectroscopy for detecting brown heart of pear. J Food Eng 83:317–323

Fumiere O, Sinnaeve G, Dardenne P (2000) Attempted authentication of cut pieces of chicken meat from certified production using near infrared spectroscopy. J Near Infrared Spectrosc 8:27–34

Golic M, Walsh KB (2006) Robustness of calibration models based on near infrared spectroscopy for the in-line grading of stonefruit for total soluble solids content. Anal Chim Acta 555:286–291

Griffiths PR (2010) Theory and instrumentation for vibrational spectroscopy. In: Chalmers JM, Griffiths P, Li Chan E (eds) Applications of vibrational spectroscopy to food science. Wiley, Chichester

Hardy J, Fanni J (1981) Application of reflection photometry to the measurement of milk coagulation. J Food Sci 46:1956–1957

He Y, Wu D, Feng SJ, Li XL (2007) Fast measurement of sugar content of yogurt using Vis/NIR-spectroscopy. Int J Food Prop 10:1–7

Heitschmidt GW, Park B, Lawrence KC, Windham WR, Smith DP (2007) Improved hyperspectral imaging system for fecal detection on poultry carcasses. Trans ASABE 50:1427–1432

Hernández-Hierro JM, García-Villanova RJ, González-Martín I (2008) Potential of near infrared spectroscopy for the analysis of mycotoxins applied to naturally contaminated red paprika found in the Spanish market. Anal Chim Acta 622:189–194

ISO (2006) 21543:2006 Milk products—guidelines for the application of near infrared spectrometry. International Organization for Standardization, Geneva

Karoui R, Dufour E, Pillonel L, Picque D, Cattenoz T, Bosset JO (2004) Determining the geographic origin of Emmental cheeses produced during winter and summer using a technique based on the concatenation of MIR and fluorescence spectroscopic data. Eur Food Res Technol 219:184–189

Karoui R, Dufour E, Pillonel L, Schaller E, Picque D, Cattenoz T, Bosset JO (2005a) The potential of combined infrared and fluorescence spectroscopies as a method of determination of the geographic origin of Emmental cheeses. Int Dairy J 15:287–298

Karoui R, Bosset JO, Mazerolles G, Kulmyrzaev A, Dufour E (2005b) Monitoring the geographic origin of both experimental French Jura hard cheeses and Swiss Gruyere and L'Etivaz PDO cheeses using mid-infrared and fluorescence spectroscopies: a preliminary investigation. Int Dairy J 15:275–286

Karoui R, Mouazen AM, Dufour É, Pillonel L, Schaller E, Baerdemaeker J, Bosset JO (2006) Chemical characterisation of European Emmental cheeses by near infrared spectroscopy using chemometric tools. Int Dairy J 16:1211–1217

Karoui R, Mazerolles G, Bosset JO, de Baerdemaeker J, Dufour E (2007a) Utilisation of mid-infrared spectroscopy for determination of the geographic origin of Gruyere PDO and L'Etivaz PDO Swiss cheeses. Food Chem 105:847–854

Karoui R, Pillonel L, Schaller E, Bosset JO, De Baerdemaeker J (2007b) Prediction of sensory attributes of European Emmental cheese using near-infrared spectroscopy: a feasibility study. Food Chem 101:1121–1129

Karoui R, De Baerdemaeker J, Dufour E (2008) A comparison and joint use of mid infrared and fluorescence spectroscopic methods for differentiating between manufacturing processes and sampling zones of ripened soft cheeses. Eur Food Res Technol 226:861–870

Kawamura S, Natsuga M, Takekura K, Itoh K (2003) Development of an automatic rice-quality inspection system. Comput Electron Agric 40:115–126

Kawamura S, Kawasaki M, Nakatsuji H, Natsuga M (2007) Near-infrared spectroscopic sensing system for online monitoring of milk quality during milking. Sens Instrum Food Qual Saf 1:37–43

Khanmohammadi M, Garmarudi AB, Ghasemi K, Garrigues S, de la Guardia M (2009) Artificial neural network for quantitative determination of total protein in yogurt by infrared spectrometry. Microchem J 91:47–52

Khurana HK, Jun S, Cho IK, Li QX (2008) Rapid determination of sugars in commercial fruit yogurts and yogurt drinks using fourier transform infrared spectroscopy and multivariate analysis. Appl Eng Agric 24:631–636

Kim J, Mowat A, Poole P, Kasabov N (2000) Linear and non-linear pattern recognition models for classification of fruit from visible-near infrared spectra. Chemom Intell Lab Syst 51:201–216

Kocaoglu-Vurma NA, Eliardi A, Drake MA, Rodriguez-Saona LE, Harper WJ (2009) Rapid profiling of Swiss cheese by attenuated total reflectance (ATR) infrared spectroscopy and descriptive sensory analysis. J Food Sci 74:S232–S239

Krivoshiev GP, Chalucova RP, Moukarev MI (2000) A possibility for elimination of the interference from the peel in nondestructive determination of the internal quality of fruit and vegetables by VIS/NIR spectroscopy. Lebensm-Wiss Technol 33:344–353

Lammertyn J, Peirs A, De Baerdemaeker J, Nicolaï B (2000) Light penetration properties of NIR radiation in fruit with respect to non-destructive quality assessment. Postharvest Biol Technol 18:121–132

Larrain M, Guesalaga AR, Agosin E (2008) A multipurpose portable instrument for determining ripeness in wine grapes using NIR spectroscopy. IEEE Trans Instrum Measurement 57:294–302

Lawrence RC, Gilles J (1980) The assessment of the potential quality of young Cheddar cheese. N Z J Dairy Sci Technol 15:1–12

Lee SJ, Jeon IJ, Harbers LH (1997) Near infrared reflectance spectroscopy for rapid analysis of curds during cheddar cheese making. J Food Sci 62:53–56

Liu YL, Chen YR, Kim MS, Chan DE, Lefcourt AM (2007) Development of simple algorithms for the detection of fecal contaminants on apples from visible/near infrared hyperspectral reflectance imaging. J Food Eng 81:412–418

Maertens K, Reyns P, Baerdemaeker JD (2004) On-line measurement of grain quality with nir technology. Trans ASAE 47:1135–1140

Manrique GD, Lajolo FM (2002) FT-IR spectroscopy as a tool for measuring degree of methyl esterification in pectins isolated from ripening papaya fruit. Postharvest Biol Technol 25:99–107

Martinsen P, Schaare P (1998) Measuring soluble solids distribution in kiwifruit using near-infrared imaging spectroscopy. Postharvest Biol Technol 14:271–281

Mateo MJ, O'Callaghan DJ, Everard CD, Fagan CC, Castillo M, Payne FA, O'Donnell CP (2009) Influence of curd cutting programme and stirring speed on the prediction of syneresis indices in cheese-making using NIR light backscatter. Lwt-Food Sci Technol 42:950–955

McGlone AV, Martinsen PJ (2004) Transmission measurements on intact apples moving at high speed. J Near Infrared Spectrosc 12:37–43

McGlone VA, Jordan RB, Martinsen PJ (2002) Vis/NIR estimation at harvest of pre- and post-storage quality indices for Royal Gala apple. Postharvest Biol Technol 25:135–144

McKenna D (2001) Measuring moisture in cheese by near infrared absorption spectroscopy. J AOAC Int 84:623–628

McMahon DJ, Brown RJ, Ernstrom CA (1984) Enzymic coagulation of milk casein micelles. J Dairy Sci 67:745–748

McQueen DH, Wilson R, Kinnunen A, Jensen EP (1995) Comparison of two infrared spectroscopic methods for cheese analysis. Talanta 42:2007–2015

Mitsumoto M, Maeda S, Mitsuhashi T, Ozawa S (1991) Near-infrared spectroscopy determination of physical and chemical characteristics in beef cuts. J Food Sci 56:1493–1496

Moros J, Inon FA, Khanmohammadi M, Garrigues S, De la Guardia M (2006) Evaluation of the application of attenuated total reflectance-Fourier transform infrared spectrometry (ATR-FT-IR) and chemometrics to the determination of nutritional parameters of yogurt samples. Anal Bioanal Chem 385:708–715

Navratil M, Cimander C, Mandenius CF (2004) On-line multisensor monitoring of yogurt and Filmjolk fermentations on production scale. J Agric Food Chem 52:415–420

Nicolaï BM, Beullens K, Bobelyn E, Peirs A, Saeys W, Theron KI, Lammertyn J (2007) Nondestructive measurement of fruit and vegetable quality by means of NIR spectroscopy: a review. Postharvest Biol Technol 46:99–118

Norris KH, Williams PC (1979) Determination of protein and moisture in hrs wheat by near-infrared reflectance spectroscopy1. Comparative-study of 12 instrumental methods. Cereal Foods World 24:459–459

Oberreuter H, Brodbeck A, von Stetten S, Goerges S, Scherer S (2003) Fourier-transform infrared (FT-IR) spectroscopy is a promising tool for monitoring the population dynamics of microorganisms in food stuff. Eur Food Res Technol 216:434–439

Ortiz C, Barreiro P, Correa E, Riquelme F, Ruiz-Altisent M (2001) PH-Postharvest technology: non-destructive Identification of woolly peaches using impact response and near-infrared spectroscopy. J Agric Eng Res 78:281–289

Park B, Chen YR, Hruschka WR, Shackelford SD, Koohmaraie M (2001) Principal component regression of near-infrared reflectance spectra for beef tenderness prediction. Trans ASAE 44:609–615

Park B, Lawrence KC, Windham WR, Smith DP (2006a) Performance of hyperspectral imaging system for poultry surface fecal contaminant detection. J Food Eng 75:340–348

Park B, Lawrence KC, Windham WR, Smith DP (2006b) Performance of supervised classification algorithms of hyperspectral imagery for identifying fecal and ingesta contaminants. Trans ASABE 49:2017–2024

Park B, Yoon SC, Lawrence KC, Windham WR (2007) Fisher linear discriminant analysis for improving fecal detection accuracy with hyperspectral images. Trans ASABE 50:2275–2283

Pasikatan MC, Dowell FE (2001) Sorting systems based on optical methods for detecting and removing seeds infested internally by insects or fungi: a review. Appl Spectrosc Rev 36:399–416

Payne FA, Hicks CL, Shen PS (1993) Predicting optimal cutting time of coagulating milk using diffuse reflectance. J Dairy Sci 76:48–61

Pearse MJ, Mackinlay AG (1989) Biochemical aspects of syneresis: a review. J Dairy Sci 72:1401–1407

Peirs A, Lammertyn J, Ooms K, Nicolal BM (2001) Prediction of the optimal picking date of different apple cultivars by means of VIS/NIR-spectroscopy. Postharvest Biol Technol 21:189–199

Pérez-Marín MD, Garrido-Varo A, Serradilla JM, Núnez N, Ares JL, Sánchez J (2001) Chemical and microbial analysis of goat's milk, cheese and whey by near infrared spectroscopy. In: Davies AMC, Cho RK (eds) Near infrared spectroscopy: proceedings of the 10th international conference. NIR Publications, West Sussex, pp 225–228

Pérez-Marín D, Sánchez M-T, Paz P, Soriano M-A, Guerrero J-E, Garrido-Varo A (2009) Non-destructive determination of quality parameters in nectarines during on-tree ripening and post-harvest storage. Postharvest Biol Technol 52:180–188

Pillonel L, Luginbuhl W, Picque D, Schaller E, Tabacchi R, Bosset JO (2003) Analytical methods for the determination of the geographic origin of Emmental cheese: mid- and near-infrared spectroscopy. Eur Food Res Technol 216:174–178
Rambla FJ, Garrigues S, de la Guardia M (1997) PLS-NIR determination of total sugar, glucose, fructose and sucrose in aqueous solutions of fruit juices. Anal Chim Acta 344:41–53
Reh C (2001) In-line and off-line FTIR measurements. In: Kress-Rogers E, Brimelow C (eds) Instrumentation and sensors for the food industry. Woodhead & CRC Press, Cambridge & Boca Raton, pp 213–232
Ren G, Chen F (1997) Determination of moisture content of ginseng by near infra-red reflectance spectroscopy. Food Chem 60:433–436
Ripoche A, Guillard AS (2001) Determination of fatty acid composition of pork fat by Fourier transform infrared spectroscopy. Meat Sci 58:299–304
Rodriguez Otero JL, Hermida M, Cepeda A (1994) Determination of fat, protein, and total solids in cheese by near-infrared reflectance spectroscopy. J AOAC Int 78:802–806
Rodriguez-Saona LE, Fry FS, McLaughlin MA, Calvey EM (2001) Rapid analysis of sugars in fruit juices by FT-NIR spectroscopy. Carbohydr Res 336:63–74
Saranwong S, Sornsrivichai J, Kawano S (2004) Prediction of ripe-stage eating quality of mango fruit from its harvest quality measured nondestructively by near infrared spectroscopy. Postharvest Biol Technol 31:137–145
Schmilovitch Ze, Mizrach A, Hoffman A, Egozi H, Fuchs Y (2000) Determination of mango physiological indices by near-infrared spectrometry. Postharvest Biol Technol 19:245–252
Scotter C (1990) Use of near infrared spectroscopy in the food industry with particular reference to its applications to on/in-line food processes. Food Control 1:142–149
Shao YN, He Y (2009) Measurement of soluble solids content and pH of yogurt using visible/near infrared spectroscopy and chemometrics. Food Bioprocess Technol 2:229–233
Shao YN, He Y, Feng SJ (2007) Measurement of yogurt internal quality through using Vis/NIR spectroscopy. Food Res Int 40:835–841
Singh PC, Bhamidipati S, Singh RK, Smith RS, Nelson PE (1996) Evaluation of in-line sensors for prediction of soluble and total solids/moisture in continuous processing of fruit juices. Food Control 7:141–148
Souto UTCP, Pontes MJC, Silva EC, Galvão RKH, Araújo MCU, Sanches FAC, Cunha FAS, Oliveira MSR (2010) UV-Vis spectrometric classification of coffees by SPA-LDA. Food Chem 119:368–371
Taifi N, Bakkali F, Faiz B, Moudden A, Maze G, D., D (2006) Characterization of the syneresis and the firmness of the milk gel using an ultrasonic technique. Meas Sci Technol 17:281–287
Tellier C, Mariette F, Guillement J-P, Marchel P (1993) Evolution of water proton muclear magnetic relaxation during milk coagulation and syneresis:structural implications. J Agric Food Chem 41:2259–2266
Togersen G, Isaksson T, Nilsen BN, Bakker EA, Hildrum KI (1999) On-line NIR analysis of fat, water and protein in industrial scale ground meat batches. Meat Sci 51:97–102
Togersen G, Arnesen JF, Nilsen BN, Hildrum KI (2003) On-line prediction of chemical composition of semi-frozen ground beef by non-invasive NIR spectroscopy. Meat Sci 63:515–523
Tsenkova R, Atanassova S, Kawano S, Toyoda K (2001) Somatic cell count determination in cow's milk by near-infrared spectroscopy: a new diagnostic tool. J Anim Sci 79:2550–2557
Ventura M, de Jager A, de Putter H, Roelofs FPMM (1998) Non-destructive determination of soluble solids in apple fruit by near infrared spectroscopy (NIRS). Postharvest Biol Technol 14:21–27
Villé H, Maes G, De Schrijver R, Spincemaille G, Rombouts G, Geers R (1995) Determination of phospholipid content of intramuscular fat by fourier transform infrared spectroscopy. J Meat Sci 41:283–291
Wang D, Ram MS, Dowell FE (2002) Classification of damaged soybean seeds using near-infrared spectroscopy. Trans ASAE 45:1943–1948
Williams PC (1979) Screening wheat for protein and hardness by near-infrared reflectance spectroscopy. Cereal Chem 56:169–172

Williams PC, Cordeiro HM (1979) Determination of protein and moisture in Hrs wheat by near-infrared reflectance spectroscopy 2. Influence of degrading factors, dockage and wheat variety. Cereal Foods World 24:460–460

Williams PC, Cordeiro HM (1981) Determination of protein and moisture in hard red spring wheat by near-infrared reflectance spectroscopy—influence of degrading factors, dockage, and wheat variety. Cereal Foods World 26:124–128

Windham WR, Lawrence KC, Park B, Buhr RJ (2003) Visible/NIR spectroscopy for characterizing fecal contamination of chicken carcasses. Trans ASAE 46:747–751

Wittrup C, Nørgaard L (1998) Rapid near infrared spectroscopic screening of chemical parameters in semi-hard cheese using chemometrics. J Dairy Sci 81:1803–1809

Woodcock T, Fagan CC, O'Donnell CP, Downey G (2008) Application of near and mid-infrared spectroscopy to determine cheese quality and authenticity. Food Bioprocess Technol 1:117–130

Chapter 5
Raman Spectroscopy

Ramazan Kizil and Joseph Irudayaraj

5.1 Introduction

Raman spectroscopy is becoming a popular analytical tool in the field of process analytical technology (PAT) as it offers both the sampling convenience of near-infrared (NIR) spectroscopy and the molecular specificity of mid-infrared (MIR) spectroscopy in a single measurement. The physical background of the Raman technique is based on inelastic scattering of light at distinct frequencies described as shifts (Raman shift) from the frequency of monochromatic incident radiation. Collecting distinct Raman shift due to light scattering, Raman spectroscopy allows probing chemical identity/structure of species in a process as a molecular fingerprinting tool. The technique is based on the laser technology and is evolving to provide better quality signal in a shorter time.

Having a unique physical basis, Raman spectroscopy is a molecule-specific, laser-excitation-dependent, sensitive and elaborate analytical tool that can be employed virtually to analysis of all kinds of food materials and ingredients. The uniqueness of Raman spectroscopy arises from the Raman effect discovered by Sir Venkata Raman in 1928. This special effect is due to the inelastic scattering of light upon interaction with matter which ends with photons having wave numbers less or greater than the monochromatic excitation at a fixed number.

Product quality and its assessment are manifest concerns of any industrial production. Having an integrated process analysis and control implementing proper analytical instrumentation will enable monitoring of the product quality throughout the whole process line. Hence, implementing PATs in manufacturing industries will

R. Kizil (✉)
Chemical Engineering Department, College of Chemical and Metallurgical Engineering,
Istanbul Technical University,
34469 Maslak, Istanbul, Turkey
e-mail: kizilr@itu.edu.tr

J. Irudayaraj
Agricultural & Biological Engineering, Purdue University,
225 S. University Street, 47907-2093 West Lafayette, IN, USA

C. P. O'Donnell et al. (eds.), *Process Analytical Technology for the Food Industry*,
Food Engineering Series, DOI 10.1007/978-1-4939-0311-5_5,

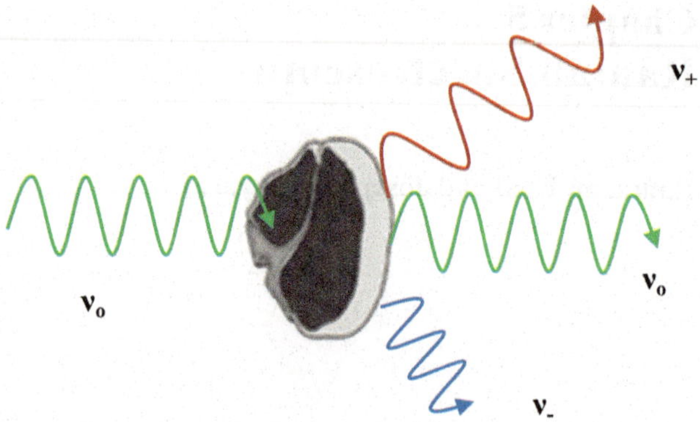

Fig. 5.1 Scattering of monochromatic light by sample; vo is the frequency of incident photon, (v+) or (v−) frequencies of the scattered photons

help optimize a process reducing production cost and increasing product quality and make the process safer and sustainable (Kueppers and Haider 2003).

Since the quality attributes of muscle foods including meat and fish are diverse and their quality loss is very fast, there is a need for inexpensive and rapid analytical tools that can be employed to quality assessment of these foods. Protein content and solubility, microbial safety, peroxide value, fatty acid composition, and texture are among important quality-related features of meat and fish. Raman as a vibrational spectroscopic technique is known to be a rapid and cost-efficient tool to elucidate quality attributes of foods at the molecular level.

Although PAT is often synonymously used with on-line analysis, it is considered as a holistic approach to process-related chemical identification and quantification (Hassel and Bowman 1998). Raman spectroscopy provides almost all possible modes of data acquisition enabling off-line, at-line, on-line, and in-line analyses in a rapid and cost-effective way. Even noncontact, noninvasive analysis can be conducted with the use of fiber-optic probes.

Raman spectroscopy is becoming one of the prominent analytical tools for identification and quantification of chemical constituents present in foodstuff. Because of its sensitivity towards the chemical composition at the molecular level, Raman technique holds a great potential for the quality assessment of food products.

The attractiveness of the Raman technique as a process analytical tool arises mainly from three distinct facts. First, the technique provides fruitful information about the chemical structure (molecular fingerprint) of virtually all kinds of samples in any physical state. Raman scattering of photons can be used to differentiate organic or inorganic functional groups in a molecule, as every chemical bond in a molecule has a specific vibrational energy. The second fact is that there is almost no need for sample preparation and no sample destruction is involved during the course of analysis. Finally, the most abundant molecule present in biological substances, water, does not interfere with the organic functional groups in Raman spectra. This feature makes Raman spectroscopy more attractive than the other spectroscopic techniques that suffer from the water/moisture-related strong signals. Raman spectroscopy is known as a nondestructive optical analytical technique whose principle is based on the inelastic scattering of photons. Light scattering phenomena by a food sample is depicted in Fig. 5.1.

5.2 Theory of Raman Spectroscopy

The first discovered scattering phenomenon, Rayleigh scattering, is related to polarizability of molecules. The extent of polarizability depends on the ability of a molecule's electrons to be distorted by an externally applied electric field due to electromagnetic radiation (Ball 2001). Rayleigh scattering of photons has the same frequency/energy with the incident light so it is considered to be elastic scattering. However, Raman scattering is a special inelastic scattering phenomenon of light interacting with a matter.

Raman scattering is inherently a weak process, because only one in every 10^6–10^8 photons scatter at a wavelength that is shifted from the incident. These scattered photons can either be "red shifted, so-called Stokes shift" or "blue shifted, so-called anti-Stokes shift" by increasing or decreasing, respectively, the ground vibrational energy level of molecules. In other words, Raman scattering produces a proportion of red shifted photons with less energy and blue shifted photons with higher energy than the incident monochromatic light. Thus, frequency shifts between the Raman scattered photons and the incident beam correspond to particular vibrational energy levels of molecules.

However, most of the scattered light has the same energy (wave number) with the incident that is known as Rayleigh scattering. Since Raman spectroscopy is based on detection of scattered light which is red or blue shifted, Rayleigh scattering has no contribution to Raman signal and it is normally rejected by a filter before entering the detector.

Despite some minor skepticism, the importance of the discovery of Raman scattering was well acknowledged by the scientific committee even before the discoverer, C. V. Raman, was awarded the Nobel Prize in 1930. In his personal letter to C. V. Raman on September 1929, Niels Bohr says: "I take this opportunity to express my most cordial congratulations to your great discovery of the new radiation phenomenon which has added so immensely to our knowledge of optics and atomic physics." Albert Einstein also admitted the importance of Raman's discovery, stating "C.V. Raman was the first to recognize and demonstrate that the energy of a photon can undergo partial transmission with matter. I still vividly recall the deep impression that this discovery made on us all." Raman scattering was also an early proof of the quantum theory of light and the photon theory postulated by Einstein.

What C. V. Raman observed in 1928 was anti-Stokes Raman scattering, yielding blue-shifted scattered photons that receive energy from bond vibration and which is quite weak. For anti-Stokes Raman scattering, the scattering photons have larger energy (lower wavelength) than the incoming photons. Although down-conversion in energy was well understood by other optical processes, such as fluorescence, at the time of discovery, this up-conversion of energy by scattering was an unidentified effect. That is why Raman called it "A New Type of Secondary Radiation" in his original paper (Raman 1928).

5.2.1 Selection Rules and Principles of Raman Technique

Although Raman scattering is used to investigate vibrational motions of molecular bonds, its principles do not apply to all types of vibrational modes. Selection rules should be considered to classify a molecular vibration as Raman or IR-active one. A vibration is said to be Raman active if it involves a change in polarizability of the molecule. However, IR-active vibrations are associated with dipole moment changes and not with the polarizability. The nature of polarizability (α) is essential for determination of the Raman activity of a vibration. Interaction of an electromagnetic field (photons) with a sample causes charge separation through alignment of the positively charged nuclei along the negative pole and electrons to the positive pole. An induced dipole moment (P) will be formed as a result of such charge separation processes. The induced dipole moment as a function of polarizability and electromagnetic field (E) is given by

$$P = \alpha . E \tag{5.1}$$

The below equation can be used to express the Raman dipole moment, incorporating the macroscopic laser electric field:

$$P = (L_{\mathrm{M}})^{1/4} \alpha . E, \tag{5.2}$$

where L_{M} is local field correction factor ($E_{\mathrm{Micro}} = (L_{\mathrm{M}})^{1/4} E_{\mathrm{Macro}}$). In vacuum and for gases, macroscopic and microscopic fields are identical (i.e., $L_{\mathrm{M}} = 1$), but L_{M} is 2.5 for water and 4 for organic solvents. Since both $\boldsymbol{P}$ and $\boldsymbol{E}$ are vectors and L_{M} is a measureable quantity (not given in the below equation), the induced dipole vector for a molecule is expressed as

$$\begin{pmatrix} P_x \\ P_y \\ P_z \end{pmatrix} = \begin{pmatrix} \alpha_{xx} & \alpha_{xy} & \alpha_{xz} \\ \alpha_{yx} & \alpha_{yy} & \alpha_{yz} \\ \alpha_{zx} & \alpha_{zy} & \alpha_{zz} \end{pmatrix} \begin{pmatrix} E_x \\ E_y \\ E_z \end{pmatrix} \tag{5.3}$$

Polarizability is a tensoral quantity and denoted by the first matrix on the right hand side of Eq. 5.3. The polarizability matrix has nine components (tensor) and a change in any of these components due to molecular vibrations can be a source of the Raman activity. The polarizability of a molecule is considered to be a constant value (α_0) at an equilibrium position (nuclear geometry) for molecules. When the molecule is free to vibrate around equilibrium position at an infinitesimal distance Δr, the instantaneous polarizability can be described by the Taylor series expansion of the components of polarizability with respect to the normal coordinates of the vibration (Ball 2001):

$$\alpha = \alpha_0 + \left(\frac{\partial \alpha}{\partial r} \right) \Delta r + \ . \ . \ . \tag{5.4}$$

where higher-order derivatives are omitted since they are several orders of magnitude smaller than the first derivative. Assuming simple harmonic motion of the vibration with a maximum amplitude of r_{max} and frequency v_{vib},

$$\Delta r = r_{max}\cos(2\pi v_{vib}t)$$

The incident light emerging from a laser source is monochromatic with a constant frequency of $v_{L,}$ induces an electromagnetic field, E, which also has harmonic behavior.

$$E = E_{max}\cos(2\pi v_{L}t) \quad (5.5)$$

where E_{max} is the maximum field. Arranging E, Δr, and α in equation 1, the induced dipole is given as

$$P = \alpha_0 E_{max}\cos(2\pi v_L t) + r_{max}E_{max}\left(\frac{\partial\alpha}{\partial r}\right)\cos(2\pi v_{vib}t)\cos(2\pi v_L t) \quad (5.6)$$

The product of cosine functions in the last equation can be rearranged yielding

$$P = \alpha_0 E_{max}\cos(2\pi v_L t) + \frac{r_{max}E_{max}}{2}\left(\frac{\partial\alpha}{\partial r}\right)\left(\cos\left(2\pi t(v_{vib}+v_L)\right)\cos\left(2\pi t(v_L - v_L)\right)\right)$$

(5.7)

Using a trigonometric identity of cosine function, the electrical dipole moment can be expressed as

$$P = P^{Rayleigh}\cos(2\pi v_L t) + P^{Raman}\left(\cos\left(2\pi t(v_L + v_{vib})\right)\cos\left(2\pi t v_L - v_{vib}\right)\right). \quad (5.8)$$

Since the scattering of a photon relies on the polarizability of molecules, the above equation can be used to define two different sources for scattering, namely Rayleigh (when the outgoing scattering has the same frequency with the incoming photon) and Raman (when scattering photons have frequencies different from that of the incoming, one as $(v_L + v_{vib})$anti-Stokes Raman scattering and the other $(v_L - v_{vib})$ Stokes Raman scattering) scattering. Equation 8 provides a gross selection rule for Raman-active modes by the classical treatment. According to the equation 8, the requirement for Rayleigh scattering is the presence of at least one nonzero component in the polarizability tensor (at the equilibrium). However, for Raman scattering, the derivative of the polarizability tensor with respect to the normal coordinate $\left(\frac{\partial\alpha}{\partial r}\right)$ must contain at least one nonzero term in any of the nine components. At this point, it will be noteworthy to define the selection rule for IR spectroscopy, since it provides complementary vibrational data to Raman spectroscopy. IR absorption emerges from vibrations as long as the derivative of the dipole moment is nonzero at the equilibrium position.

5.2.2 *Quantum Mechanical Description of Raman Scattering Effect*

The quantum mechanical representation of the effect of light (electromagnetic radiation) energy absorption at the molecular level can be schematically illustrated by the Jablonski diagram using the concepts of quantization of the electronic and vibrational energy levels. Electronic states are aligned vertically by increasing energy and grouped horizontally by spin multiplicity in the Jablonski diagram. A molecule at rest has two electronic states, namely the ground and excited, and each state has its own vibrational energy planes as illustrated in Fig. 5.2. In the ground state, the electrons occupy the lowest energy state as dictated by the Pauli exclusion principle. The ground state is named as singlet state S_0, at which the total spin is zero. The electronic excited state is the condition of a nucleus (or molecule) generated upon gaining energy (at room temperature) as a result of absorption of electromagnetic radiation or inelastic collisions with other particles. The excited state has higher energy than the ground state and is composed of a singlet state (S_1) with a total spin of zero and a triplet state (T_1) with a total spin of one. The higher energy electronic states can be denoted by S_2, T_2, etc., if needed. In addition to electronic states, internal atomic vibrations are considered as minor perturbation to the electronic state. Since the energy level of vibrational states is remarkably smaller (nearly 0.2 eV for most organic molecules) than that of the electronic states, vibrational states are regarded as the substructure of the electronic states with narrow energy spacing. The energy spacing between electronic states typically lies at the ultraviolet (UV) and visible (Vis) level of the electromagnetic spectrum (200–700 nm), whereas the energy spacing for vibronic bands is in the NIR region (800–1104 nm).

Excitation of a molecule by electromagnetic radiation prompts electrons to a higher electronic or vibrational energy level. After excitation, electrons return to the ground state by two distinct mechanisms, radiative and nonradiative (heat) energy loss. When light sources operating at the IR region are used to excite a molecule, the energy to be absorbed will be just enough to change its vibrational energy level at the ground electronic state. The change in the vibronic energy level of molecules is known as absorption (IR absorption). Excitation at UV or Vis regions of electromagnetic spectrum can bring electrons to an energy level different from the ground state. A virtual electronic state is generally used to describe energy levels which lie between the ground and first excited electronic states. Absorption of Vis or UV radiation can excite molecules from a vibrational level in the ground electronic state to another vibrational level in the electronic excited state, usually the first excited singlet state S_1.

Considering a transition from a lower to a higher vibrational level in the excited state upon optical energy absorption, three different but competing paths will be involved to release this excess internal energy, taking the electrons back to the lowest vibrational energy status. These paths are heat loss (nonradiative decay), molecular decomposition (photochemical reaction), and luminescence (radiative decay). The energy release by molecular decomposition results in the formation of a new molecule. Radiative decay is indicated by a return to the ground state through emission

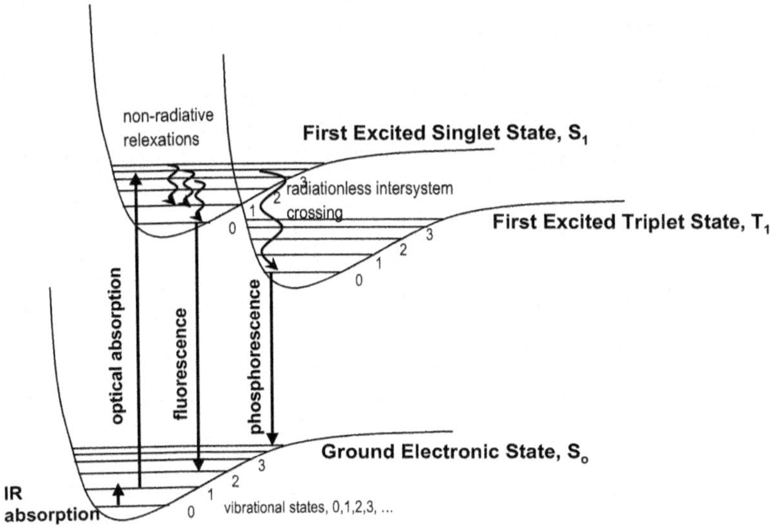

Fig. 5.2 Quantization of vibrational and electronic levels of molecules upon light interaction

of a photon. The transition phenomenon from a singlet excited state to a singlet ground state is called "fluorescence" with lifetimes on the order of nanoseconds or less. When relaxation into an excited triplet state occurs followed by the emission of a photon, the radiative phenomenon is called "phosphorescence" and exhibits a lifetime ranging from milliseconds to seconds (Lakowicz 1999).

Scattering of light either by Rayleigh or Raman phenomena is an instantaneous two-photon process which does not require any excitations corresponding to any electronic or vibronic transitions of the molecule. This implies that Raman scattering signals can be obtained using excitations at any frequency. However, the signal intensity is proportional to one-fourth power of the wavelength of laser excitation. For Raman spectroscopy, excitation sources operating at UV, Vis, or IR (usually NIR) can be utilized. At the standard temperature conditions, the Boltzmann distribution of vibrational states impose that most molecules are present in their ground vibrational state. In either Rayleigh or Raman scattering, molecules start and finish in the electronic ground state. Figure 5.3 is used to illustrate changes in the quantized energy of molecules during scattering process in the simplified (diatomic energy level) Jablonski diagram.

Since light energy comes in discrete quanta known as photons, whose energy is directly proportional to wavelength, the charged particles in molecules will not able to absorb the light permanently. Therefore, the charged particles will soon reemit the light. Upon interacting with light, the charged particles begin to shift into a new quantum state, a "virtual" state. This virtual electronic state may or may not be permanently allowed. The virtual state lies below the excited state, allowing the electron to reside in the real excited state for a certain time allowed by the uncertainty theorem (Pitt et al. 2005). The simultaneous absorption of an incident photon

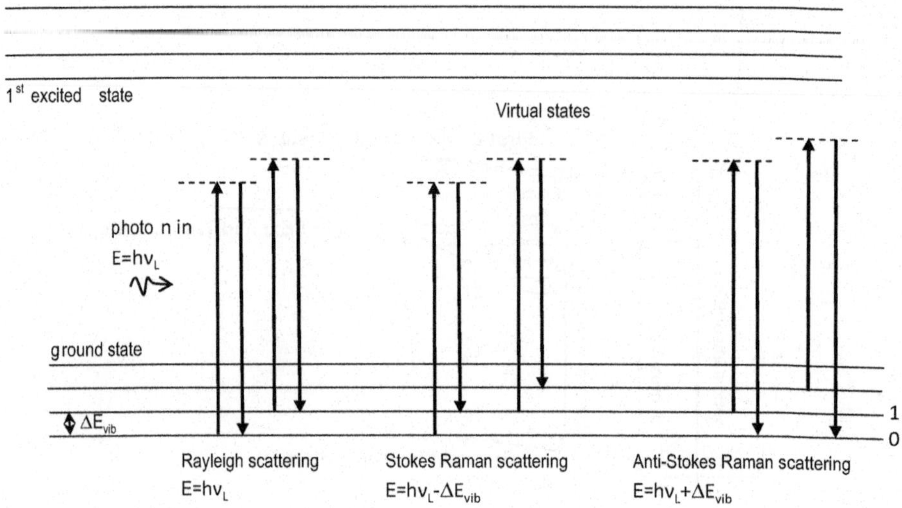

Fig. 5.3 Energy quantization for Rayleigh and Raman (Stokes and Anti-Stokes Raman scattering)

with energy of $h\nu_L$ and emission of scattered photons at different energies are given by: Rayleigh scattered photon $h\nu_L$, Stokes Raman scattered photon $h\nu_L - \Delta E_{vib}$, and anti-Stokes Raman scattering $h\nu_L + \Delta E_{vib}$. The Rayleigh scattering is the dominating process, since most photons scatter in this elastic (energy conserved) way. The process starting from the ground vibrational state (0) by absorption of energy and the promotion to a higher-energy excited vibrational state (1) is known as Stokes scattering in which the photon scatters inelastically at lower energy (shifted wavelength towards the red end of the spectrum). For this type of scattering, the energy difference between the incident and scattered photons corresponds to the energy of vibrational modes. Since the population state of a molecule is principally in its ground vibrational state at room temperature, Stokes scattering is the significant Raman scattering effect for experiments at room conditions. Whereas anti-Stokes process starts at a vibrationally excited state (1) that is thermally populated and ends with a return the ground vibrational state (0). The relative intensities of these two Raman processes rely on the population of the various states of the molecule which is characterized by the Boltzmann equation (Long 2002). However, anti-Stokes scattering will become significant compared to Stokes scattering at high temperatures. The difference in intensities of Raman bands in Stokes and anti-Stokes scattering can also be used to measure temperature. The positions of Stokes and anti-Stokes scattering lines present in Raman spectra are characteristic of materials that can be used for composition/structure identification.

Comparing the intensities of scattering processes, Rayleigh scattering is 10^3–10^4 times weaker than the incident light; at the same time, Raman scattering is 10^3–10^6 times weaker than Rayleigh scattering. Magnitude of Raman shifts is independent of the wavelength of excitation.

5.2.3 Special Raman Techniques for Signal Intensity Improvement

Conventional Raman scattering is a linear two-photon spontaneous process in which only a tiny fraction of the incoming light (incident) exchanges quanta of vibrational and/or rotational energy with the molecule under investigation and scatters as incoherent Stokes or anti-Stokes frequencies. Hence, inherently, conventional Raman is hampered by limited sensitivity. The signal intensity of classical Raman spectroscopy depends only on the fourth power of the frequency of the light used to induce the Raman effect. When the incident radiation matches an electronic transition of the irradiated molecule, a fraction of Raman-active vibrational modes of the molecule are enhanced, presenting from two to six orders of magnitude enhancement more than conventional (dispersive) Raman (McCreery 2000) enhancement.

The resonance effect can be used to selectively observe a molecule that possesses an absorption transition (electronic) in a complex medium when the absorption energy matches with the energy of the incoming photons. If a chemical group of a molecule is not involved in electronic transition following irradiation, there will be no resonance effect for that chemical group and no resonance Raman will be detected for this case. The most common resonance enhancement is Franck–Condon enhancement. This enhancement is related to expansion of a molecule along a component of the normal coordinate of a vibrational motion upon an electronic excitation. The more the molecule involves in expansion, the larger the enhancement factor.

Resonance enhancement does not necessarily start at a particular wavelength. Raman signal enhancement is also observed when the exciting laser is even a few hundred wave numbers below the electronic transition of a molecule. This is known as pre-resonance enhancement. The theory of resonance enhancement is beyond the scope of this study.

Surface-enhanced Raman scattering (SERS) is a special signal enhancement technique in which metallic surfaces (Au and Ag surfaces mostly) are utilized as host for absorption of the molecule of interest. The Raman signal enhances enormously at the metal–molecule interface. Because of the large signal amplification, SERS has been extensively applied to ultrasensitive analytical applications. The SERS enhancement involves chemical mechanism through the charge transfer between metal and molecule under test and also the electromagnetic mechanism coupling of the incident and scattered fields to the surface plasmons (collective excitations of the conduction electrons in metal).

5.2.4 Nonclassical (Nonlinear) Raman Effects

Linear Raman spectroscopy is a sensitive and unique analytical tool, but it often suffers from the fluorescence effect that may swamp the entire Raman signal of most biological samples. Generally, the nonlinear Raman technique is more sensitive than the linear Raman technique and the fluorescence effect can be effectively discriminated by the nonclassical Raman effect (Borman 1982). Important examples of these nonlinear Raman effects are stimulated Raman scattering (SRS), stimulated Raman scattering (SRC), stimulated Raman gain (SRG), hyper-Raman scattering (HRS), inverse Raman scattering (IRS), coherent anti-Stokes Raman spectroscopy (CARS), and coherent Stokes Raman Spectroscopy (CSRA).

SRS is a nonlinear signal amplification technique that may be used to enhance the signal as high as 10^{13}. The SRS was accidentally discovered when nitrobenzene cell was used as Q-switch of a ruby laser. In SRS phenomenon, light wave at a frequency of one of the Stokes-shifted photons is also incident on the sample being investigated simultaneously with a laser line at a defined frequency (Laubereau 1982). So, a significant enhancement in Raman signal at a frequency corresponding to the strongest Raman transition can be observed. However, this technique has not found much applicability as a chemical analytical tool; rather, it is often utilized for extending the tuning ranges of the existing lasers.

SRG and IRS are closely related nonlinear techniques. The former involves stimulated gain at a Stokes-shifted frequency and the latter is stimulated loss at an anti-Stokes-shifted frequency. IRS can also be called stimulated Raman loss. The stimulated gain is thought to be an induced emission at a Stokes scattering frequency and is achieved by directing two laser lines, which are a pump laser at frequency v_p and a probe laser, to the molecule being analyzed. The details of these techniques can be found elsewhere (Carreira and Horovitz 1982; Laubereau 1982). The coherent Raman output from nonlinear Raman techniques is highly direction dependent so that it can be easily separated from incoherent emissions, such as fluorescence. Compared to SRG, IRS provides better fluorescence rejection, since the signal beam is upshifted away from intense fluorescence emission.

Among nonlinear Raman techniques, CARS draw the most attention due to its extensive application in biological or chemical analysis. In CARS, two laser sources, a pump and probe laser, are mixed up and as a result of this a coherent beam at a third frequency is generated. Both CARS and CSRS involve vibrational coherences that can operate at electronically resonant conditions exhibiting resonance enhancement. Like other coherent techniques, the directional nature of CARS and CSRS avoids fluorescence interference in the spectra.

HRS was first reported by Terhune and coworkers in 1965 using a pulsed ruby laser as the excitation source. A small fraction of this incident energy was converted into scattered radiation shifted slightly from the second harmonic of the ruby laser. The intensity of scattered photons was proportional to the square of the laser intensity. HRS is a two-photon excited scattering phenomenon which provides complementary information to that of linear Raman and IR spectroscopes. The hyper Raman effect relies on the hyperpolarizability which arises from the quadratic electric field of the laser. Although it can reveal vibrational modes that cannot be recorded by classical Raman or IR spectroscopy, HRS yields extremely weak signals with scattering cross sections 35 orders of magnitude smaller than one-photon excited classical Raman and 15 orders of magnitude less significant than two-photon excited nonlinear Raman techniques (Kneipp et al. 2006).

5.3 Raman Instrumentation

A classical Raman instrumentation can be divided into three basic components:

- Laser source and excitation and optics
- Spectrometer and interface elements
- Detector and collecting optics

Although Raman instrumentation is divided into three main parts, there is a strong connection between the light source being used and the detecting system because of the operating region limitations in the detector technology. The Raman effect was first demonstrated by the discoverer using filtered sunlight as the monochromatic source of photons, a colored filter as the monochromator, and the human eye as the detector. The first easy-to-use Raman system, Cary 81 Raman spectrometer, was delivered to research laboratories in 1954; the system consisted of an excitation source of 3 kW helical Hg-arc, a Czerny-Littrow double monochromator with 1200 grooves per mm grating as the spectrograph, and a photomultiplier tube (PMT) as the detector (Adar et al. 2007). Following these early stages of the discovery, instrumentation was first adapted to laser technology and semiconductor-based charge-coupled device (CCD) detectors.

5.3.1 Raman Lasers

Lasers are used to excite molecules to generate Raman scattered photons out of samples which can be in solid (crystalline), liquid, or gaseous states. The lasers are coupled with spectrographs and then signal is transmitted to a detector. This brings inherently weak optical throughputs and necessities utilizing a high-power laser in Raman instrumentation. The use of lasers as a Raman source was first suggested in 1961 by the inventor of the laser, Charles H. Townes, who worked as a professor of physics at Columbia University, MIT, and UC-Berkeley, and as a consultant to Bell Labs (Townes 1961). The theory of Raman and phonon maser was presented by Townes and students (Chiao et al. 1964). Weber and Porto (1965) conducted first laser-based Raman experiments for gases using a continuous wave (CW) He–Ne laser. The involvement of laser in Raman experiments then was followed using other lasers such as the argon and krypton ion lasers which are expensive and require a supply of cooling water (Pitt et al. 2005).

The advances in spectrograph design enhanced the optical throughput especially after 1990 which was an important milestone for the use of lower-power lasers (typically 10–50 mW) in Raman instrumentation (Pitt et al. 2005). Well-collimated light at high-power output and thermal stabilization are required criteria for a laser to be implemented in fiber-optic-based industrial process control. New solid-state lasers meet these criteria and are attractive for industrial applications requiring rugged and portable systems. It is important to note that all laser operations and fiber optics must have safety-appropriate switches and operational warning signals.

The need for integrating lasers operating at NIR region (Nd-YAG laser, 1064 nm wavelength) within a Raman system has emerged as a necessity to avoid fluorescence effects that is frequently encountered in conventional Raman experiments of biological samples with Vis region excitation sources (Barbillat and da Silva 1997). This integration is usually not useful unless the system operates in conjunction with interferometric multiplexing detection.

There are lasers operating at UV, Vis, or NIR regions that can be used as the excitation source in Raman instrumentation. Lasers providing UV emission lines between 229 and 264 nm are typically larger in size and expensive than other lasers. Although laser line in UV region can bring thermal damages to samples, the sensitivity is automatically increased due to the fourth-power wavelength dependence of Raman scattering intensity. The selection for the laser source might seem simple and is mainly based on the wavelength selection and nominal output power of the selected laser line. Most Raman spectrometers use CW lasers, as long as wavelength and output stability are maintained. This is because Raman scattered signal sent to the detector is directly related to the power quantitatively serious problem and wavelength reference for Raman shift calculation of the laser. When Raman spectroscopy is involved in a continuous on-line process monitoring, system should provide stable peak intensities with minimal drift or fluctuations. The Raman laser line must be coupled to systems with detectors sensitive to Raman scattering being generated. For a typical Raman instrumentation, lasers operating at all Vis and UV wavelengths are currently coupled to multichannel detectors, which are generally CCDs, and NIR region lasers except 830 nm excitation are being used with germanium (Ge) or indium gallium arsenide (InGaAs) single-channel detectors along with the Fourier transform Raman (FT-Raman) configuration that provides multiplexing (Adar 2001).

5.3.2 *Spectrometer (Spectrograph)*

Spectrographs are the first instruments used to measure wavelengths, which still hold an important position in spectroscopy, particularly when equipped with optical multichannel analyzers (Laser Spectroscopy book). Spectrographs are optical instruments that form images from light that passes the entrance slit for slivering of the light then the images are laterally separated for different wavelengths (λ) (spectral lines) of the incident radiation. This lateral dispersion is achieved either by spectral dispersion in prisms or by diffraction on plane or concave reflection gratings. The spectral lines are identified by wavelength when captured by the detector.

The spectrometer design is essentially based on the detector being used. When the signal is detected with a PMT, it is required to transfer a tight focus to the exit slit of spectrometer to maintain optical spectral resolution. A spectrometer should minimize the tray light in the optical system. Czerny–Turner and Sergent-Rozey are two most common spectrometer designs; one uses slits along the horizontal trace and the other has the slits long the vertical line. The Czerney–Turner design essentially consists of two gratings; the first is focusing optics and three adjustable slits to filter out the Rayleigh scattered light and the second has gratings disperse the Raman light.

When lasers were first introduced to Raman instrumentation in 1970s, the spectrometers of that period were based on either double or triple monochromator designs. The first monochromator of double system and the first two of the triple monochromator system were used to filter out the stray light.

5.3.3 Detectors

Multiplexed fashion behavior like the photographic plate as well as electronic response in a wide dynamic range with high signal to noise characteristic, digital data storage, and manipulation are the main requirements for a detector to be employed in a Raman system.

The early multichannel Raman detectors were the intensified photodiode array and imaging PMT having good signal quality but limited dynamic range capture (Adar et al. 2007). These detectors have currently been replaced by the high-sensitivity CCD array detectors known as the CCD camera technology, delivering the speed and dynamic range benefits to multichannel detection. CCD detector is a two-dimensional rectangular array composed of photosensitive elements with a typical resolution of 1024×256 pixels and operation rate of 50 kHz. During data collection, the horizontal rows of CCD are used to register wavelength, while the column elements are binned to record the intensity of scattered light at each wavelength. There are three different configurations available for the current CCD technology; front illuminated, thinned back-illuminated, and front or back deep depletion.

5.3.4 Base Raman Instrument Technology

Based on the signal detection technology, Raman system can be divided into two subgroups as dispersive Raman and interferometric FT-Raman instruments. Dispersive systems operate mainly at the Vis and UV region and should include components that provide adequate resolution to measure the spectral lines and removal of the interference and stray light from the laser line. The resolution varies with the wavelength region and the dispersion of the grating within the spectral region of interest. As the resolution increases, the spectral/wavelength range coverage of the Raman system decreases. The most common grating density for the scanning grating operation in the Vis region is between 1000 and 2000 lines per mm. The early dispersive Raman instruments involved either double or triple monochromators but with the advent of holographic notch filters (HNFs) for filtering out the laser line the need for a sophisticated spectrograph in the instrumentation was lifted. The current technology enables both scanning and static grating mode of operations with array detectors such as CCD camera.

In a Raman experiment, a laser source is used to shine the incident light on the sample. Then inelastically scattered photons are collected through collection optics and filtered out from the cosmic rays and other interfering photons, such as

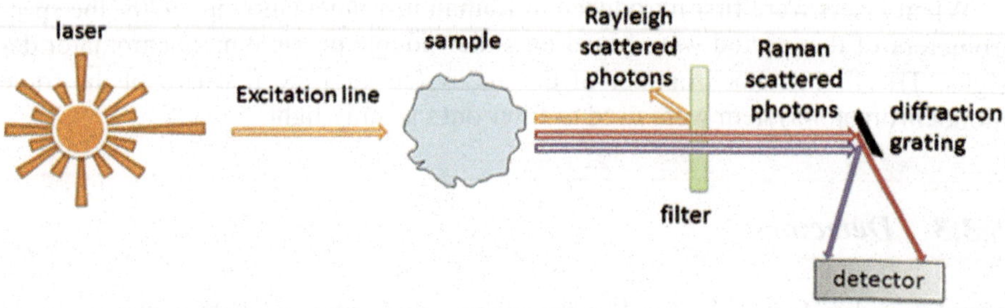

Fig. 5.4 A typical Raman experiment setup used to collect inelastically scattered light

Rayleigh scattering. The final step in the experiment is detection of the scattered photons using a suitable detector. A PMT (in old technology) or CCD camera is employed to record Raman signals when UV and Vis sources are involved in excitation (Fig. 5.4). InGaAs and Ge semiconductor detectors, the former operates at the room temperature but the latter must be cooled by liquid nitrogen, are needed if the excitation is done with a NIR range operating laser.

Filtering can be done using HNFs. A grating system separates photons by their wave numbers and Raman signal is recorded by a CCD camera. In a research grade dispersive Raman spectrometer, the sample is analyzed through a microscope.

FT-Raman configuration is specially designed to collect fluorescence-free and wavelength-stable spectra from a wide range of samples spanning from crystals to biological tissues (Chase 1987). A typical FT-Raman spectroscopy instrumentation is composed of an excitation source, a neodymium-doped yttrium aluminum garnet (Nd:YAG) laser which operates at NIR region (1064 nm), a sampling unit with proper collection optics (90 or 180° in general), an interferometer comprising a fixed and moving mirrors, and a detecting unit. A He–Ne laser (633-nm "red" excitation) is made collinear to the excitation laser, to aid obtaining the most proper optical alignment. In addition to guiding the Nd:YAG laser, the He–Ne laser in FT-Raman instrumentation serves as the reference laser for the operation of the interferometer. A line filter helps filtering plasma lines and the light is focused on a sample sitting in a sample compartment. In FT-Raman experiments, there is no need for sample preparation; the signal can be collected virtually from food samples at any physical state. The scattered light is then collected using either 180 or 90 back-scattering geometry and passes through a dielectric filter to cutoff the 1064- and 633-nm laser lines before entering the detector which can be nitrogen cooled Ge or room temperature InGaAs.

5.3.5 Process Raman Measurements with Fiber-optic Probes

It is convent to use a flow system to provide an on-line line sampling for Raman measurements when the analyte is a fluid. Continuous pumping of fluidic samples from a process vessel to a simple glass tube (vial) in the sample compartment of

Raman spectrometer will be enough to acquire Raman signals during any timeframe of the process, because excitation line can penetrate a couple of centimeters in the glass without any interference. Since samples can be probed through fine glass containers, Raman spectroscopy offers a great flexibility compared to MIR and NIR approaches. For a Raman experiment, path length is of minor consideration; however, optical assembly must be designed to ensure that the feature of interest matches the beam and focal point for reproducibility of measurements (Smith and Dent 2005).

Nonetheless, flow cells may not satisfy the real process conditions, such as temperature, and create lag from the real time due to transportation of samples from the process to the spectrometer. Delivering the incident light and conducting the scattered radiation to a specific destination through fiber-optic probes provides a convenient means of adoption for on-line measurements in the field of spectroscopy, eliminating the time delay due to transportation of the sample from one medium to the other. Raman spectroscopy becomes more sampling flexible when the system is integrated with a fiber-optic probe.

A fiber-optic probe to be coupled to a Raman spectrometer has to perform both excitation and collection at the same time. The challenges of using a fiber-optic probe for Raman experiments include contraction of the laser beam into the delivery fiber optic (generally one fiber is used to deliver incident light), the probe design architecture, and contamination protection and coupling of the collecting fiber optics to the detector. The laser line is connected to fiber-optic probe using a coupler that attaches the fiber directly to output of the laser. A simple lens in the coupler can be used to inject the laser line into the fiber, precluding the misalignment. Although a number of different fiber-optic designs are available, two design architecture is predominantly used. The first one is n-around-1 design and the second one is coaxially filtered probe design. The n-around-1 probe design uses only one fiber for the delivery of the incident radiation and the rest of fiber for collection of scattered light.

5.4 PAT Applications of Raman Spectroscopy in Food Industry

PAT applications using Raman spectroscopy starts always with reliable attribution of the scattering bands/modes (features or markers) represented in the spectra to the sample's chemical functional groups. Determination of the functional groups in the spectra is a critical step to identify the sample being investigated. A Raman spectroscopic analysis begins with the identification of a sample from its molecular fingerprints in the spectral form and continues with monitoring of alterations in band positions and intensities of some characteristic chemical functional groups, such as C–C, C–H, C=O, S–S, etc. Monitoring the spectral fingerprints, an analyzer can probe the chemical nature of a system under investigation during processing. Spectral information provides a good overview about the chemical structure (identity), conformational state of macromolecules such as proteins, level of saturation,

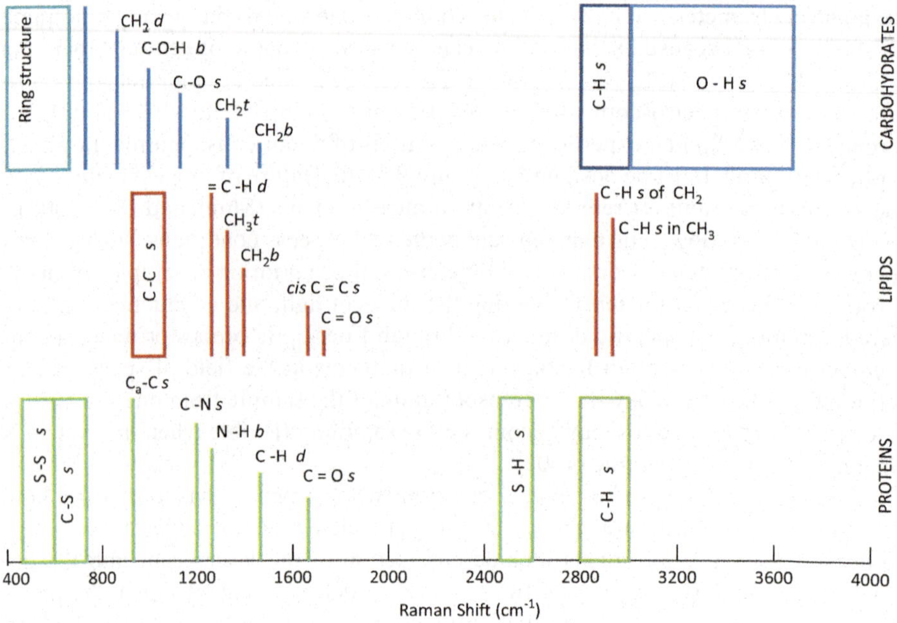

Fig. 5.5 Raman chart for major food components, carbohydrates (*top*), lipids (*middle*), and protein Raman peaks are given at the *bottom*. *s*: stretch *d*: deformation *b*: bend *t*: twist

degree of crystallinity or polymorphism, and inter- and intramolecular interactions in the system of interest.

Attribution of Raman scattering modes to corresponding chemical functional groups is a key to most spectroscopic analysis. Since Raman technique is a well-established molecular technique, assignments of chemical functional groups to most Raman modes are available through Raman atlases or references. Using such references, even a novice of Raman spectroscopy can make successful band/peak assignments, matching the spectral features with the chemical structure. In addition to the direct vibrational-mode-related qualitative analysis, experienced users can predict conformational states of proteins, changes in the level of unsaturation of lipids, and decomposition of polysaccharides from their Raman responses. Such information offered by Raman spectroscopy may provide useful insights on the assessment of process-related changes in a food system. The chart given in Fig. 5.5 summarizes significant Raman modes due to molecular stretch, bend, and other vibrations for major food components; proteins, fats, and carbohydrates.

Raman spectroscopy has been shown to provide promising benefits for the quality assessment of food products. Authentication of foods in a short time (in several minutes) and nondestructively using Raman and other vibrational spectroscopic techniques satisfy the current needs for quality screen of food products off-line. However, on-line monitor of foods or food processing using Raman spectroscopy is limited partly due to selectivity limitation of the technique or the lack of calibration. The current PAT using Raman spectroscopy in food science and industry is mainly concentrated on the assessment of the quality of the final food product. Although Raman spectroscopy can be successfully employed to monitor physicochemical properties of food products during food processing, the complexity of most food systems precludes the industrial-scale application of the technique for on-line monitoring of product quality. Because the technique is very sensitive to detect a change in the composition or structure of macromolecules such as proteins, it is not selective enough to probe only an interested change. The interested change cannot be directly evaluated from the information-rich spectral information, rather special data pretreatments are often necessary to draw a conclusion about the process. In addition, the product regulations are not very strict as it is for the pharmaceutical industry that may tolerate processing inefficiencies to a degree. However, deployment of Raman spectroscopy to on-line monitoring of the properties or the quality indices of food products during manufacturing or processing becomes an attractive alternative with the improvements in the instrumentation enabling fast and high-quality data collection and advances in quantification of spectral information through chemometrics.

Raman systems with Vis-range excitation should operate under dark ambient condition to prevent interference from outside light during process monitoring. However, laser sources operating at NIR field do not require dark condition for recording ambient noise- and fluorescence-free signal.

Comparing to MIR spectroscopy, Raman measurements provide spectral features not interfered by the water (moisture)-related vibrational modes which are particularly important for the analyses of food and biological materials. In addition, Raman spectroscopy offers sampling flexibility of NIR spectroscopy, making measurements available for the samples in a glass vial, or the laser line can be extended to remote a sample using a fiber-optic probe.

Carbohydrates show strong Raman responses in the fingerprint region due to various stretch-, bend-, twist-, and deformation-related vibrations of carbon atoms making covalent bonding with H, C, and O atoms (Fig. 5.5). The most distinct Raman bands for polysaccharides arise from the skeletal mode vibrations of the reign structure recorded below 800 cm^{-1} (Fig. 5.6). The Raman peak at 478 cm^{-1} can directly be attributed to the polymeric glucose chain. This Raman mode can also provide quantitative information about the degree of polymerization of polysaccharides (Celedon and Aguilera 2002). During the gelatinization process of starches, the Raman features due to crystalline phase almost disappear, whereas the O–H stretch mode becomes apparent as the uptake of water into starch granules becomes significant. Although the signal quality for starch gels was not good enough with the dispersive Raman setup used at that time, Kim et al. (1989) proposed a molecular-based mechanism for the gelatinization process of starch, screening the Raman responses of starches and their gel forms using a Vis excitation.

Utilizing an FT-Raman spectrometer with NIR excitation for starch gel analysis, the signal quality of gelatinized polysaccharides was noticeably improved, diminishing the fluorescence effect. Using an FT-Raman configuration, Schuster et al. (2000) reported on-line monitoring of both gelatinization of potato starch and the subsequent enzymatic hydrolysis of the gel first to dextrin (liquefaction) and then to

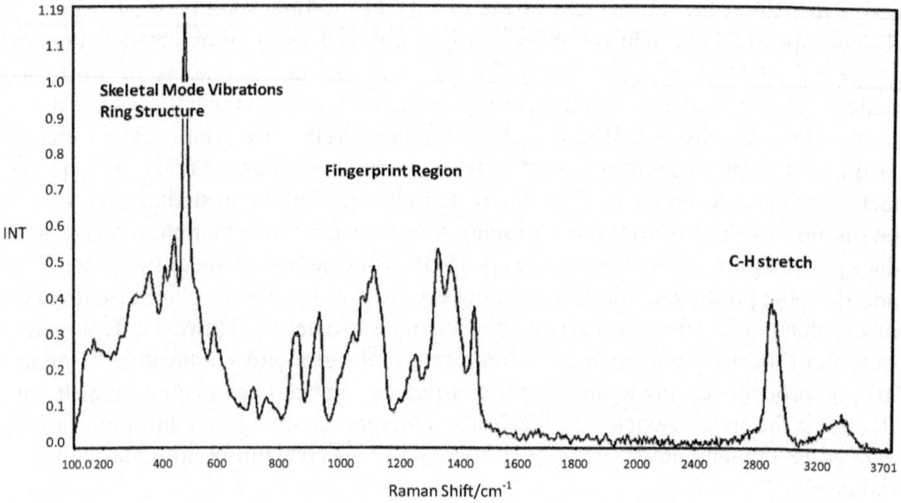

Fig. 5.6 FT-Raman spectra of granular starch

glucose (saccharification). The conventional process monitoring approach for starch gelatinization and liquefaction is based on viscosity measurements. The progress of saccharification is traditionally assessed by determination of dextrose equivalents through wet chemical techniques. With this novel spectrometric approach, Schuster et al. (2000) studied on-line monitoring of the gelatinization, liquefaction, and saccharification processes separately using a flow-through cell attached to the spectrometer. The flow cell was made up of a plastic material with a small CaF_2 window. The backside of the cell was coated with palladium mirror so that better-quality Raman scattering could be acquired by backscattering. Circulating 0.4 ml/min mixture from an experimental size (10 ml) gelatinization vessel to the flow cell, gelatinization at various temperatures were monitored by continuously collecting spectra, then liquefaction was monitored at 50 °C by adding α-amylase enzyme into the gelatinization reactor. Finally, the enzymatic saccharification of 50 g/l dextrin was monitored in a different batch, hydrolyzing dextrin to the simple sugar glucose. The chemical complexity of the reaction mixture makes the use of advanced data-processing techniques necessary to extract useful quantitative information from Raman data which is overlapped by the solvent and interfered by the byproducts. The same reaction system nowadays can be monitored by means of a fiber-optic probe attached to the Raman spectrometer and Raman scattering can be measured that can be used to collect through a thin transparent glass window of the reactor or the optical probe itself can be directly immersed in the mixture.

The swelling and gelatinization processes of starches at elevated temperatures result in apparent decrease in the intensity of all granular starch Raman bands except the water-uptake-related O–H bend (1633 cm^{-1}) and stretch (3213 cm^{-1}) modes. Monitoring changes in the relative intensity of major potato starch gel bands in the temperature range between 50 and 85 °C for 30 min, Schuster et al. (2000) draw an

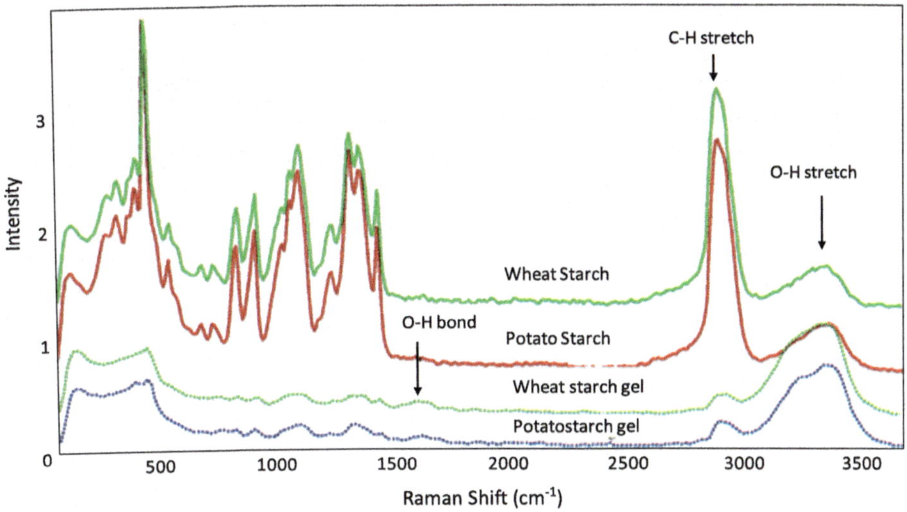

Fig. 5.7 FT-Raman spectra of different starches and their gels. (Kizil and Irudayaraj 2006)

attention to the break of intra- and intermolecular hydrogen bonds and formation of new hydrogen bonds with free water uptake upon the swelling of starch particles.

Although the signal quality gets poor due to optical transitions in the starch-water dispersion during gelatinization, Raman spectroscopy provides clear evidences for the heat-induced deformation in the structure of starch. As shown in Fig. 5.7, granular starch gives strong skeletal mode (ring structure)-related Raman responses but they diminish drastically upon gelatinization (Kizil and Irudayaraj 2006). The skeletal mode C–C stretch was shown to be useful for determination of the loss of birefringence in starch granules as a function of temperature (Celedon and Aguilera 2002).

Monitoring enzymatic hydrolysis reaction of pre-gelatinized potato starch, Schuster et al. (2000) have shown that the process brings three remarkable spectral changes. The first one is diminishing of all Raman modes originating from the skeletal vibrations (below 800 cm^{-1}) of granular starch. The second process-related change is an apparent wavelength shift of Raman mode at 947 cm^{-1} due to the cleavage of glycosidic linkages of the polysaccharide by the enzymatic activity. The last one is intensifying and broadening of the O–H-bend-related Raman peaks around 1633 cm^{-1}.

FT-Raman monitoring of the subsequent enzymatic reaction (saccharification) in which dextrins produced by liquefaction is converted into dextrose results in total disappearance of the skeletal responses of granular starch (480 and 735 cm^{-1}) in the Raman spectra and new bands appeared as the reaction proceeds (Fig. 5.8). Saccharification-related Raman bands in the low wavelength region were detected at 426 and 519 cm^{-1} and noted as Raman markers for the reaction. The enzymatic breakdown of the α,1–4 glycosidic linkage was monitored as a peak shift from

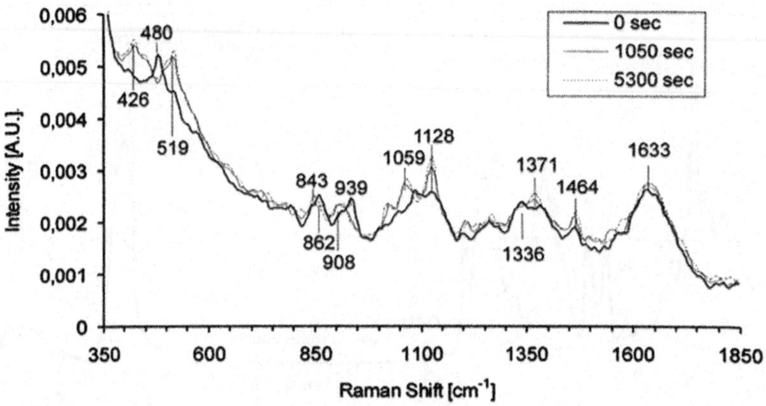

Fig. 5.8 FT-Raman monitoring of saccharification. (Schuster et al. 2000)

934 cm^{-1} to the left. Since glucose is the final product, the process results in the formation of remarkable C–O (1128 cm^{-1}) and C–O–H (1059 cm^{-1})-related Raman peaks.

Schuster et al. (2000) demonstrated the potential of Raman spectroscopy in on-line monitoring of an industrially important process for glucose production. They showed that the structure of materials used in the process can be determined and reactions can be characterized by the wavelength shifts, disappearance of polysaccharide Raman peaks, or appearance of new bands due to chemical conversion. This study infers that on-line monitoring of such a production scheme by means of Raman spectroscopy can provide a good control over the process, improving the product quality and yield through each reaction.

Since fat and oil do not mix with water and other hydrophilic constituents in food, Raman spectroscopic studies of lipids offer easy measurement advantage. The study of oils with Raman spectroscopy became more accessible after the development of instrumentation which enables rejection of Rayleigh line with holographic filtering, high-sensitivity CCD detectors, and reduced fluorescence through the use of NIR-region excitation lasers. The high detection sensitivity of Raman towards π-bonds present in oils and fats makes it an attractive analytical tool to predict the level saturation and determine isomerization ratio of lipids.

Compositional analysis offered by Raman spectroscopy makes this technique very useful for determination of the nutritional and dietary value of food products. The fatty acid compositions of adipose tissue from chicken, beef, pork, and lamb can be predicted using even a home-built Raman spectrometer system operating with a 785-nm laser excitation at nearly 100-mW laser power and detecting the signal using a spectrograph coupled to a liquid-nitrogen-cooled CCD detector (Beattie et al. 2006). The Raman signal accumulated during 60 s of data acquisition period in the 270 and 1900 cm^{-1} Raman shift region was used to probe C–C, C=C, C–O, and C=O stretch modes and C–H bend vibrations of lipids (Fig. 5.9). Table 5.1 presents tentative assignment of bands in the Raman spectra of adipose tissue as presented

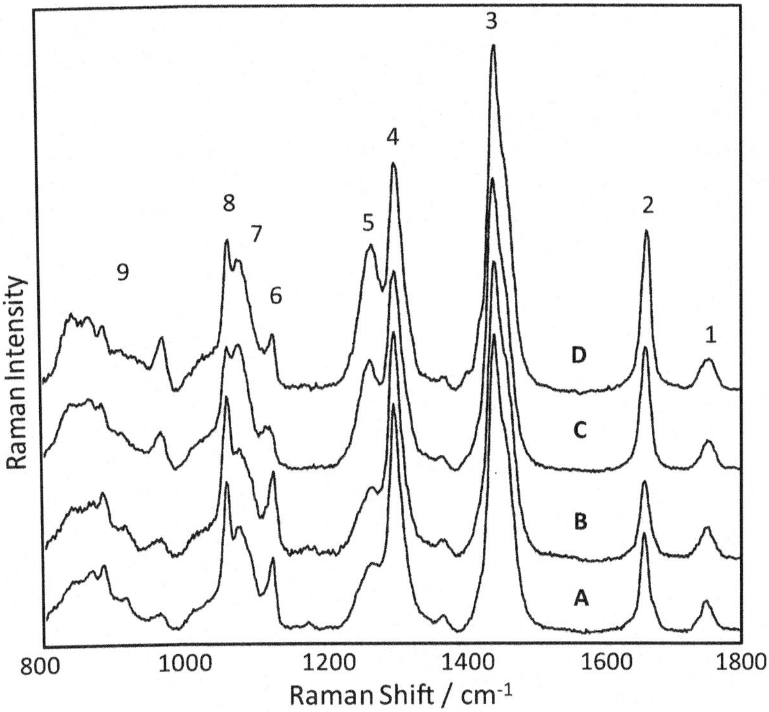

Fig. 5.9 Normalized average Raman spectra collected from A- beef, B- lamb, C- chicken, and D- pork adipose tissue. The numbering of the bands refers to tentative band assignments given in Table 5.1. (Beattie et al. 2006)

Table 5.1 Tentative assignment of bands in the Raman spectra of adipose tissue. (Modified from Beattie et al. 2006)

Band #	Band position range (cm^{-1})	Tentative assignment
1	1730–1750	C=O carbonyl stretch
2	1650–1680	*cis* (1650–1660, *trans* (1660–1680) C=C olefinic stretch
3	1400–1500	(CH_2) methylene scissor deformation
4	1295–1305	(CH_2) methylene twist deformation
5	1250–1280	(=CH2) in-plane *cis* olefinic H bend
6	1100–1135	(C–C) in-phase aliphatic C–C stretch all *trans*
7	1080–1090	(C–C) aliphatic C–C stretch all *gauche*
8	1060–1065	(C–C) aliphatic out of phase stretch all *trans*
9	800–920	Combination of (C_1–C_2) stretch, (CH_3) rocking, C–O stretch

by Beattie et al. (2006). Multivariate calibration models were developed for the estimation of the bulk properties of adipose tissue and a detailed fatty acid profiling by partial least squares (PLS) regression provided prediction accuracy as good as other spectroscopic techniques such as NIR and Fourier transform infrared (FTIR). These results suggest that the approach can be considered as a secondary standard for the determination of chemical properties of animal fat. Such a spectroscopic approach, however, can be put to a practical on-line application, in which noncontact and rapid measurements with no sample preparation necessity can be maintained.

Raman spectroscopic information can also allow classification of fats and oils (adipose tissue samples) different from their botanical or animal origins, applying a multivariate discrimination tool to specific Raman data. Baeten et al. (1998) studied classification of 38 oils and fats from 21 different sources based on the level of unsaturation, applying principal component analysis (PCA) to Raman data acquired with a FT-Raman spectrometer. Beattie et al. (2007) studied classification of adipose tissues from four different animal sources using a dispersive Raman system operating with a 785 nm a diode laser which produces less fluorescence background comparing to visible laser lines.

For the classification of adipose tissues from their Raman responses, a discriminative multivariate statistical tool was employed. Raman spectra were collected from 255 different samples, of which 102 were used to build a discrimiation model while 153 were used for testing. To reduce the computational complexity in the analysis, PLS or PCA was applied to Raman spectral information, reducing the data to a manageable size. Finally, the latent variables obtained from the data reduction technique were transferred to a nonsupervised (nondirected) discrimination tool called partial least squares discriminant analysis (PLSDA). Figure 5.10 shows the separation of Raman observations obtained from four different sources. The circles depict the 90 % confidence intervals for the statistical grouping decision. Based on the Fig. 5.10, most of the lamb and beef adipose tissue samples are well separated and grouped into specific spaces defined by two discriminant score variates, discriminant score (DS) 1 and 2. Only few beef (marked by an arrow) and lamb tissue samples were represented out of their groups described by the 90 % confidence intervals. However, the discriminant analysis was not able make a good separation for porcine and chicken adipose tissues, so both samples were clustered in the same group. The results show that Raman spectroscopy can be effectively used to differentiate ruminant adipose tissues from the nonruminant ones.

The potential of Raman spectroscopy for rapid and nondestructive prediction of omega-6 and omega-3 fatty acids in pork adipose tissue was demonstrated utilizing a contact fiber-optic probe (Olsen et al. 2008). The Raman system for this study consists of a NIR region operating laser for excitation, a ball-type fiber-optic probe with a 6-mm-diameter spherical sapphire lens as the sampling device and an electrothermal-cooled CCD as the detector. Raman spectra were collected from both the dorsal and ventral sides of cube-like cut adipose tissues by direct contacting the probe with the sample at 20 °C and from melted fat around 50 °C by immersing the probe into the melt.

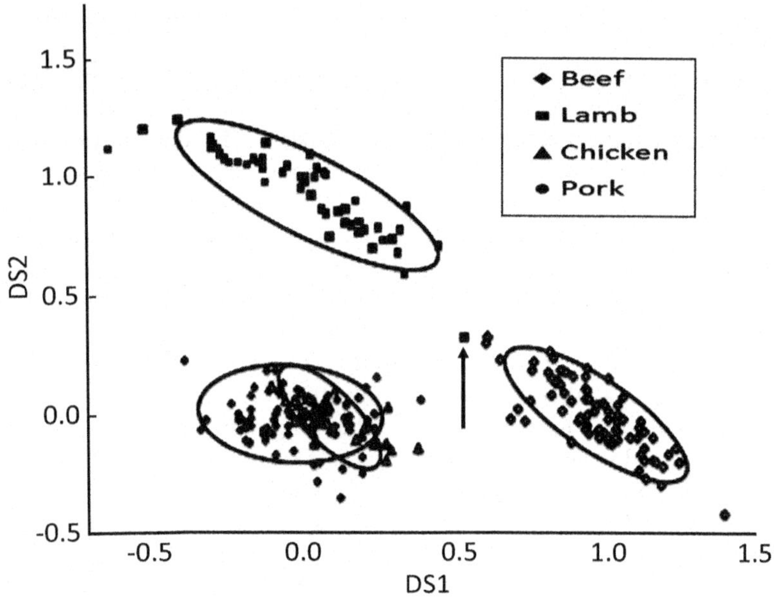

Fig. 5.10 Classification of adipose tissues from four animal sources using PLSDA applied to dispersive Raman results. (Beattie 2007)

The acquisition time to collect Raman spectra was 20 s with the CCD detector. The Raman spectra from the melted samples were compared with FTIR results. For the analysis of omega-6 and omega-3 fatty acids, double bond structures of C=C are important. Raman spectroscopy gives free distinct peaks related to the HC=CH vibrations at 3009, 1655, and 1263 cm^{-1}.

The band at 3009 cm^{-1} is apparent in both Raman and FTIR spectra and attributed to the asymmetric stretch of *cis*-configured C=C bonds that is often used as an indicator of degrees of unsaturation. The peak at 1655 cm^{-1} appears to be one of the prominent bands in Raman spectra of oils and fats, whereas it has very weak IR absorption. This peak is a characteristic of *cis*-configured C=C vibration of olefins and shifts between 1670 and 1650 cm^{-1} depending on various factors. This peak serves as a basis to predict the iodine value of margarines. The symmetric rocking of unconjugated *cis*-configured double bond scatter at 1263 cm^{-1}. This Raman peak is free from interference of any *trans* bond vibrations and can be used to estimate the level of *cis* unsaturation of lipids.

Raman spectroscopic investigations, which offer in situ nondestructive analysis with sampling convention of fiber-optic probes, can be adapted to rapid yet accurate prediction of omega fatty acid contents of adipose tissues, making measurements possible directly from carcasses and their cuts. This approach is expected to bring a positive effect in marketing of meat products. Raman spectroscopy with fiber-optic probing and a suitable regression method can be employed to on-line sort of the carcasses or cuts based on the content of omega-3 and omega-6 fatty acids in slaughterhouses (Olsen et al. 2008).

Raman spectroscopy was shown to be a rapid and nondestructive quality assessment tool for the analysis of edible oils and fats. Although during the 1980s and 1990s the FT-Raman setup was the primary choice for the analysis of fats and oils, which is a very powerful technique for overcoming the fluorescence effect, recent advances in Raman instrumentation made it possible to study oils and fats or high-lipid-containing samples with a good signal quality in a shorter time with a dispersive Raman system. The botanical origin (Baeten et al. 1998; Yang et al. 2005), authenticity (Yang and Irudayaraj 2000; El-Abbasy et al. 2009), and degree of unsaturation (Barthus and Poppi 2001; Silveira et al. 2009) of edible oils and fats can be determined using chemical information obtained from selected Raman bands. For example, adulteration of olive oil with cheap vegetable oils such as sunflower, olive pomace, or hazelnut oil, which has long been a problem in the olive oil market, was studied to determine the amount of the adulterant, treating the spectral data by means of a regression tool such as PCA or PLS. Comparing the result of NIR, FTIR, and FT-Raman methods, the best standard error of prediction of the correlation was obtained from FT-Raman data as low as 1.72 % for pomace oil adulteration (Yang and Irudayaraj 2001). The detection limit for the sunflower oil in olive oil was determined as below as 500 ppm by using dispersive Raman spectroscopy with Vis range excitation, that actually is not a practical adulteration ratio for economical purposes but important for the trace element analysis (El-Abassy et al. 2009). FT-Raman spectroscopy was also demonstrated to be an effective tool for determination of conjugated linoleic acids in cow milk fat within the range of 0.56–4.70 % (Bernuy et al. 2008).

Oxidation is a major chemical deterioration process for lipids. Lipid oxidation is a complex free-radical chain reaction, causing a variety of chemical and physical changes in lipids. The products of lipid oxidation can be identified using gas or liquid chromatography systems. Among the lipid oxidation products, aldehydes, which are responsible for the development of rancidity, can be determined through specific wet chemical techniques such as anisidine and 2-thiobarbituric acid (TBA) value analyses. The progress of the lipid oxidation reaction can be studied by chemical parameters such as peroxide value and UV absorbance measurements at 232 and 270 nm. Muik et al. (2005) showed that the oxidation process of edible oils can be monitored using FT-Raman spectroscopy, providing information about the extent of oxidation and the products. Formation of aldehydes, isomerizations of *cis* and *trans* double bonds, and conjugation of double bonds were determined screening the C–C stretch region of oil's Raman spectra. Anisidine value and spectrophotometric response at 270 nm as the primary standards can be correlated with Raman results to determine the extent of oxidation.

Proteins are one of the major components of nutritionally valuable foods, such as animal-based foods and some plant-based food like cereals and nuts. Proteins are composed of amino acids that are connected by peptide linkages. The structure of a protein is determined by the linear assortment of amino acid backbone chained via peptide linkages and progress through ordered arrangements (secondary, tertiary, and quaternary structures) to form the final molecular shape. There is a close relationship between the structure and function for protein molecules. The myofibrillar proteins are the main contributors to the textural and functional properties of muscle foods. The functional properties such as foamability, water-holding capacity, solubility, gel-forming ability, emulsifying activity and stability and viscosity, and the textural properties of high-protein-containing foods depend to a great extent on the structure of their protein content. Changes in the structure of proteins can happen through reversible or irreversible processes. Once the structure of a protein changes irreversibly, that protein will no longer exhibit its inherent functional properties possessed in its ordered form. Elucidation of protein structure using spectroscopic or other optical techniques as well as x-ray crystallography has always been of prime interest in biological fields since functional properties and chemical nature of a protein can be predicted from its structure.

Monitoring food processing operations using rapid and nondestructive Raman spectroscopic technique, which provides fruitful information about the chemical structure of proteins at the molecular level, provides a convenient and accurate means of on-line detection of food quality as well as prediction of the functional and textural properties of the final processed product.

Raman spectra contain plenty of information about protein backbone structure and side chains in the form of well-separated bands/peaks with varying intensities at distinct wave numbers. Protein Raman spectra contain nearly 30 peaks arising mainly from the vibrational modes of the backbone structure and some contributions from vibrations of amino acid side chains (Thomas 2002).

Raman peaks of proteins in the 500–1750 cm^{-1} region are attributed to the vibrations of the backbone structure (amide bond linked chain) and amino acid side changes as well as ringed structures of particular amino acids, such as tyrosine "tyr," phenylalanine "phe," and tryptophan "trp." The Raman bands attributed to sulfur groups arise from the C–S and S–S stretch modes recorded below 700 cm^{-1}. Since the disulfide bridge is structure stabilizing bonding, the S–S (508 cm^{-1}) vibration modes are useful for determination of protein structure.

Proteins are composed of amino acids that are connected through covalent linkages, the peptide (amide) bonds. The linear assortment of amino acids via peptide bonds is known as protein backbone and the sequence of amino acids along the backbone determines the primary structure of proteins. Hence, length of a protein backbone structure can be represented by the number of peptide bonds covalently attached. Peptide bonds involve in the formation of hydrogen bonds between the water molecules and it also makes intramolecular hydrogen bonding with proteins own N–H and C=O groups. Hydrogen bonding is a weak force but important for stabilization of the protein structure along with other molecular interactions. Therefore, it is important to probe backbone amide and sulfur group vibrations to understand how protein structure changes. Secondary structure is ordered arrangement (conformation) of amino acids by means of a variety of molecular interactions in localized regions. The secondary structure of a protein can be predicted from the molecular vibrations of the amide group. These vibrations are C=O-, N–H-, and

C–N-related in-plane vibrations that appear in five different regions having different names for each region. These are amide I, II, III, and amide A and B regions.

Among the amide-related vibrational modes, the most prominent one is the amide I envelope appearing between 1590 and 1720 cm^{-1} of protein Raman spectra. The amide I envelope is closely associated with the carbonyl (C=O) groups (nearly 80 % of all vibrational contribution) that serve as acceptors for the formation of hydrogen bonding in the backbone. Therefore, this Raman feature can be used to predict the secondary structure of proteins. Because the amide I region can be decomposed into various component peaks and most of these component peaks correspond to helical, turn, random coil, and sheet motifs at specific wavelengths. However, some of the component bands may arise from amino acid side chain and unordered structures. The component peaks in the amide I region are determined through curve-fitting algorithms using Gaussian or Lorentzian functions/shapes. Various mathematical approximations can be utilized to obtain the best fit to the amide I band with a series of component peaks. The peaks of the component bands obtained by curve fitting technique correspond to designated structural motifs, helix, turn, and sheet. Raman band associated with the α-helical content of protein appears in the interval of 1650–1658 cm^{-1} in the amide I region. Proteins with high helical content generally give Raman amide I response centered at 1650 cm^{-1}. This is clearly observed in the Raman spectra of muscle tissue samples from fresh hake fish (Fig. 5.11). Other prominent protein motifs, such as *β*-sheets give a rise to Raman signals in the range of 1665–1680 cm^{-1}, and random coil structures exhibit Raman bands at the interval of 1660–665 cm^{-1}.

Although freezing is a useful food-preserving technique, which provides considerable increase in the storage life of a food product, it may bring detrimental changes in the textural and functional properties by changing the structure of proteins. Freezing of fish products tends to cause formation of high molecular weight aggregates of proteins, which mainly contain myofibrillar proteins, myosin, and actin. Raman spectroscopic investigations have indicated that freezing-related changes in protein structure can be elucidated through the analysis of the amide I and III bands and the nature of hydrophobic interaction of proteins can be probed from the C–H stretch (2800–3000 cm^{-1}; Herrero 2008). Careche and Li-Chan (1997) showed that myosin isolated from cod underwent structural modifications associated with a loss in the helical content due to freezing at −20 °C and the protein involved in more hydrophobic interactions after frozen storage or formaldehyde addition. Structural changes in actomyosin from ling cod during cold storage were also determined in the presence of various cryoprotectants (Sultanbawa and Li-Chan 2001). Low-temperature gelation process of protein extracts from underutilized fish species for making surimi, which is utilized as imitation shell fish in the Western world, causes unfolding of helical motifs and progress with cross-linking of the protein network (Ogawa et al. 1999). In addition to myofibrillar proteins, structure of connective tissue from fresh and thawed cod fish can also be studied using Raman spectroscopy. Badii and Howell (2003) reported cold storage, formaldehyde, and fish oil addition effects on the structure of fish collagen. They showed that the amide I band centered at 1660 cm^{-1} along with secondary structure stabilizing bonds and interaction-re-

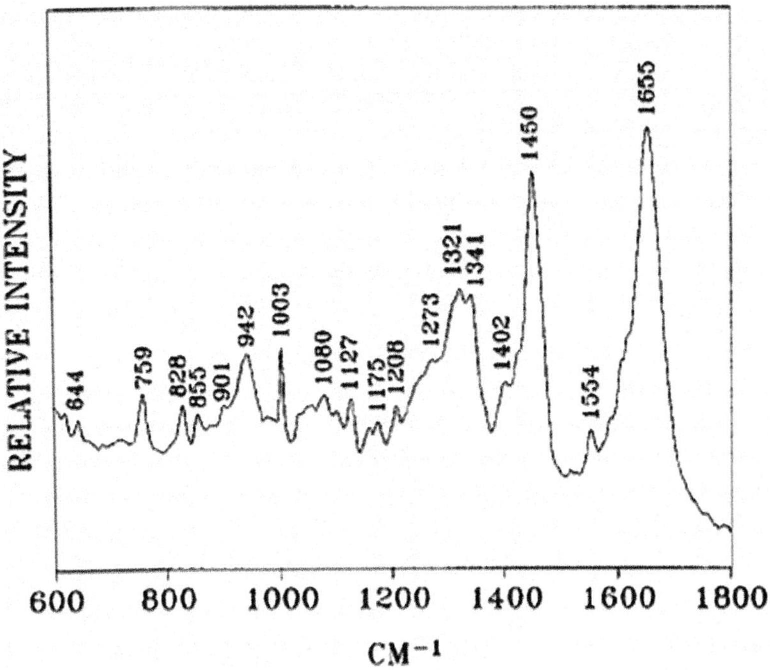

Fig. 5.11 Raman peaks of fresh muscle tissue of hake, showing a prominent peak at 1665 cm^{-1} is due to helical content. (Careche et al. 1999)

lated Raman modes are prone to changes in the presence of fish oil and formaldehyde. Raman spectra show chemical treatment effects on the structure of fish collagen particularly at the amide I band centered at 1660 cm^{-1}, CH_3 symmetric stretch (937 cm^{-1}), Phe ring bend (1034 cm^{-1}), protein backbone CN stretch (1128 cm^{-1}), and aliphatic CH3 asymmetric rock (1160 cm^{-1}) as well as Trp bands at 1554 and 1451 cm^{-1} regions. Changes in the intensity and shape of Raman scattering bands due to the hydrophobic amino acids such as Phe and Trp suggest that some amino acid residues along the backbone get buried or exposed to the solvent upon chemical treatments. It is inferred that the structure of fish collagen and changes in the hydrophobic residues such as Phe and Trp can be monitored by Raman spectroscopy. These spectroscopically determined parameters can provide useful insights about quality improvement of stored food. It is also reported that protein Raman bands in 876–951 cm^{-1} and 3071–3128 cm^{-1} region can also be used to predict water-holding capacity of fresh porcine meat (Pedersen et al. 2003).

Although all these studies provide valuable information about the secondary structure and hydrogen bonding assembly of food proteins, the analyses were performed on proteins isolated and purified from a variety of fish species, not directly recording responses from the food (fish) itself. The chemical composition of food such as meat and fish is complicated and includes various organic compounds all contributing to the Raman spectra. Raman spectra of complex food have often overlapping peaks and the dynamic nature of food may create artifact effect when analyzing the conformational state of proteins. For this reason, a good sampling strategy, posttreatment of data, and the choice of Raman instrumentation type are critical factors to obtain reliable results about the structure and composition of food samples.

There are a limited number of Raman spectroscopic studies that investigate food quality directly recording relevant quality measures from the food itself. Marquardt and Wold (2004) studied the potential of rapid quality screening of fish by a dispersive Raman system operating at 785-nm excitation line. The Raman system utilized in this study is capable of measuring a selected region along the fish samples by means of a fiber-optic head and a microscope objective. Coupling the probe head with a microscope objective, the Raman system can provide real-time screening of fish samples. The quality of fish is determined by attributes including chemical composition, color, freshness, and texture. Since Raman spectroscopy provides molecular level insight about the chemical composition of fish when good signal-to-noise ratio is attained and bands are well separated, the quality measure related to the chemical composition can be well determined. Collagen and fat composition was investigated to correlate quality of fish with the chemical composition in this study. Apart from the freshness, color for some fish species, for example salmon, is the main factor that effects the consumer perception. The pink color of salmon fish is attributed to deposition of carotenoids astaxanthin and canthaxantin. Collagen is protein that significantly contributes to the texture of fish. Fat is another key quality effecting parameter which prominently influences the nutritional, sensory, and processing properties of fatty fishes like salmon. The authors attempted to determine quality of different fishes using Raman spectra measured from the intact fish fillets of species with eight different content of carotenoids, collagen, or fat without performing a reference measurement. Analyzing simultaneously the deposition of carotenoids in salmon muscle, collagen type (I and V), and fats, Marquardt and Wold (2004) showed the potential of Raman spectroscopy for rapid quality screening of fish species. In the analysis, it was critical to apply a data posttreatment routine to remove fluoresce background from Raman spectra. A fourth-order polynomial function was fitted to the raw spectra and then subtracted from the raw spectrum to eliminate fluorescence background intensities in the Raman spectra. In addition to the fluorescence interference, spectral overlap of chemical components can also occur when a complex sample such as muscle tissue is analyzed.

The collagen type I was differentiated from type V whose fluorescence was recorded to be half of the type I. The carotenoids in salmon was detected at 1005, 1159, and 1518 cm^{-1} wave numbers. Based on the carotenoids and fat-related Raman responses, white fish samples were discriminated using PCA. The heme group of myoglobin can also show resonance enhancement. Monitoring the structural changes of myoglobin in pressure-treated porcine meat, meat quality was studied by directly collecting spectra from meat samples by a resonance Raman spectrometer operating at 413 nm excitation laser source.

Prediction of the sensory quality of meat and meat products has long been interest to food industry since the sensory quality attributes predominantly influence the

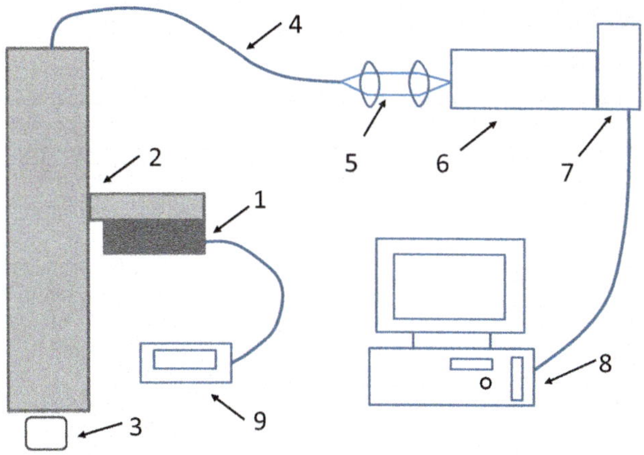

Fig. 5.12 Raman experimental setup employed to in situ investigation of meat. (*1*) laser module, (*2*) Raman optical bench, (*3*) meat sample, (*4*) optical fiber, (*5*) launch optics, (*6*) spectrometer, (*7*) CCD camera, (*8*) computer, (*9*) laser driver. (Sowoidnich et al. 2010)

consumer's perception. Consumer perception is generally measured through sensory panels as a time-consuming and destructive quality-assurance method in meat industry. The prediction of consumer response using an instrumental technique that offers fast and nondestructive testing and sensitivity provides an alternative approach for the assessment of meat quality. Such an instrumental technique can be a promising tool of quality assessment for the food-processing industry and even for the inspectors and consumers. Although NIR and also MIR spectroscopy have been successfully employed to prediction of meat quality and compared with the results of shear force value, which is the primary instrumental measure for the meat's tenderness, the more elaborate and moisture insensitive vibrational spectroscopic tool, Raman spectroscopy has not yet fully been implemented in this area. Only a few preliminary investigations for the prediction of the sensory attributes of meat and meat products were performed by Raman spectroscopy (Beattie et al. 2004; Brondum et al. 2000)

In the analysis of complex food system, proper selection of instrumentation to reduce optical interferences other than Raman scattering, the resonance effect forming lasing, filtering the signal, and experience of multivariate data treatment improve the quality of Raman measurements and provide better interpretations of the results for quality assessment purposes. For example, resonance Raman spectroscopy was employed to monitor progress of lipid oxidation in mechanically separated turkey over the oxidative bleaching of β-carotene (Kathirvel et al. 2008). This study showed the potential of resonance enhancement of β-carotene peaks as the marker in meat samples for determination of the extent of lipid oxidation in turkey meat. As Raman instrumentation technology evolves, new Raman probe systems are being developed and coupled to Raman spectrometers that can make on-line investigation of food attractive. A prototype Raman probe attached to a customized 671-nm microsystem diode laser was used for in situ investigation of meat spoilage (Sowoidnich et al. 2010). The experimental setup is illustrated in Fig. 5.12 and the system was used to monitor biochemical changes in pork *longissimus dorsi* due to storage.

References

Adar F (2001) Evolution and revolution of Raman instrumentation—application of available technologies to spectroscopy and microscopy. In Lewis IR, Edwards HGM (eds) Handbook of Raman spectroscopy. Marcell Dekker, New York, pp 11–40

Adar F, Delhaye M, DaSilva E (2007) Evolution of instrumentation for detection of the Raman effect as driven by available technologies and by developing applications. J Chem Educ 84(1):50–60

Badii F, Howell NK (2003) Elucidation of the effect of formaldehyde and lipids on collagen from frozen cod by Raman spectroscopy and differential scanning calorimetry. J Agric Food Chem 51:1440–1446

Baeten V, Hourant P, Morales MT, Aparicio R (1998) Oil and fat classification by FT-Raman spectroscopy. J Agric Food Chem 46(7):2638–2646

Ball DW (2001) Theory of Raman spectroscopy. Spectroscopy 16(11):32–34

Barbillat J, da Silva E (1997) Near infra-red Raman spectroscopy with dispersive instruments and multichannel detection. Spectrochim ACTA Part A Mol Biomol Spectrosc 53(13):2411–2422

Barthus RC, Poppi RJ (2001) Determination of the total unsaturation in vegetable oils by Fourier transform Raman spectroscopy and multivariate calibration. Vib Spectrosc 26(1):99–105

Beattie RJ, Bell SJ, Farmer LJ, Moss BW, Desmond PD (2004) Preliminary investigation of the application of Raman spectroscopy to the prediction of the sensory quality of beef silverside. Meat Sci 66(4):903–913

Beattie JR, Bell SEJ, Borgaard C, Fearon A, Moss BW (2006) Prediction of adipose tissue composition using Raman spectroscopy: average properties and individual fatty acids. Lipids 41:287–293

Beattie JR, Bell SEJ, Borgaard C, Fearon A, Moss BW (2007) Classification of adipose tissue species using Raman spectroscopy. Lipids 42:679–685

Bernuy B, Meurens M, Mignolet E, Larondelle Y (2008) Performance comparison of UV and FT-Raman spectroscopy in the determination of conjugated linoleic acids in cow milk fat. J Agric Food Chem 56(4):1159–1163

Borman S (1982) On-linear Raman-spectroscopy. Anal Chem 54(9):A021

Brondum J, Byrne DV, Bak LS, Bertelsen G, Engelsen SB (2000) Warmed-over flavour in porcine meat—a combined spectroscopic, sensory and chemometric study. Meat Sci 54(1):83–95

Careche M, Li-Chan ECY (1997) Structural changes in cod myosin after modification with formaldehyde or frozen storage. J Food Sci 62(4):717–723

Careche A, Herrero AM, Rodriguez-Casado A, Del Mazo ML, Carmona P (1999) Structural changes of Hake (*Merluccius merluccius* L.) fillets: effects of freezing and frozen storage. J Agric Food Chem 47:952–957

Carreira LA, Horovitz ML (1982) Resonance Raman gain and Raman loss spectroscopy. In: Kiefer W, Long DA (eds) Nonlinear Raman spectroscopy and its chemical applications. NATO Advanced Study Institutes Series. Series C: Mathematical and Physical Sciences. R. Reidel Publishing Company, Dordrecht, The Netherlands

Celedon A, Aguilera JM (2002) Applications of microprobe Raman spectroscopy in food science. Food Sci Technol Int 8(2):101–108

Chase B (1987) Fourier-transform Raman-spectroscopy. Anal Chem 59(14):A881

Chiao RY, Townes CH, Stoicheff BP (1964) Stimulated Brillouin Scattering and Coherent Generation of Intense Hypersonic Waves. Phys Rev Lett 12:592–595

El-Abbasy RM, Donfack P, Materny A (2009) Visible Raman spectroscopy for the discrimination of olive oils from different vegetable oils and the detection of adulteration. J Raman Spectrosc 40(9):1284–1289

Hassel CD, Bowman EM (1998) Process analytical chemistry for spectroscopists. Appl Spectrosc 52(1):18A–29A

Herrero AM (2008) Raman spectroscopy for monitoring protein structure in muscle food systems. Crit Rev Food Sci Nutr 48(6):512–523

Kathirvel P, Ermakov IV, Gellermann W, Mai J, Richards MP (2008) Resonance Raman monitoring of lipid oxidation in muscle foods. Int J Food Sci Technol 43(11):2095–2099
Kim IH, Yeh AI, Zhao BL, Wang SS (1989) Gelatinization kinetics of starch by using Raman-spectroscopy. Biotechnol Progr 5(4):172–174
Kizil R, Irudayaraj J (2006) Discrimination of irradiated starch gels using FT-Raman spectroscopy and chemometrics. J Agric Food Chem 54:13–18
Kneipp J, Kneipp H, Kneipp K (2006) Two-photon vibrational spectroscopy for biosciences based on surface-enhanced hyper-Raman scattering. Proc Natl Acad Sci U S A 103(46):17149–17153
Kueppers S, Haider M (2003) Process analytical chemistry—future trends in industry. Anal Bioanal Chem 376:313–315
Lakowicz JR (1999) Principles of fluorescence spectroscopy, 2nd edn. Kluwer/Plenum, New York
Laubereau A (1982) Stimulated Raman scattering. In: Kiefer W, Long DA (eds) Nonlinear Raman spectroscopy and its chemical applications. NATO Advanced Study Institutes Series. Series C: Mathematical and Physical Sciences. R. Reidel Publishing Company, Dordrecht, The Netherlands
Long DA (2002) The Raman effect: A unified treatment of the theory and Raman scattering by molecules. John Wiley & Sons Ltd., Chichester, UK
Marquardt BJ, Wold JP (2004). Raman analysis of fish: a potential for rapid quality screening. Lebensm-Wiss Technol 37:1–8
McCreery RL (2000) Raman spectroscopy for chemical analysis. Wiley-Interscience, New York, NY USA
Muik B, Lendl B, Molina-Diaz A, Ayora-Canada MJ (2005) Direct monitoring of lipid oxidation in edible oils by Fourier transform Raman spectroscopy. Chem Phys Lipid 134(2):173–182
Ogawa M, Nakamura S, Horimoto Y, An H, Tsuchiya T, Nakai S (1999) Raman spectroscopic study of changes in fish actomyosin during setting. J Agric Food Chem 47(8):3309–3318
Olsen EF, Rukke EO, Egelandsdal B, Isakson T (2008) Determination of omega-3 and omega-6 fatty acids in pork adipose tissue with nondestructive Raman and Fourier transform infrared spectroscopy. Appl Spectrosc 62(9):968–974
Pedersen DK, Morel S, Andersen HJ, Engelsen SB (2003) Early prediction of water-holding capacity in meat by multivariate vibrational spectroscopy. Meat Sci 65(1):581–592
Pitt GD, Batchelder DN, Bennett R, Bormett RW, Hayward IP, Smith BJE, Williams KPJ, Yang YY, Baldwin KJ, Webster S (2005) Engineering aspects and applications of the new Raman instrumentation. IEE Proc Sci Measurement Technol 152(6):241–318
Raman CV, Krishnan KS (1928) A new type of secondary radiation. Nature 121:501–502
Schuster KC, Ehmoser H, Gapes JR, Lendl B (2000) On-line FT-Raman spectroscopic monitoring of starch gelatinization and enzyme catalysed starch hydrolysis. Vib Spectrosc 22:181–190
Silveira FL, Silveira L, Villaverde AB, Pacheco MTT, Pasqualucci CA (2009) Use of dispersive raman spectroscopy in the determination of unsaturated fat in commercial edible oil- and fat-containing industrialized foods. Instrum Sci Technol 38(1):107–123
Smith E, Dent G (2005) Modern Raman spectroscopy—a practical approach. Wiley, Chichester, pp 210
Sowoidnich K, Schimidt H, Maiwald M, Sumf B, Kronfeldt HT (2010) Application of diode-laser Raman spectroscopy for in situ investigation of meat spoilage. Food Bioprocess Technol 3:878–882
Sultanbawa Y, Li-Chan ECY (2001) Structural changes in natural actomyosin and surimi from ling cod (*Ophiodon elongatus*) during-frozen storage in the absence or presence of cryoprotectants. J Agric Food Chem 49(10):4716–4725
Thomas GJ (2002) New structural insights from Raman Spectroscopy of proteins and their assemblies. Biopolymers 67(4–5):214–225
Townes CH (1961) Advances in quantum electronics. Columbia University Press, New York, pp 3–11 (Singer JR (ed))
Weber A, Porto SPS (1965) He-Ne laser as a light source for high-resolution Raman spectroscopy. J Opt Soc Am 55:1033–1034

Yang H, Irudayaraj J (2000) Rapid determination of vitamin c by NIR, MIR and FT-Raman techniques. J Pharm Pharmacol 54(9):1247–1255
Yang H, Irudayaraj J (2001) Comparison of near-infrared, Fourier transform-infrared, and Fourier transform-Raman methods for determining olive pomace oil adulteration in extra virgin olive oil. J Am Oil Chem Soc 78(9):889–895.
Yang H, Irudayaraj J, Paradkar MM (2005) Discriminant analysis of edible oils and fats by FTIR, FT-NIR and FT-Raman spectroscopy. Food Chem 93(1):25–32

Chapter 6
Magnetic Resonance Imaging and Nuclear Magnetic Resonance Spectroscopy

Michael J. McCarthy and Kathryn L. McCarthy

6.1 Introduction

Nuclear magnetic resonance (NMR) and magnetic resonance imaging (MRI)-based sensors provide a wealth of information concerning the properties of a material. Examples of the type of information that can be obtained from magnetic resonance (MR)-based measurements are shown in Fig. 6.1. Figure 6.1a is a plot correlating the MR proton signal to moisture content in navy beans, while Fig. 6.1b shows variations in internal properties of Roma tomatoes as a function of maturity. MR-derived information is complementary to information obtained from other spectroscopy-based sensors because the information content is obtained from the entire sample. In contrast to MR, the information content from many other technologies (e.g., optical spectroscopy) often provides information primarily from the material surface or near-surface region. MR-based measurements have additional features, making them an attractive measurement technique to employ in process analytical technology (PAT). The MR signal is directly proportional to the number of nuclei in a specific sample volume and is linear from the detection limits of ~ 10 ppt to 100 % (Skloss et al. 1994). Highly specific chemical information may be obtained since spectra can be recorded from only one nucleus (^{1}H, ^{31}P, ^{23}Na, or ^{13}C) or a combination of nuclei. Multiple types of information, including chemical, physical state, sample internal structure at a range of length scales, pH, and temperature, may be measured. MR measurements can be made rapidly and at a speed compatible with most food processing lines. From the reliability standpoint, an MR spectrometer has no moving parts and needs minimal maintenance.

Incorporating an MR-based sensor into a process requires consideration of the influence of motion, temperature, and environmental factors. The influence of motion impacts the MR signal through changes in intensity, changes in effective relaxation times, and through limiting the time, a sample is in the measurement zone.

M. J. McCarthy (✉) · K. L. McCarthy
Department of Food Science and Technology, University of California,
Davis, CA 95616, USA
e-mail: mjmccarthy@ucdavis.edu

C. P. O'Donnell et al. (eds.), *Process Analytical Technology for the Food Industry*, Food Engineering Series, DOI 10.1007/978-1-4939-0311-5_6,

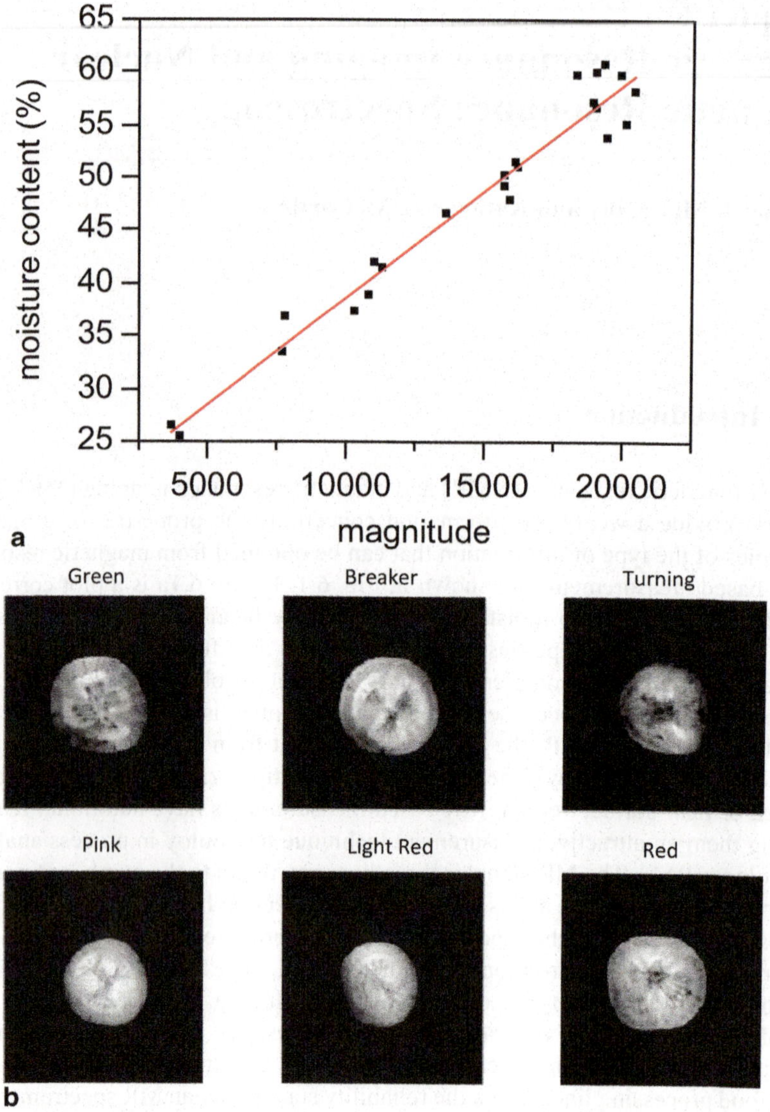

Fig. 6.1 **a** Magnetic resonance calibration curve for determining the moisture content of navy beans. **b** Proton magnetic resonance images of tomatoes demonstrating differences in maturity

Temperature changes or variations impact the signal and the phase of components so that care must be taken to ensure appropriate sample temperature control and history for many measurements (e.g., solid–liquid ratio in cocoa butter). Environmentally, the magnet and spectrometer electronics will generally need to be temperature controlled and the magnet protected from unintentional insertion of ferrous metals. All of these considerations can be addressed and successful implementations of MR-based process sensors can be achieved.

One of the earliest implementations involved the measurement of proton spectra from a flowing process stream (Nelson et al. 1960). A 30-MHz system was employed and a sample spectra was recorded every 6 s from a 3-mm-diameter glass tube. This was demonstrated in a pilot plant process. Applications noted in the literature related to foods include the measurement of viscosity (Snoddy 1993), fat content in meat (McDonald 1995), and moisture content in foods (Tellier and Mariette 1995). An MRI system was employed by Surrey Medical Imaging Systems to detect microbial spoilage in aseptically packaged infant formula and intravenous food packages. The system was based on a whole-body permanent magnet initially designed for clinical application. The magnet was coupled to a conveyor belt and the entire case of product was imaged to detect microbial contamination (McCarthy and Bobroff 2000). A higher-field superconducting magnet was employed in Japan for measuring the quality of watermelons for several years.

While NMR and MRI have been used for process control in food production, the applications remain limited in number of systems. This is in contrast to the petroleum refinery industry, where many NMR spectroscopy systems have been implemented to improve yield, and the polymer industry, where NMR is used for measuring melt index, xylene solubles, and other quality parameters. As the value of food products increase and the cost of MR sensors decrease, one can anticipate greater use of MR-based sensors in food production.

6.2 Theory of Nuclear Magnetic Resonance

MR is a phenomenon that occurs between atomic particles and an external magnetic field. The atomic particles responsible for this interaction are the electrons and the nucleus. The interaction between the atomic particles and an external magnetic field is similar to what happens when iron filings are placed near a bar magnet. The filings become oriented and a magnetic field is induced in the metal. However, unlike the filings, the physical orientation of the atomic particles is not altered. At most common magnetic field strengths, only the magnetic moment of the atomic particles is induced. The phenomenon of resonance is observed in these systems because they absorb and emit energy at specific frequencies. The specific frequency depends on the individual atomic particle and the strength of the applied magnetic field.

Work discussed in this chapter will focus on using the nucleus as the atomic particle, and in this case, the phenomenon is referred to as *nuclear magnetic resonance*. Common nuclei with magnetic moments include ^{1}H, ^{31}P, ^{15}N, and ^{23}Na. The most commonly studied nuclei in food systems is the ^{1}H. In both medical and food applications of NMR, the technique is often referred to simply as MR. The term *nuclear* is omitted so that patients/consumers will not confuse this technique with nuclear procedures that use radioactive materials. MR is a safe, experimental procedure and does not harm or alter the sample, the operator, or the environment (McCarthy 1994).

When an ensemble of spins is placed in a magnetic field $\mathbf{B} = (0, 0, B_0)$, the ensemble magnetization vector $\mathbf{M}$ precesses around this field at the Larmor frequency $\omega_0 = \gamma B_0$. Resonance phenomenon results upon application of a transverse radio frequency (RF) field $\mathbf{B}_1$ oscillating at the same frequency ω_0. At resonance, the magnetization simultaneously precesses around the longitudinal field B_0 at ω_0 and around the RF field B_1 at ω_1. A short burst of a resonant RF field is known in NMR as the RF pulse. If the duration of the RF pulse is t, then the magnetization will rotate by an angle $\omega_1 t$ about the direction $\mathbf{B}_1$ in the rotating frame.

The detection of the NMR signal is governed by Faraday's law and depends on the motion of the magnetization vector. Suppose a receiver RF coil is placed around the sample with its symmetry axis transverse to the polarizing field B_0, then a RF pulse will produce transverse magnetization precessing at the Larmor frequency ω_0 which will induce an oscillatory electromotive force (emf) at the same frequency ω_0. Therefore, the primary NMR signal is measured in the time domain as an oscillating, decaying emf (known as the free induction decay (FID) signal). For instance, for a single pulse experiment, where a 90° RF pulse ($\omega_1 t = \pi/2$) is applied to the equilibrium spin magnetization $(0, 0, M_0)$, the magnetization at time t after the RF pulse is $\mathbf{M}(t) = (M_0 \cos\omega_0 t \exp(-t/T_2), M_0 \sin\omega_0 t \exp(-t/T_2), 0)$, or in complex number notation $\mathbf{M}(t) = M_0 \exp(i\omega_0 t) \exp(-t/T_2)$. The time constant T_2 characterizes the transverse (spin–spin) relaxation, which has to do with interactions between spins of the ensemble. Another time constant, important in MR applications, is longitudinal (spin–lattice) relaxation time T_1. This time parameter characterizes how fast the thermal equilibrium of the spin system is restored (for instance, after application of the RF pulse) and depends on the exchange of energy between the spin system and the lattice (this term refers to the surrounding thermal reservoir).

In conventional NMR spectroscopy, the spectrum of nuclear precession frequencies provides information about the chemical environment of the spins, and therefore, it is important to remove the inhomogeneities in the B_0 field prior to each experiment (this removal is achieved by careful adjustment of the currents in the magnet shim coils). However, if a profile of magnetic field is deliberately varied linearly across the sample, the Larmor frequencies of the sample spins will show the same spatial dependence. The linearly varying field is known as a field gradient, which is created by means of specially designed gradient coils. Since the gradient fields are much smaller than the field B_0, the Larmor frequencies are affected only by gradient field components $\mathbf{G}$ parallel to $\mathbf{B}_0$. The Larmor frequency at the point $\mathbf{r}$ then can be written as:

$$\omega(\mathbf{r}) = \gamma(B_0 + \mathbf{G}\cdot\mathbf{r}). \tag{6.1}$$

This linear relation between the Larmor frequency and the spin location, $\mathbf{r}$, expresses the fundamental idea of MRI (Callaghan 1991). The integrated NMR signal from a sample may be written as:

$$S(t) = \iiint \rho(\mathbf{r}) \exp(i\gamma\mathbf{G}\cdot\mathbf{r})\, d\mathbf{r} \tag{6.2}$$

where **r** is a position of the volume element d**r** and $\rho(\mathbf{r})$ is the local spin density. The signal is so defined by taking into account that, relative to the reference resonant frequency $\omega_0 = \gamma B_0$, heterodyne detection provides the signal frequency offsets in the audio range. The effects of transverse relaxation are not taken into account, since the dephasing due to the spread in $\gamma\mathbf{G}\cdot\mathbf{r}$ is much more rapid. The concept of a reciprocal space vector, **k**, may be introduced:

$$\mathbf{k} = \gamma \mathbf{G} t / 2\pi. \tag{6.3}$$

The fundamental relationship of MRI is based on the concept of the Fourier transform and states that the signal $S(\mathbf{k})$ and the spin density $\rho(\mathbf{r})$ are mutually conjugate. Therefore, $S(\mathbf{k})$ is measured in the time domain and the Fourier transform yields $\rho(\mathbf{r})$ in the frequency domain:

$$S(t) = \iiint \rho(\mathbf{r}) \exp(-i2\pi\, \mathbf{k}\cdot\mathbf{r}) d\mathbf{r}. \tag{6.4}$$

In general, the phase accumulation at the location **r** during the application of the time-dependent gradient $\mathbf{G}(t)$ can be written as:

$$\phi(t) = \int (\omega(\mathbf{r}(t) - \omega_0)\, dt = \gamma \int \mathbf{G}(t)\cdot\mathbf{r} dt. \tag{6.5}$$

For the uniaxial flow of fluid along the z-direction, $\mathbf{r} = (x, y, z(t))$. The pulsed gradient spin echo (PGSE) pulse sequence can be employed to measure the velocity in one spatial dimension. PGSE uses two distinct rectangular shape phase encoding gradient pulses of the same duration τ, inducing phase accumulations ϕ_1 and ϕ_2:

$$\phi_1 = \gamma z_1 \int G_z(t)\, dt = \gamma z_1 G_z \tau. \tag{6.6}$$

$$\phi_2 = \gamma z_2 \int G_z(t)\, dt = \gamma z_2 G_z \tau. \tag{6.7}$$

If the first phase accumulation can be inverted by a 180° RF pulse, then the net phase is $\phi = \phi_2 - \phi_1 = \gamma G_z \Delta z \tau$, thus encoding the displacement of the fluid, Δz, over the flow time T (the time between the velocity encoding gradient pulses).

The MR signal will be:

$$S(k_x, q_x) = \iint P(\Delta z; x; T) \rho(x) \exp(-i2\pi\, k_x x) \exp(-i2\pi\, q_z \Delta z)\, dk_x d\Delta z \tag{6.8}$$

where $P(\Delta z; x; T)$ is the probability density for a fluid element at radial position x to displace for a distance Δz within the flow time T. The values k_x and q_z are given by $(2\pi)^{-1}\gamma G_x t$ and $(2\pi)^{-1}\gamma G_z \tau$. The inverse Fourier transform is needed for reconstruction and yields a map of $P(\Delta z; x; T)\, \rho(x)$ for each radial position x:

$$P(\Delta z; x; T) \rho(x) = \iint S(k_x, q_x) \exp(-i2\pi\, k_x x) \exp(-i2\pi\, q_z \Delta z)\, dk_x d\Delta z. \tag{6.9}$$

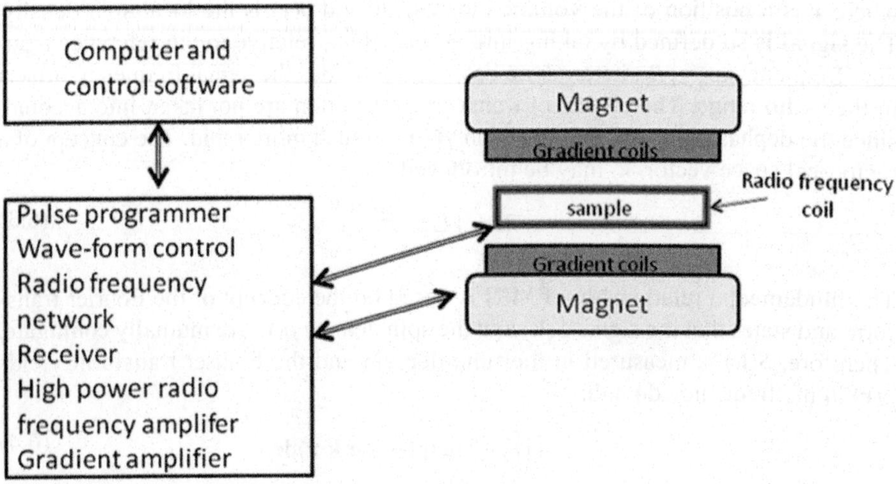

Fig. 6.2 Major components of an NMR/MRI spectrometer

If the flow of fluid is in steady state, division of the fluid displacement Δz by the flow time T yields velocity and the resulting image is referred as either the position-displacement conditional probability density or the velocity profile. The PGSE method is also routinely used for measurements of the coefficients of molecular diffusion, D. The method uses the dependence of the echo signal on the amplitude of the encoding pulse gradients. The echo signal will be attenuated when the spins move in an uncorrelated fashion (for instance, due to self-diffusion), since the rephasing of the echo is not complete. The amplitude of the echo can be expressed as:

$$E(b) = \exp(-bD) \tag{6.10}$$

where b is proportional to the amplitude and duration of the gradients. Thus, the parameter b can be varied by changing the amplitude of the gradient, and the diffusion coefficient D can be found from the plot of $E(b)$ versus b.

6.3 Magnetic Resonance Equipment

NMR equipment includes a computer with control software, RF electronics, a magnet, and a RF coil. The spectrometer is used to apply energy to the sample and record the decay of energy from the sample. The magnet is used to polarize the spins in the sample and the RF probe is used to couple the sample to the spectrometer. MRI spectrometers additionally include magnetic field gradient coils which are used to induce linear gradients in the applied main magnetic field for spatially localizing the signal. A diagram of the basic components utilized to construct an MR sensor is shown in Fig. 6.2.

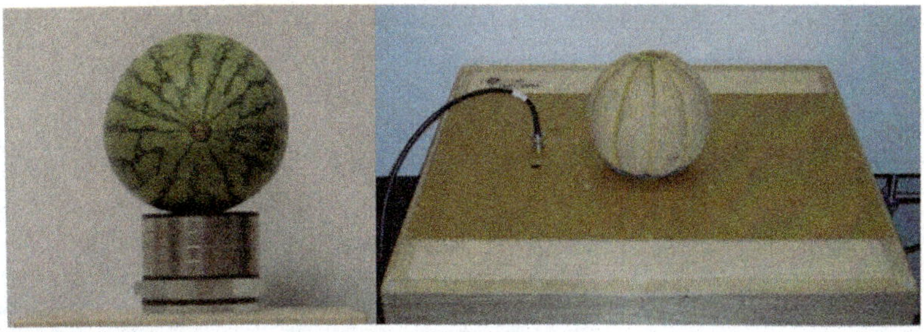

Fig. 6.3 Single-sided magnet configurations. (Photos courtesy of ABQMR Inc.)

The user interface is the control software and computer. The primary tasks for the software are control of data acquisition, processing of the data, and data storage. A variety of options exist from very sophisticated programming interfaces to simple one-button operation. One of the challenges in implementation of NMR or MRI sensors for control applications is the writing of specific software for the application and integration of the information into control of the production process. This is especially true in the case of application to food systems where the range of quality factors is extensive (e.g., from viscosity control to sugar acid ratios in fresh fruit).

Control of the RF electronics is well advanced. The RF electronics are generally referred to as the console or spectrometer. The spectrometer includes RF electronics to generate well-defined RF pulses, a high-power RF transmitter, electronics to control magnetic field shim coils, an analog to digital converter, and a low-noise amplifier to increase the signals from the sample. MRI systems also include linear gradient amplifiers to provide high-current gradient waveforms that coupled with gradient coils inside the magnet cavity produce linear variations in the main magnetic field. Spectrometers can vary from single-board computers to multi-board systems. Shown in Fig. 6.3 is a picture of a two-board system that includes one board built around a digital signal processing chip and one board that is a high-power RF amplifier. These were powered by batteries, controlled by a portable computer, and coupled to a small permanent magnet.

Magnets used for NMR and MRI can be based on permanent magnet materials, electromagnets, or superconducting magnets. For utilization in a production environment, permanent magnets are generally the preferred option. Electromagnets require very stable constant current supplies and cooling water. Superconducting magnets require cryogenic gases to cool the magnet and vacuum jacketing to maintain magnet temperature. Permanent magnets require, at most, thermal blankets to keep them at a constant temperature.

Permanent magnets exist in a large number of different configurations. These configurations include single-sided, where the field and signal are acquired at a distance from the magnet surface as shown in Fig. 6.3, where a watermelon and cantaloupe are shown positioned for measurement on such magnets. Alternate configurations are based on cylindrical cavities like a Halbach design with the sample in the center of the cylinder (Fig. 6.4) and have two plates where the sample is located

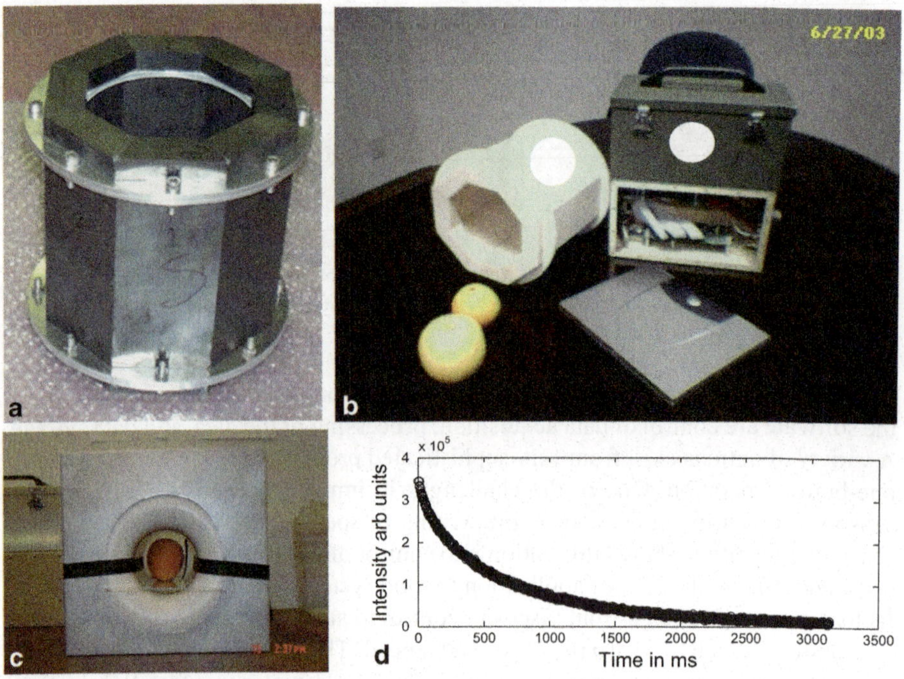

Fig. 6.4 **a** The Halbach magnet just after manufacture. **b** The prototype design for a field portable NMR to measure freeze damage in navel oranges (*1*) is the spectrometer that consists of two PC boards, one a spectrometer and the other a radio frequency amplifier (*2*), is the conceptual design for the Halbach magnet shown in **a**. **c** The magnet enclosed in protective housing with an orange sample. **d** Experimental data for an orange recorded using this system

between the plates. Permanent magnets are made of an assembly of individual magnetized bricks. Materials that are used to fabricate the bricks include alnico alloys or rare-earth alloys. The bricks are often coupled to an iron yoke (shown in Fig. 6.3) or face plate to increase the magnetic field strength/uniformity. Field strength for these types of magnets range from 0.05 to 2.35 T (proton resonance frequencies from 2 to 100 MHz). For example, the strength of the magnet shown in Fig. 6.3 is 0.04 T with a homogeneity of 2500 ppm over a volume of 0.11 m sphere. A prototype developed for sorting fruit is shown installed in a conveyor sorting citrus fruit (Fig. 6.5). This magnet has a field strength of 1 T and a homogeneity of better than 70 ppm over 0.06 by 0.09 m elliptical imaging volume. Almost all past and current applications of MR for process control have utilized permanent magnets.

Development of NMR and MRI hardware is occurring at a rapid pace. There are many small companies that are developing unique magnets and unique small portable spectrometers. All of which can be combined into a process sensor system. The equipment and magnet costs have been reduced significantly in the past 15 years. The range of costs for these types of sensors is from tens of thousands of US$ to >500,000 US$ depending upon magnetic size and system configuration.

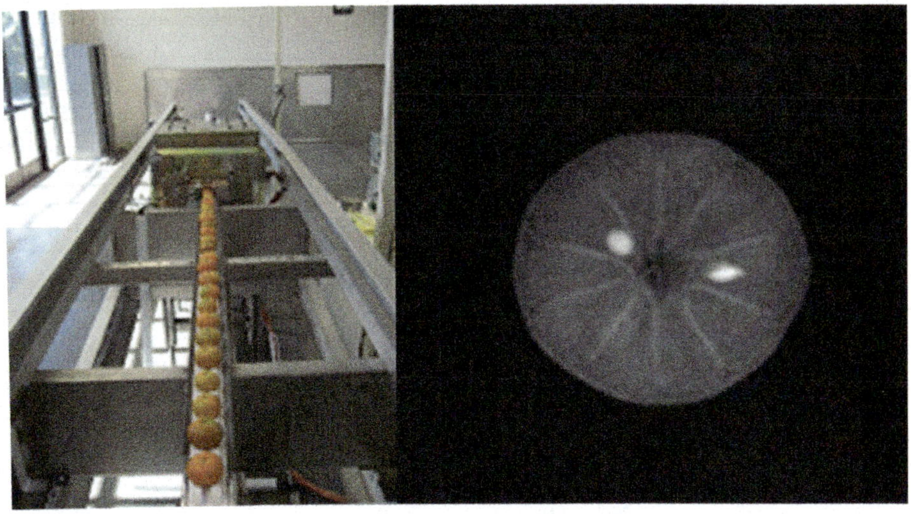

Fig. 6.5 A magnetic resonance imaging system coupled with a fruit conveyor used for detecting seeds in citrus fruit. An MRI taken with the system of a clementine fruit with two seeds is shown on the *right*. (Photo and image courtesy of Uri Rapoport, Aspect-AI Ltd.)

6.4 Applications

The range of food properties that can be measured using NMR and MRI are extensive. Chemical properties that can be measured include component concentrations, reaction rates, and exchange rates. Physical properties that can be measured are diffusion coefficients, thermal properties, rheological properties, and phase behavior. The structure of a food material can be measured from the pore size or droplet size to insect damage and layer thicknesses. This range of potential measurements has resulted in a very diverse array of applications to process control and quality assurance in the food industry using MR.

6.4.1 Measurements of Component Concentration

The measurement and control of component concentrations are critical to the quality, profitability, and safety of many food products, and NMR/MRI can be used to make these measurements. Most of the MR systems applied to measuring component concentrations have been permanent magnet based low-resolution systems. The term low-resolution refers to the homogeneity and/or strength of the magnetic field. Either one or both of these features prevent chemical shift interactions from being resolved. These systems are good for measuring the signal magnitude, spin–spin relaxation, spin–lattice relaxation, and diffusion behavior. The MR system shown in Fig. 6.4 is an example of a low-resolution spectrometer.

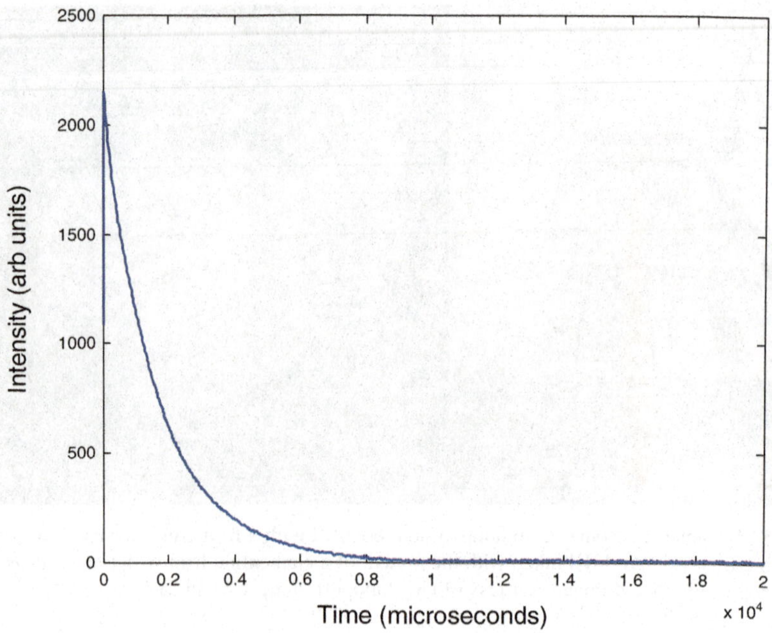

Fig. 6.6 Free induction decay of moist navy beans, data taken at 1-Tesla field strength

Two groups made significant contributions to the early application of MR for moisture measurement: Tri-Valley Research and The Southwest Research Institute (McDonald 1995). These groups developed measurements based on analysis of the NMR signal in the time domain. The simplest approach used by these two groups was to measure the FID (signal after one 90° RF pulse). The FID from moist navy beans is shown in Fig. 6.6. The intensity of the signal at the beginning of the decay is proportional to the moisture content in this sample. A series of different bean samples were soaked in water as a function of time and then data was recorded as in Fig. 6.6. The initial signal magnitudes were then paired with moisture content measured using an oven-drying technique, and the plot in Fig. 6.1a was generated. This type of measurement can be implemented in-line or on-line to measure moisture content prior to adding sauce to beans, permitting the final product to have a more consistent bean to sauce ratio.

This development of these type of measurements has continued, especially noteworthy have been the accomplishments of the Minispec Division of Bruker Optik GmbH which has applied low-resolution time-domain NMR to measure the moisture and/or fat content of a wide variety of food materials, including cereal, bread, cookies, beans, lentils, powders, flavors, flour, cocoa powder, and chocolate (Todt et al. 2006). These techniques are standard practice in many quality assurance laboratories and have been recognized as International Standard Methods (e.g., ISO 829 International Standard 1991).

6.4.2 *Rheology*

The goal of most rheological measurements is to develop quantitative relationships between deformation and force and to connect these to structure and composition of a material. From a process monitoring and control standpoint, once these relationships are developed, the rheological measurements are used as a tool to provide information about the operating status of a unit operation (Arola et al. 1997). This section describes the development and application of an MRI-based viscometer that meets the needs of industry for rapid characterization of complex fluids, in particular foods. Specifically, the goal of the work has been to provide a means to monitor rheological properties in-line or on-line in a food factory setting.

This type of viscometer is based on the use of MRI to directly measure velocity profiles in flowing pipe systems by using either a time-of-flight technique (e.g., McCarthy et al. 1992a) or a displacement phase-encoded imaging (e.g., Seymour et al. 1993). Although, ultimately, the displacement phase-encoded imaging is the more developed technique, the first work to report the combined use of tube viscometry and flow visualization utilized the time-of-flight method (McCarthy et al. 1992b). The importance of the work lies in the recognition that a "conventional" tube viscometer, as described by Steffe (1996), can be successfully used as an MRI flow system. For Newtonian and power law fluids, the rheological parameters were obtained by pressure drop measurements at multiple flow rates to create a rheogram that reflected wall stresses and wall shear rate values. Good agreement was observed when comparing the theoretical velocity profiles using the rheological parameters to the MRI velocity images. Shortly thereafter, researchers recognized the potential to obtain rheological data from a single velocity profile.

6.4.2.1 Single Point Measurements and Fluid Characterization

The *single point* method was first introduced in 1994 (Powell et al. 1994) and described in detail in a 1996 patent, based on work carried out in M. McCarthy's research group (Maneval et al. 1996). For a single velocity image in pipe flow, data are obtained over a range of shear rate (e.g., velocity gradient), which ranges from the maximum shear rate at the pipe wall to a minimum at the pipe center.

For this *single point* method to be successful, a well-defined flow field must be established. To evaluate shear viscosity in tube (or pipe) flow, an incompressible fluid undergoes steady pressure-driven flow in the laminar regime. The conservation of linear momentum, which equates pressure forces to viscous forces, provides the relationship between the shear stress, σ, and the tube radius, r:

$$\sigma(r) = \frac{-(\Delta P)}{2L} r, \tag{6.11}$$

where ΔP is the pressure drop over the tube length L. In this method, the shear rate, γ, is obtained at the same radial position using the velocity profile. The expression for the shear rate in tube flow is:

$$\dot{\gamma}(r) = \left|\frac{dv(r)}{dr}\right|, \tag{6.12}$$

where v is the axial velocity. Using Eqs. 6.11 and 6.12, the apparent viscosity η is determined by:

$$\eta(r) = \frac{\sigma(r)}{\dot{\gamma}(r)}. \tag{6.13}$$

Alternately, the shear stress can be plotted as a function of shear rate to determine rheological parameters for a specific model, e.g., Newtonian and power laws.

Initial publications that described the *single point* method illustrated proof of concept by presenting rheograms and/or spectra without fully characterizing the test fluids (Powell et al. 1994; Seymour et al. 1995; Arola et al. 1997). The most complete characterization from the mid-1990s was the evaluation of 0.2 to 1.0% aqueous polyacrylamide solutions by Li and McCarthy (1995). These solutions were modeled as shear-thinning power law fluids over a range of flow rates by evaluating an MRI velocity image and the simultaneous pressure drop. Flow behavior indices (n), but not consistency indices (K), were given in a table for all model fluids at all volumetric flow rates. A rheogram that compared MRI data and conventional rotational data for 1% aqueous polyacrylamide solution illustrated excellent agreement between the two methods (Li and McCarthy 1995). In 1999, Sadikin characterized skim milk concentrates and coffee concentrates as Newtonian fluids (n=1). These results are given in Table 6.1, in terms of the more general Herschel–Bulkley model:

$$\sigma = \sigma_0 + K\dot{\gamma}^n, \tag{6.14}$$

where σ_0 is the yield stress, K is the consistency index, and n is the flow behavior index. The shear rate ranges given in Table 6.1 were based on four to five flow rates for each concentrate; the coefficient of determination values (R^2) were greater than 0.97 for all curve fits. These viscosity values are relevant to the spray drying of milk and coffee products. For many Newtonian materials like skim milk concentrate or coffee concentrate, the spin–spin relaxation time is often proportional to viscosity; however, this relationship can fail at higher solids levels.

6.4.2.2 Direct Measure of Yield Stress and Slip Velocity

A consistent theme of the MRI-based viscometer work has been to extend the shear rate range and the accuracy of the measurement technique while attempting to anticipate an economically feasible cost unit for actual plant applications (Arola et al.

Table 6.1 Skim milk concentrate and coffee concentrate characterized as Newtonian fluids by MRI-based viscometer

Fluid	Model	Parameters, SI units			Shear rate range (s^{-1})	T (°C)
		K (Pas^n)	n	σ_0 (Pa)		
Skim milk concentrate						Ambient
26.5 Brix	Newtonian	9.3×10^{-3}	1	–	4–61	
31.9 Brix	Newtonian	13.7×10^{-3}	1	–	5–77	
36.1 Brix	Newtonian	33.4×10^{-3}	1	–	3–72	
Coffee concentrate						Ambient
36.6 Brix	Newtonian	8.1×10^{-3}	1	–	5–93	
41.8 Brix	Newtonian	14.7×10^{-3}	1	–	6–120	
52.5 Brix	Newtonian	61.4×10^{-3}	1	–	6–117	

1997; Choi et al. 2002). As velocity and radial resolution improved, more subtle aspects of the flow imaging were revealed. McCarthy and McCarthy (1995) reported yield stress measurements (σ_0) for tomato concentrates (5.5–12 °Brix) that were obtained directly from the blunt center region characterized by R_0:

$$\sigma_0 = \frac{(\Delta P)R_0}{2L}. \tag{6.15}$$

In addition, nonzero wall velocity was quantified for fluid suspensions that have been described as exhibiting "slip" behavior. Slip occurs when a thin layer of fluid, having a viscosity lower than the bulk fluid, forms at the wall of the viscometric device. This may occur in fruit and vegetable suspensions and high-fat foods. In conventional tube viscometry, an additional term is added to the volumetric flow rate and an effective slip velocity is found by acquiring data using multiple tube radii (Steffe 1996). In contrast, the slip velocity is directly measureable from the MR velocity image.

Lee et al. (2002) characterized tomato concentrates that ranged in soluble solids from 6–16 °Brix over the temperature range of 20–110 °C. This work was the first to characterize fluid foods that were not at ambient temperature and that exhibited wall slip by the MRI-based viscometer. After correction for the wall slip, the fluids were modeled as Herschel–Bulkley fluids. The flow behavior indices were not significantly affected by temperature and the average flow behavior index ($\overline{n}$) for each concentration was used in the model (Eq. 6.14), where the consistency index was given by an Arrhenius-type relationship:

$$K = K_T \exp(E_a / RT), \tag{6.16}$$

where K_T is a constant and E_a/R is the activation energy term, with T in units of absolute temperature. The rheological information was incorporated into a residence time distribution model to assess the uniformity of thermal processing of these commercially important products.

Recognizing the heat sensitivity of many foods, researchers designed single-pass flow systems to replace the closed-loop circulation designs of the previously described studies. Yoon and McCarthy (2002) characterized yogurt flowing at 25 and 35 °C as Herschel–Bulkley fluids after correcting for wall slip. Processed cheese and processed cheese spread at 85 °C were modeled as power law fluids; no wall slip was evident for those fluids, though fat content was between 21 and 28 % (wb) (Yoon and McCarthy 2003). After correcting for wall slip, Wichchukit et al. (2005) characterized molten milk chocolate at 42 °C by the Casson model, frequently applied to fluid chocolate in the confectionery industry:

$$\sigma^{0.5} = \sigma_0^{\ 0.5} + (\eta_{CA}\dot{\gamma})^{0.5}, \tag{6.17}$$

where σ_0 is the Casson yield stress and η_{CA} is the Casson viscosity. The Casson plastic viscosity and yield stress values of chocolate are affected by emulsifier level, among other factors, and was documented in this work. The deviation between the MRI-based viscometer and rotational viscometry results was due to uncorrected wall slip during the rotational measurement. The MRI-based viscometer Casson parameters were incorporated into an unsteady state mass balance to predict film thickness during the enrobing process, which is an important consideration in process control of these products.

6.4.2.3 Development of Graphical User Interface

Sadikin (1999) recognized the need to implement a means to generate and display data to individuals with little or no knowledge of computer programming and for process control applications. As part of her Master of Science thesis project, she developed a graphical user interface (GUI) in MatLab (The Mathworks, Inc., Natick, Mass., USA). The GUI facilitates interacting with the computer code without programming commands and represents a significant step toward implementing the MRI-based viscometry in a factory setting. The idea was to process the data from the MRI measurement and to construct the rheogram immediately after the flow imaging data acquisition was complete. The four most common constitutive equations were programmed: Newtonian, power law, Bingham plastic, and Herschel–Bulkley. The program has four major parts: reconstruction of the velocity image, generation of the velocity profile, determination of the shear rate range, and application of constitutive models. More extensive discussion is given in Choi et al. (2005). Current and ongoing modifications to the GUI code include automating the selection of an MR velocity image, incorporating a symmetry check of the velocity profile, improving the reliability of the yield stress estimate, and achieving a specified shear rate range.

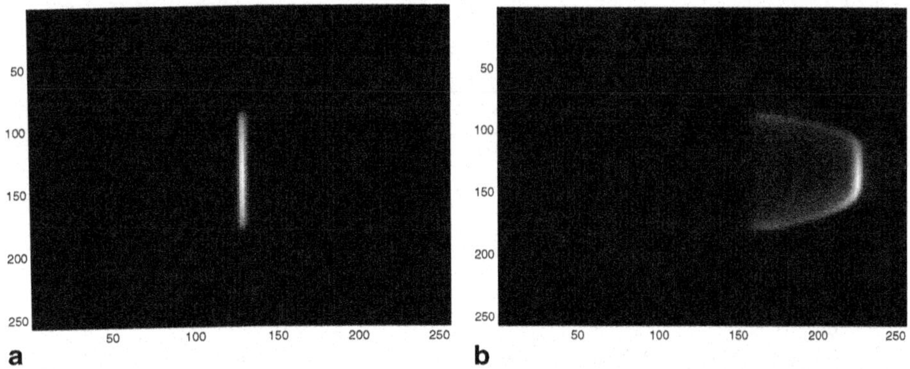

Fig. 6.7 MR flow image of tomato concentrate at 9 °Brix for no flow conditions (**a**) and at a volumetric flow rate of 4.5 L/min (**b**). The blend is 80 % H-8 and 20 % H-1

6.4.2.4 Example of Factory Application

The MRI technique and hardware for the MRI-based viscometer have advanced to the point of permitting real-time process control in a factory. An example application is given in McCarthy and McCarthy (2009) for the blending of tomato pastes, packed at different Bostwick readings, for use in tomato ketchup production. The in-line viscosity measurements of 12 °Brix tomato concentrates provided by the MRI-based viscometer were correlated to the final ketchup quality, as measured by the Bostwick consistometer. The Bostwick consistometer, which measures the extent of flow in a trough, has remained an integral part of assessing the consistency of tomato products in the factory. Although a slow and laborious quality assurance test, US Department of Agriculture (USDA) ketchup grades are identified by the Bostwick measurement. The 12 °Brix tomato concentrate blends were characterized as Herschel–Bulkley fluids. The in-line rheology measurements of the intermediate product (the blends) were correlated to final product Bostwick measurements by identifying a range of blends that yielded the consistency required for grade A ketchup.

A striking aspect of the work was the ability to distinguish between tomato concentrates at the same soluble solids level (e.g., °Brix). Although not reported in McCarthy and McCarthy (2009), 9 °Brix blends were also prepared from the concentrates packed at 23 °Brix (designated as H-1) and at 35 °Brix (designated as H-8). Figure 6.7a illustrates an MR image acquired under no flow conditions to illustrate the proton signal at the zero axial velocity position (pixel 129 on the x-axis), across the diameter of the pipe (pixel range 78–172 on the y-axis). At a volumetric flow rate of 4.5 L/min in a 1.9 cm ID pipe, the velocity image is shown in Fig. 6.7b. The offset from pixel 129 (x-axis) was due to wall slip; the slip velocity was 13.7 cm/s at an average fluid velocity of 26.2 cm/s. The apparent viscosities of the six blends at a soluble solids level of 9 °Brix is shown in Fig. 6.8. All blends were modeled as

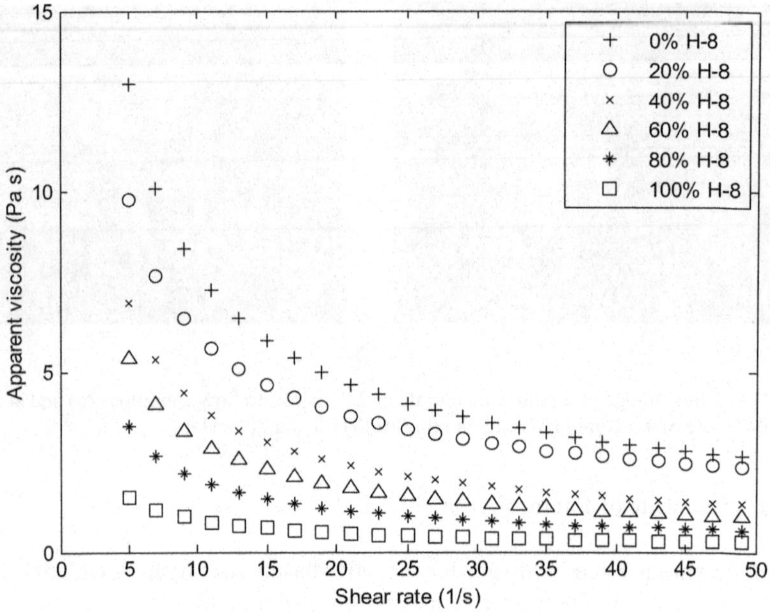

Fig. 6.8 Apparent viscosities of six tomato concentrate blends at a soluble solids level of 9 °Brix. All blends were modeled as Hershel–Bulkley fluids

Herschel–Bulkley fluids; different blends at the same soluble solids level are easily distinguishable.

6.4.3 *Fruit Quality*

Evaluation of the quality of fruit is primarily based on surface appearance and selecting a subsample of a lot for destructive testing. This subsample approach is suboptimal because of the range of individual variations within the batch yield many fruit that differ from the desired quality specifications. Ideally, the internal chemical and structural features of each and every fruit would be measured. High quality fruit should have the appropriate maturity, texture, chemical composition, structure, as well as the absence of defects (e.g., bruises, browning, mold, insect damage). There are a number of potential technologies that can be used to measure fruit quality, including infrared spectroscopy, mid-infrared spectroscopy, Raman spectroscopy, ultrasound, dielectric imaging, x-ray imaging, and MR techniques. A sensor system for any fresh fruit will most likely be constructed using two or more of these technologies to achieve an evaluation of the most important quality factors. MRI will often be one of the technologies of choice because MRI can be readily implemented to measure a wide range of fruit quality factors.

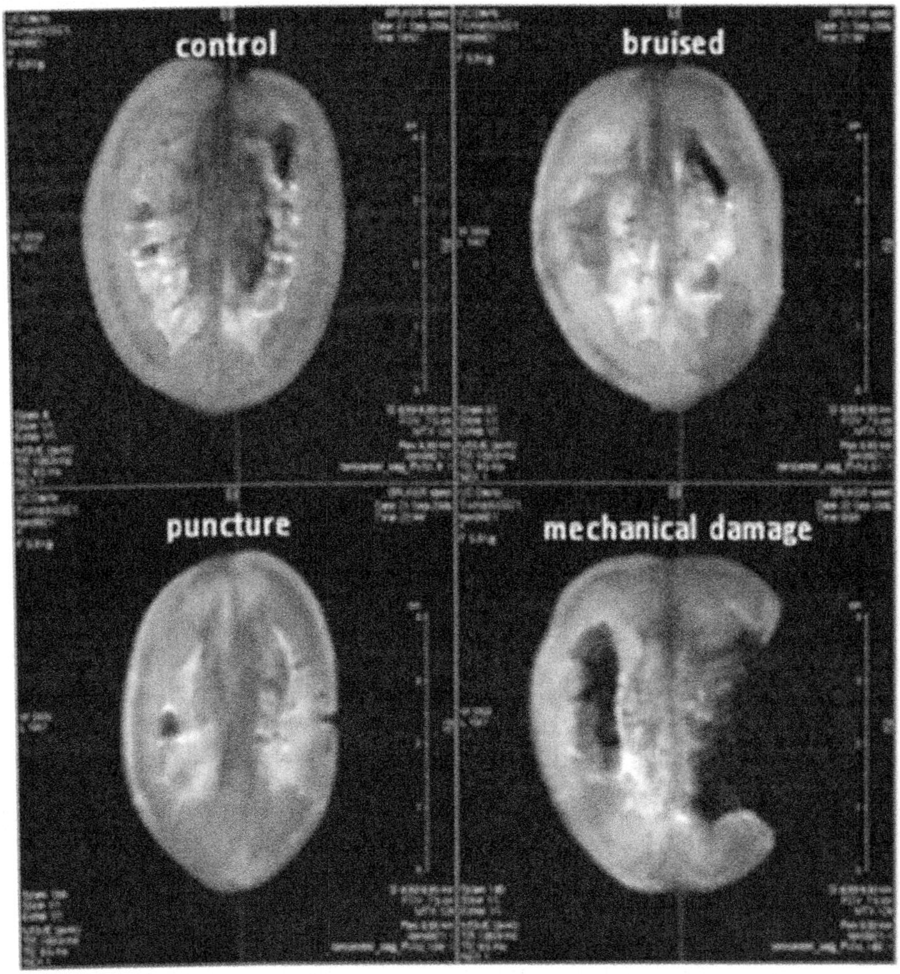

Fig. 6.9 Mechanical damage in Roma processing tomatoes

MRI can be used to measure chemical properties, such as °Brix, °Brix/acid ratio, moisture content, oil content, and maturity, and structural features, such as browning, rot, insect damage, bruises, and presence of foreign material. A review by Hills and Clark (2003) of the applications and status of using NMR/MRI to measure the quality of horticultural products provides an excellent summary of the potential and current state-of-the-art to measure fruit internal quality factors (Hills and Clark 2003). Over 30 different fruits and vegetables have been examined for internal quality factors. An example of the type of images that can be acquired at lines speed are shown in Fig. 6.9 for processing tomatoes. The results to date for applying NMR and MRI to detect the internal quality of fruits and vegetables are extensive and impressive. These applications include coupling a conveyor with research grade MR

Fig. 6.10 One-Tesla field strength magnet for MRI sorting of citrus fruit being installed in the packing line. The three tubes extending from the magnet will each have a lane of fruit feeding into the system. The three lanes will be imaged simultaneously. One image from the system is shown on the *right*. (Photo and image courtesy of Uri Rapoport Aspect AI Ltd., Netanya, Israel)

systems (Chen et al. 1996; Kim et al. 1999; Kim and McCarthy 2006; Hernandez et al. 2005; Hernandez-Sanchez et al. 2006; Zion et al. 1994, 1997). The prototype conveyor and imaging system shown in Fig. 6.5 was recently used by Kim et al. (2008) to develop algorithms to detect seeds and freeze damage in clementine fruit. The prototype has been redesigned and enlarged for application in a production sorting line, and the system is shown in Fig. 6.10 during installation.

Utilization of MR for sorting has involved development of hardware and software. For applications like sorting citrus by number of seeds, the software development is straightforward and can employ traditional image processing (Kim et al. 2008; Barreiro et al. 2008) or multivariate image analysis techniques (Milczarek et al. 2009). Software based on either approach is very effective and generally results in greater than 90% correct classification for both seedless and seeded fruit. For other quality attributes, the development of quantitative measures can be difficult since the definitions involve phrases like "substantially free of a specific defect" or "the evaluation of a defect is subjective." Hence, prior to implementing a sensor, a quantitative scale for the quality factor needs to be developed. Consider quantitative evaluation of the extent of freeze damage in citrus fruit.

The approach to detect freeze damage in California-grown citrus fruit is currently destructive, and only a few fruit from a lot are examined. The sample fruits are investigated for determination of freeze damage by using a segment cut. The segment cut proceeds by removing both the stem and blossom end of the fruit such that

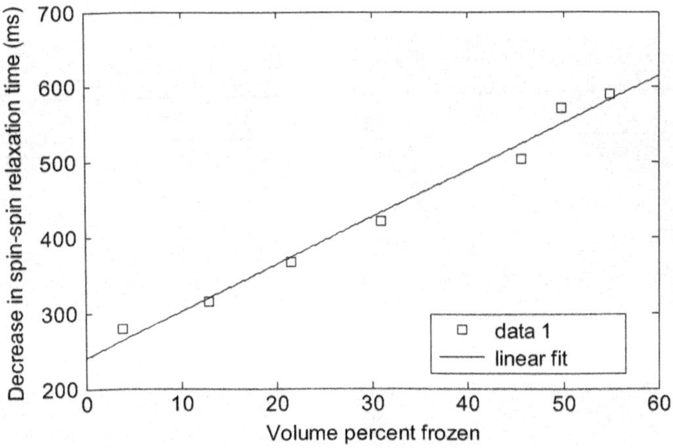

Fig. 6.11 Change in proton spin–spin relaxation time in a whole orange as a function of volume percent of segments frozen

a center section of fruit remains, that is, 1 to 1.5 in. in width. This center segment of the fruit is carefully opened to show the segments (opened from one cut, though only the peel). The segments are then inspected, and an orange must show damage to the entire length of both sides of two segments to be classified as freeze damaged. Damage is a water-soaked appearance or evidence of previous water-soaking or the presence of crystals. The tolerance in a lot of fruit for freeze damage is 15 %. This definition of freeze damage is useful; however, it is difficult to translate to a quantitative scale for spectroscopic measurement.

Development of an MRI-based sensor to detect percentage of freeze damage in a citrus fruit needs to yield results that are comparable to the destructive segment cut method or at least translatable. The impact of freeze damage on fruit is well known; cell membrane and hence fruit structure are altered. In citrus, this results in a decrease in the proton spin–spin relaxation time (Gambhir et al. 2005), and by weighing the image intensity by spin–spin relaxation time, regions within the fruit that are freeze damaged can be detected. The relationship can be quantified by constructing a plot of T_2 compared to percent of the orange frozen as shown in Fig. 6.11.

The difficulty becomes relating percent frozen in an orange to the results of the segment cut test. This is currently an open question, and the comparison is at best qualitative between NMR/MRI results and the segment cut. A relationship is most useful during the first week after a freeze event since the oranges that have been freeze damaged begin to lose moisture and begin to form voids. The progression of moisture loss and void formation in a freeze damaged orange is shown in Fig. 6.12 as a function of the number of days after the freeze event. A central plane in the orange perpendicular to the stem blossom axis approximately 1 cm thick is used to follow the loss of moisture. Voids in the oranges begin to be easily detected at about 1 week after the freeze and increase significantly as storage time is extended to approximately 1 month. The orange in Fig. 6.12 was held in cold storage and imaged

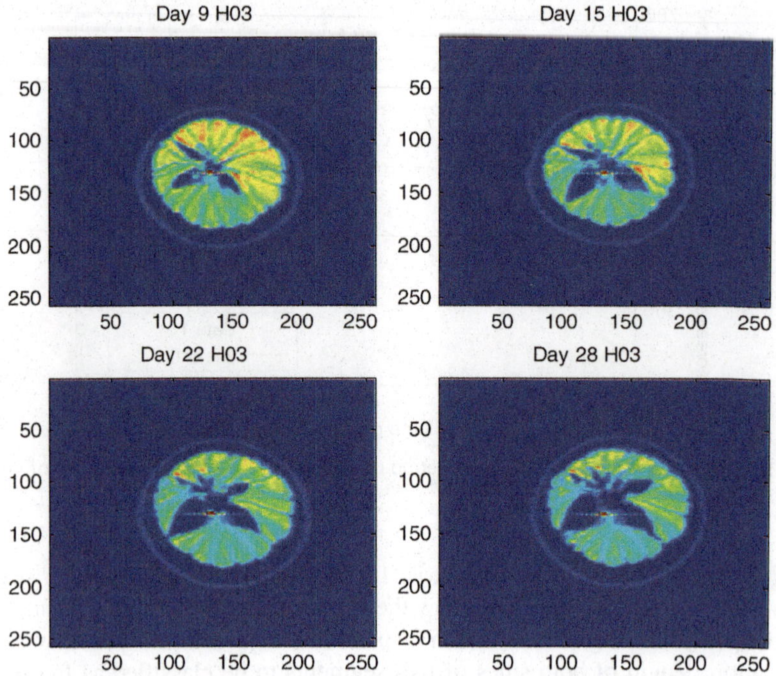

Fig. 6.12 Development of voids in whole navel oranges (variety Washington) after being frozen in an orchard during the winter of 2007 in California

approximately once a week. Once the voids become significant, freeze damage can be detected using a density measurement and an MRI would not be required, prior to void formation. NMR/MRI is the only nondestructive method currently available to quantify freeze damage.

The challenge to compare the measurement of a spectroscopic technique to a visually determined quality factor is not unique to NMR/MRI. Evaluation of fruit visual, as well as structural features, is important to the consumer at purchase. Flavor, texture, and aroma become most important at consumption. Yet relating consumer preference and subjective evaluation to quantitative measures of fruit attributes is a challenge that will involve considerable work and development. It is important to note, though, that purchasers of fruit are now requiring attributes of lots that cannot easily be verified for each and every fruit (e.g., completely seedless mandarins). This economic driving force for continued development of quantitative evaluation of fruit quality should result in significant development of quantitative scales for fruit attributes and the need for advanced sensor systems.

Acknowledgments We appreciate the assistance of R. R. Milczarek, S. McCarthy, and L. Zhang in preparation of the figures for this chapter. We are indebted to Uri Rapoport of Aspect AI Ltd. and Eiichi Fukushima of ABQMR Inc., for sharing with us photographs and data from current PAT applications of NMR/MRI. Work shown in this chapter was supported by ConAgra Foods Inc., Paramount Citrus Association, and the Center for Process Analytical Chemistry at the University of Washington, Seattle, WA, USA.

References

Arola DF, Barrall GA, Powell RL, McCarthy KL, McCarthy MJ (1997) Use of nuclear magnetic resonance imaging as a viscometer for process monitoring. Chem Eng Sci 52(13):2049–2057

Barreiro P, Zheng C, Sun D-W, Hernandez-Sanchez N, Perez-Sanchez JM, Ruiz-Cabello J (2008) Non-destructive seed detection in mandarins: Comparison of automatic threshold methods in FLASH and COMSPIRA MRIs. Postharvest Biol Technol 47:189–198

Callaghan PT (1991) Principles of nuclear magnetic resonance microscopy. Clarendon, Oxford

Chen P, McCarthy MJ, Kim S-M, Zion B (1996) Development of a high-speed NMR technique for sensing maturity of avocados. Trans ASAE 39(6):2205–2209

Choi YJ, McCarthy KL, McCarthy MJ (2002) Tomographic techniques for measuring fluid flow properties. J Food Sci 67(7):2718–2724

Choi YJ, McCarthy KL, McCarthy MJ (2005) A MATLAB graphical user interface program for tomographic viscometer data processing. Comput Electron Agric 47:59–67

Gambhir PN, Choi YJ, Slaughter DC, Thompson JF, McCarthy MJ (2005) Proton spin-spin relaxation time of peel and flesh of Navel orange varieties exposed to freezing temperature. J Sci Food Agric 85:2482–2486

Hernandez-Sanchez N, Barreiro P, Ruiz-Cabello P (2006) On-line identification of seeds in mandarins with magnetic resonance imaging. Biosyst Eng 95(4):529–536

Hills BP, Clark CJ (2003) Quality assessment of horticultural products by NMR. Ann Rep NMR Spectrosc 50:75–120

Lee Y, Bobroff S, McCarthy KL (2002) Rheological characterization of tomato concentrates and the effect on uniformity of processing. Chem Eng Commun 189(3):339–351

Li T-Q, McCarthy KL (1995) Pipe flow of aqueous polyacrylamide solutions studied by means of nuclear magnetic resonance imaging. J Nonnewton Fluid Mech 57(2):155–175

Kim S-M, McCarthy MJ (2006) Analysis of characteristics of in-line magnetic resonance sensor. Key Eng Mat 321-323:1221–1224

Kim S-M, Chen P, McCarthy MJ, Zion B (1999) Fruit internal quality evaluation using on-line nuclear magnetic resonance sensors. J Agric Eng Res 74(3):293–301

Kim SM, Milczarek RR, McCarthy MJ (2008) Fast detection of seeds and freeze damage of mandarins using magnetic resonance imaging. Mod Phys Lett B 22(11):941–946

Maneval JE, McCarthy KL, McCarthy MJ, Powell RL (1996) Nuclear magnetic resonance imaging rheometer, U.S. Patent No. 5532593

McCarthy MJ (1994) Magnetic resonance imaging in foods. Chapman and Hall, New York

McCarthy MJ, Bobroff S (2000) Nuclear magnetic resonance and magnetic resonance imaging for process analysis. In: Meyers RA (ed) Encyclopedia of analytical chemistry. Wiley, Chichester, pp 8265–8281

McCarthy KL, McCarthy MJ (1995) Yield stress measurements of tomato concentrates using magnetic resonance imaging. In: Narsimhan G, Okos MR, Lombardo S (eds) Advances in food engineering: proceedings of the 4th conference of food engineering. Purdue Research Foundation, West Lafayette, pp 86–90

McCarthy KL, McCarthy MJ (2009) Relationship between in-line viscosity and Bostwick measurement during ketchup production. J Food Sci 74(6):E291–E297

McCarthy MJ, Maneval JE, Powell RL (1992a) Structure/property measurements using magnetic resonance spectroscopy and imaging. In: Singh RP, Wirakartakusumah MA (eds) Advances in food engineering. CRC Press, Boca Raton

McCarthy KL, Kauten RJ, McCarthy MJ, Steffe JF (1992b) Flow profiles in a tube rheometer using magnetic resonance imaging. J Food Eng 16(1/2):109–125

McDonald PJ (1995) The use of nuclear magnetic resonance for on line process control and quality assurance. In: Gaonkar AG (ed) Food processing recent developments. Elsevier, Oxford

Milczarek RR, Saltveit ME, McCarthy MJ (2009) Assessment of tomato pericarp mechanical damage using multivariate analysis of magnetic resonance images. Postharvest Biol Technol 52:189–195

Nelson FA, Reilly CA, Savage WE (1960) Process monitor using high resolution nuclear magnetic resonance. Ind Eng Chem 52(6):487–489

Powell RL, Maneval JE, Seymour JD, McCarthy KL, McCarthy MJ (1994) Nuclear magnetic resonance imaging for viscosity measurements. J Rheol 38(5):1465–1470

Sadikin S (1999) Viscometric measurement by nuclear magnetic resonance imaging. Master of science in engineering. University of California, Davis, CA. 208 pp

Seymour JD, Maneval JE, McCarthy KL, McCarthy MJ, Powell RL (1993) NMR velocity phase encoded measurements of fibrous suspensions. Phys Fluids A Fluid Dyn 5(11):3010–3012

Seymour JD, Maneval JE, McCarthy KL, Powell RL, McCarthy MJ (1995) Rheological characterization of fluids using NMR velocity spectrum measurements. J Texture Stud 26(1):89–101

Skloss TW, Kim AJ, Haw JF (1994) High resolution NMR process analyzer for oxygenates in gasoline. Anal Chem 66(4):536–542

Snoddy ML (1993) The potential of process NMR on flowing streams. Spectroscopy 8(3):41–47

Steffe JF (1996) Rheological methods in food process engineering, 2nd edn. Freeman Press, East Lansing

Tellier C, Mariette F (1995) On-line applications in food science. Ann Rep NMR Spectrosc 31:105–122

Todt H, Guthausen G, Burk W, Schmalbein D, Kamlowski A (2006) Water/moisture and fat analysis by time-domain NMR. Food Chem 96:436–440

Wichchukit S, McCarthy MJ, McCarthy KL (2005) Flow behavior of milk chocolate melt and the application to coating flow. J Food Sci 70(3):E165–E171

Yoon WB, McCarthy KL (2002) Rheology of yogurt during pipe flow as characterized by magnetic resonance imaging. J Texture Stud 33:431–444

Yoon WB, McCarthy KL (2003) Flow behavior of processed cheese melts. J Food Process Eng 26:559–576

Zion B, McCarthy MJ, Chen P (1994) Real-time detection of pits in processed cherries by magnetic resonance projections. Food Sci Technol LEB 27(5):457–462

Zion B, Kim S-M, McCarthy MJ, Chen P (1997) Detection of pits in olives under motion by nuclear magnetic resonance. J Sci Food Agric 75(4):496–502

Chapter 7
Computer Vision

Cheng-Jin Du and Qiaofen Cheng

List of Abbreviations

3-D	Three-dimensional
ANN	Artificial neural network
CCD	Charge coupled device
CMOS	Complementary metal oxide silicon
DBC	Differential box counting
FD	Fractal dimension
GLCM	Grey-level co-occurrence matrix
HSV	Hue, saturation, and value
IMF	Intramuscular fat
KFCM	Kernel fuzzy c-means
MRI	Magnetic resonance imaging
RGB	Red, green, and blue
RLM	Run length matrix
SC	Statistical classification
SVM	Support vector machine
WT	Wavelet transform

C.-J. Du (✉)
Warwick Systems Biology Centre, University of Warwick,
Coventry House, CV4 7AL Coventry, UK
e-mail: c.du@warwick.ac.uk

Q. Cheng
Department of Food and Nutritional Sciences, Whiteknights,
RG6 6AP Reading, UK
e-mail: q.cheng@reading.ac.uk

C. P. O'Donnell et al. (eds.), *Process Analytical Technology for the Food Industry*,
Food Engineering Series, DOI 10.1007/978-1-4939-0311-5_7,

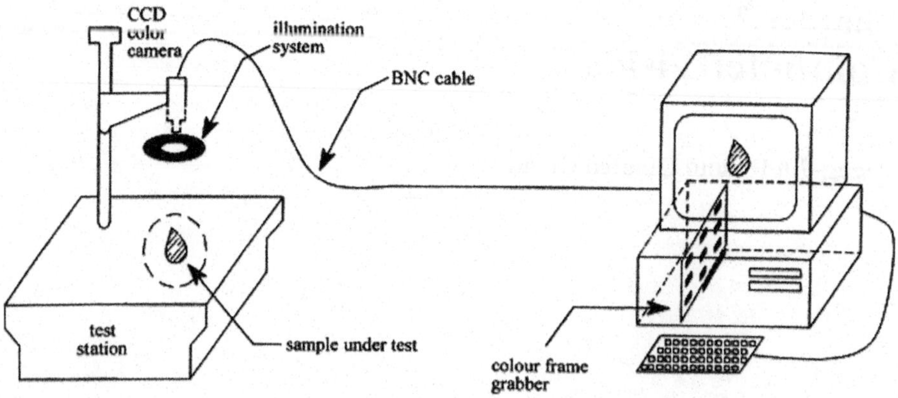

Fig. 7.1 A typical computer vision system. (Reprinted from Abdullah et al. (2006) with permission from Elsevier)

7.1 Introduction

Computer vision is the science and technology that enables programming a computer to simulate the physiological processes behind visual perception in humans, and to clone human behaviour of performance in colour, content, shape, and texture inspection. The history of computer vision can be traced back to the 1960s. Backed by the powerful learning systems, computer vision provides a mechanism in which human thinking process is simulated artificially and can help humans in making complicated judgments accurately, quickly, and very consistently over a long period (Abdullah et al. 2004). It is able to provide a rapid, consistent, and objective inspection tool for quality assurance and process control with a level of sensitivity and objectivity that humans cannot match.

The application potential of computer vision to the food industry has long been recognized (Tillett 1990). The food industry ranks among the top ten industries using computer vision technology (Gunasekaran 1996). By applying computer vision for automatic quality evaluation and process control, production speed and efficiency can be improved in addition to the increased assurance accuracy, with an accompanying reduction in production costs (Sun and Brosnan 2003). It could offer flexibility in application and be reasonable substitutes for the human visual decision-making process. Recently, computer vision techniques have been developed rapidly, and its hardware is relatively inexpensive and easy to use. Computers are by magnitudes faster and images are in better quality. Subsequently, computer vision plays a more and more important role in the food quality assurance and process control by maintaining accuracy and consistency while eliminating the subjectivity of manual inspections.

The organization of a computer vision system is highly application dependent. Typically, it is based on a computer endowed with image processing algorithms in connection with various instrumentations as shown in Fig. 7.1. This chapter is to introduce the instrumentation and the image processing techniques used in computer vision, illustrate their role in the food quality assurance and process control, outline the applications, and discuss the challenges encountered in the food industry.

7.2 Instrumentation

The instrumentation set-up of a typical computer vision system consists of an illumination device, a camera, a frame-grabber, and a computer. Illumination is an important prerequisite of image acquisition. The quality of captured images can be greatly affected by the lighting condition. Choosing the right lighting configuration is a prerequisite to high-quality image acquisition. A high-quality image can help to reduce the time and complexity of subsequent image processing steps. By enhancing image contrast, a well-designed illumination system can improve the accuracy and lead to successful image analysis (Gunasekaran 1996).

There is no general guideline for choosing the right lighting strategy. Different application may require different illumination strategy. Novini (1990) reported that most lighting arrangement could be grouped as one of followings: front lighting, back lighting, and structured lighting. Care must be taken when acquiring images with reflection, which can be found even using the back lighting. In order to control the reflection, it is especially important to understand what happens when light hits the surface. Besides, the choice of illuminant itself is also a key factor of image quality. A white fluorescent bulb with high frequency is a popular choice for many computer vision systems (Pedreschi et al. 2006; Pandit et al. 2007).

The camera and frame-grabber are the two elements responsible for the capture of an image in digital form. The camera converts photons to electrical signals, and the frame-grabber then digitizes these signals to give a stream of data or image. During the past decades, considerable amount of research effort has been directed at developing techniques for image acquisition. There are various types of light-sensitive cameras, such as the complementary metal oxide silicon (CMOS) cameras and charge-coupled device (CCD) cameras. Compared with CMOS cameras, CCD cameras have less noise, higher sensitivity, and a greater dynamic range. Therefore, it is frequently employed by computer vision systems for food quality assurance and process control.

CCD cameras can convert light into electrical charges and create high-quality, low-noise images with lots of pixels and excellent light sensitivity, which are free of geometric distortion and highly linear in their response to light. Recently, fishery, fruit, grain, meat, vegetable, and other food quality assurance issues have provided many actual and potential applications of the CCD camera. Among the applications, CCD cameras are widely used for quality classification, physical characteristic detection, and property estimation of food products. In some cases, it is difficult to evaluate food quality in the spectral region typically used. Through the use of different filters fitted to CCD cameras, analysis of images from selected spectral regions can be performed.

In addition to CCD cameras, there is a growing interest in using other image acquisition techniques for food quality assurance and process control applications. These are technologies such as magnetic resonance imaging (MRI; see Chap. 6), thermal imaging (see Chap. 8), hyperspectral imaging (see Chap. 9), ultrasound (see Chap. 10), and X-ray (see Chap. 12).

7.3 Image Processing

Image processing is the core of a computer vision system. Once the image is acquired, a preprocessing step will normally be carried out to obtain an enhanced image. After that, the image is segmented into disjoint and nonoverlapping regions, each of which typically corresponds to one object. The characteristics of these objects can then be measured, such as size, shape, colour, and texture. The objects are finally identified by classifying them into different groups.

7.3.1 Image Preprocessing

The captured image is subject to various types of noises, which may degrade the image quality. In order to improve the quality of an image, operations need to be performed on it to remove or decrease degradations suffered by the image during its acquisition. The purpose of preprocessing is to suppress unwilling distortions or enhance some image features that are important for further processing, and create a more suitable image than the original one for a specific application. Two types of image preprocessing approaches can be identified for food quality assurance and process control, i.e. spatial domain and transform domain methods.

7.3.1.1 Spatial Domain Methods

Spatial domain methods operate directly on the pixels of an image. Mean filtering is one of the most commonly used spatial domain methods for reducing noise in images. Each pixel value is replaced by the mean of its neighbourhood pixels:

$$I_{\mathrm{out}}(m,n)=\frac{1}{N}\sum\sum I_{in}(m-i,n-j), \qquad (i,j)\in W, \tag{7.1}$$

where W is the neighbourhood around the pixel (m, n), e.g. a 3×3 square, N is the total number of pixels in the neighbourhood W. The matrix formed by the weights $1/N$ is called a kernel. Quite often, the weights are assigned unequal values such as Gaussian kernel. Unlike the uniformly weight of mean filtering, Gaussian filter assigns the weight more towards the value of the central pixels.

One of the main problems for mean filtering is that it is unable to preserve the edges of an image well. When the kernel straddles on an edge, it will interpolate new values for pixels on the edge and thus blur the edge. In order to avoid edge blurring, one way is to use a median filter, which allows edges to be preserved while filtering out unwanted noises. It replaces the output pixel with the median of its neighbouring pixel values instead of a weighted sum of those values. Another way to smooth an image while preserving edges is to use the bilateral filter proposed by Tomasi and Manduchi (1998). The implementation of bilateral filter is noniterative, local, and simple where the output denoised image $\mathbf{I}_{out}(\mathbf{x})$ is a weighted average of the input noisy image $\mathbf{I}_{in}(\mathbf{x})$. It combines colours based on both their geometric closeness (the spatial domain $\mathcal{S}$) and their photometric similarity (the range domain $\mathcal{R}$). The weight assigned to each neighbour decreases with both the distance in the spatial domain $\mathcal{S}$ and the distance in the range domain $\mathcal{R}$. The pixels at larger distances are assigned smaller weight, and the larger the pixels value difference, the smaller the pixels contribute to the weight. In practice, two Gaussian functions are used as the decreasing functions for the spatial and range domain respectively. A bilateral filter applies to an input image $\mathbf{I}_{in}(\mathbf{x})$ and produces an output image $\mathbf{I}_{out}(\mathbf{x})$ as follows:

$$\mathbf{I}_{out}(\mathbf{x}) = \Gamma^{-1}(\mathbf{x}) \int \exp\left\{-\frac{|\xi - \mathbf{x}|}{2\sigma_s^2}\right\} \exp\left\{-\frac{|\mathbf{I}_{in}(\xi) - \mathbf{I}_{in}(\mathbf{x})|}{2\sigma_r^2}\right\} \times \mathbf{I}_{in}(\xi) d\xi, \quad (7.2)$$

where $\Gamma(\mathbf{x})$ is a normalization term:

$$\Gamma(\mathbf{x}) = \int \exp\left\{-\frac{|\xi - \mathbf{x}|}{2\sigma_s^2}\right\} \exp\left\{-\frac{|\mathbf{I}_{in}(\xi) - \mathbf{I}_{in}(\mathbf{x})|}{2\sigma_r^2}\right\} d\xi$$

The first Gaussian function measures the geometric closeness between the neighbourhood centre $\mathbf{x}$ and a nearby point ξ, where σ_s controls the extent of the spatial neighbourhood used to filter a pixel. The parameter σ_r of the second Gaussian function controls the discrimination power of an adjacent pixel because of the intensity value difference.

7.3.1.2 Transform Domain Methods

In many cases, transform domain methods are more effective than their spatial domain counterparts because noise can be more easily separated from the objects in the transform domain, where an image is expressed as a combination of a set of basic signals, known as the basis functions. In the process of transform domain methods, the input image firstly undergoes a forward transform, such as Fourier and wavelet, resulting in an array of transform coefficients. Then operations are carried out on the coefficients. Finally, an inverse transform is performed to reconstruct the processed coefficients.

The Fourier transform decomposes an image into sinusoidal signals with different periods, which describe the spatial frequencies in an image. In the Fourier domain image, low- and medium-frequency components correspond to smooth regions or large structures and image features, while high-frequency components are dominated by noise. A low-pass filter can be used to denoise an image by suppressing all frequencies higher than the cut-off frequency C_0 while leaving smaller frequencies unchanged:

$$H(u,v) = \begin{cases} 1 & if \ \mathrm{D(u,v)} \le C_0 \\ 0 & if \ \mathrm{D(u,v)} > C_0 \end{cases}, \tag{7.3}$$

where $H(u,v)$ is the filter function and $D(u,v) = \sqrt{u^2 + v^2}$ is the distance of a point from the origin in the Fourier domain. The drawback of this filter function is to cause ringing artefacts in the filtered spatial domain image. Better results can be achieved with the Butterworth filter of *n*th order

$$H(u,v) = \frac{1}{1 + [D(u,v)/C]_0^{2n}}. \tag{7.4}$$

The Fourier transform is an analysis of global frequency content in the image, which is not suitable for the analysis to be localized in the spatial domain. In contrast, wavelet transform (WT) captures both frequency and location information, and is more suitable for such application. As a result, WT provides a better tool to analyse edge regions in the image, where the signals are transient and time-variant (nonstationary). There are many types of wavelets available, among which a translation-invariant WT is more appropriate for image preprocessing. Normally, small WT coefficients correspond to noise. An image can be enhanced by reducing those coefficients to a value near zero.

7.3.2 *Image Segmentation*

Image segmentation partitions an image into its constituent objects, which is a challenging task because of the richness of visual information in food images. The techniques of image segmentation developed for food quality assurance can be divided into four different philosophical approaches, i.e. thresholding-based, region-based, boundary-based, and classification-based segmentations.

7.3.2.1 Thresholding-Based Segmentation

Thresholding-based segmentation is a particularly effective technique for scenes containing solid objects resting upon a contrasting background, which distinguishes the object from the remaining part of an image with an optimal value. The key of this method is the selection of threshold value. No universal methodology for threshold selection works on all kinds of images. A variety of techniques have been proposed to set the threshold value under different circumstances. The simple way is to set the threshold values as the dips of the grey-level histogram if two or more distinct modes exist. When the histogram is noisy, the locations of dips might be obscure and unreliable. This can be overcome by smoothing the histogram before trying to find separate modes. However, smoothing might shift the position of the dips in the histogram when two peaks are unequal in size.

The isodata algorithm proposed by Ridler and Calvard (1978) is an iterative threshold selection technique. A threshold value T^0 is initialized as the midway between the maximum and minimum grey level. Next, the means μ_F^0 and μ_B^0 of the foreground and background pixels are calculated, respectively. A new threshold value $T^1 = \mu_F^0 + \mu_B^0 / 2$ is then obtained. This process is repeated until the threshold value no longer changes.

Another way to look at the problem is to consider the values in the two regions as two classes. The optimum threshold separates those two classes in a way that their combined spread (within-class variance) is minimal (Otsu 1979). Define the within-class variance as the weighted sum of each class variances:

$$\sigma_w^2(T) = S_F(T)\sigma_F^2(T) + S_B(T)\sigma_B^2(T), \tag{7.5}$$

where $S_F(T) = \sum_{i=0}^{T-1} p(i)$ and $S_B(T) = \sum_{i=T}^{N-1} p(i)$ are the probabilities of the foreground and background, respectively, $\sigma_F^2(T)$ and $\sigma_B^2(T)$ are their variances. Minimizing the within-class variance is the same as maximizing the between-class variance, which is

$$\sigma_b^2(T) = \sigma^2 - \sigma_w^2(T) = S_F(T)S_B(T)\left[\mu_F(T) - \mu_B(T)\right]^2. \tag{7.6}$$

Thus, the optimal threshold can be obtained by iteratively updating the class probabilities S_i and class means μ_i in turn.

7.3.2.2 Region-Based Segmentation

Thresholding-based segmentation methods only take into account the distribution of grey levels without considering any spatial information. Region-based methods exploit spatial context by grouping adjacent pixels or small regions together into larger regions, which can be divided into two basic classes: region growing-and-merging and region splitting-and-merging. The former is a bottom-up method that groups pixels or subregions into larger regions according to a set of homogeneity criteria; and the latter is a top-down method that successively divides an image into smaller and smaller regions until certain criteria are satisfied. Region-based algorithms are computationally more expensive than those simpler techniques, e.g. thresholding-based segmentation, but region-based segmentation is able to utilize several image properties directly and simultaneously determine the final boundary location. It shows the greatest promise in the segmentation of food products because strong a priori knowledge is normally not available.

An example of region-growing methods is the vector confidence connected algorithm (Yoo et al. 2002). A seed point should be supplied for the initialization of the algorithm. This point can be calculated as the position with the median pixel value of the object. The vector mean $\mathbf{m}_p$ and the covariance matrix $\mathbf{C}_p$ of a small neighbourhood region around the seed point are then computed. The membership of a pixel $\mathbf{p}$ to the region is measured using the Mahalanobis distance D as follows:

$$D = \sqrt{(\mathbf{p}-\mathbf{m}_p)^T \mathbf{C}_p^{-1} (\mathbf{p}-\mathbf{m}_p)}. \quad (7.7)$$

Taking distribution of the points (correlations) into account, Mahalanobis distance is a very useful way of determining the "similarity" of a set of values from an "unknown" sample to a set of values measured from a collection of "known" samples. Besides the grey level, other image properties can be easily incorporated into the membership function, such as position, texture, and colour.

Region-splitting methods commonly use a data structure called quadtree, each node of which corresponds to a square-shaped region. The root node represents the whole image, while the leaf nodes represent a coherent region. As quadtrees impose one type of regular decomposition into an image, a merging process after each split must be applied to compare adjacent regions of four nodes with a common parent and merge them if they obey the criterion. The drawback of this method is that it is difficult to decide where to make the partition.

7.3.2.3 Boundary-Based Segmentation

The thresholding approach accomplishes segmentation by partitioning the image into sets of interior and exterior points. By contrast, boundary-based approaches attempt to find the edges directly by their high gradient magnitudes. The first step of a boundary-based segmentation method is to detect edges of an image. Edges are those pixels that have sharp grey-level changes or discontinuities, which can be detected by looking for the maximum in the first derivative or zero-crossings in the second derivative of the image. The detected edges can then be used to establish the boundaries of objects.

The detected edge points seldom form closed connected boundaries required for image segmentation. Hence, it is usually required to link those edges, which can be accomplished by searching a neighbourhood around an endpoint for other endpoints and then filling in boundary pixels as required to connect them. The application of the boundary-based segmentation is limited because completed boundaries are difficult and sometimes impossible to trace in some food images.

Due to the complex nature of food images, the segmentation approaches described above are often found only partly suitable for performing such task. For some specific application, the performance could be improved by combining those

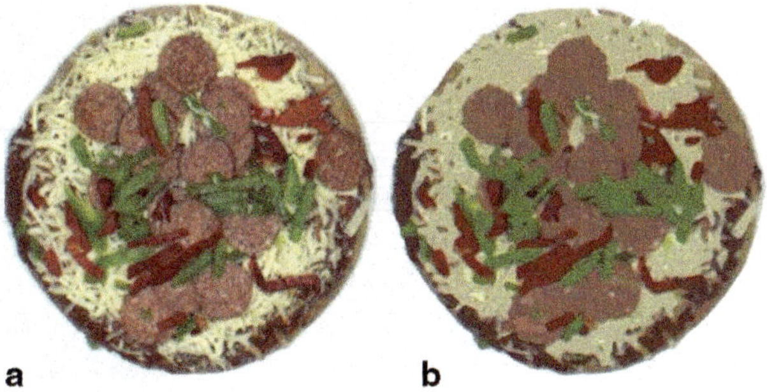

Fig. 7.2 An example of combined method for pizza-topping image segmentation. **a** An example pizza image. **b** The result of image segmentation. (Reprinted from Sun and Du (2004) with permission from Elsevier)

traditional methods. To partition a pizza-topping image into homogeneous regions automatically, Sun and Du (2004) developed a region growing-and-merging method known as "stick growing and merging", which employs the traditional region-based segmentation as a dominant method and combines the strengths of both thresholding- and edge-based segmentation techniques. The algorithm consists of four major steps: stick initialisation, stick merging, subregion merging, and boundary modification. It is started from an initial decomposition of the image into small sticks and non-sticks. The small sticks are merged to obtain the initial subregions on the basis of homogeneity criteria. Then smaller subregions with only one stick are merged into larger subregions and subsequently all subregions are merged into regions according to the criteria. Finally, non-sticks and separate small sticks are merged and the degree of boundary roughness is reduced by boundary modification. Figure 7.2b shows the segmented results of a pizza-topping image, which includes ham, red and green peppers, cheese shreds, and tomato sauce. The original image, Fig. 7.2a, is complex for several reasons: the inhomogeneous character inside the natural foods, object overlapping, shadows, and light reflection.

7.3.2.4 Classification-Based Segmentation

Classification-based methods attempt to assign each pixel to different objects based on classification techniques, including supervised and unsupervised methods. Supervised methods require training data to be specified to train a classifier, while unsupervised methods learn a classification directly from the data, i.e. no training data required. Given the complex nature of food images, unsupervised techniques appear to be the preferable solution for reliably and consistently separating an image into parts of interest without human intervention.

As one of the most effective unsupervised methods, the kernel fuzzy c-means (KFCM) algorithm has been used successfully for image segmentation. The major advantage is that each image pixel has a membership grade indicating its belongingness degree to each cluster by the introduction of fuzziness. This makes KFCM more robust by retaining more information from the original image than the crisp segmentation methods, such as k-means, which assigns the image pixel to only one

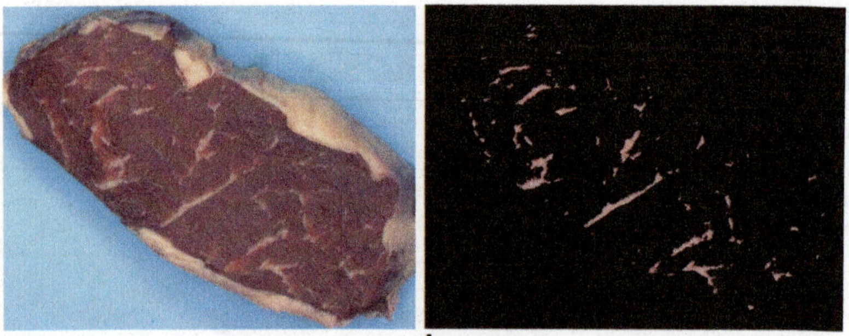

Fig. 7.3 An illustration for the classification-based segmentation method. **a** An example of beef-steak image. **b** Marbling image obtained via the kernel fuzzy c-means method

of the clusters. Furthermore, by mapping the feature vectors to a higher dimensional space via the kernel trick (Boser et al. 1992), it is able to separate the image pixels non-linearly.

Suppose that an image has a size of n pixels, and each pixel is denoted by a vector $\mathbf{x}_k$, the KFCM algorithm can be described as the task of finding l cluster centroids so that the following objective function is minimized:

$$J_{KFCM}(\mathbf{U},\mathbf{C}) = \sum_{i=1}^{l}\sum_{k=1}^{n} u_{ik}{}^{w}(K(\mathbf{x}_k,\mathbf{x}_k) - 2K(\mathbf{x}_k,\mathbf{c}_i) + K(\mathbf{c}_i,\mathbf{c}_i)), \tag{7.8}$$

where $\mathbf{U}$ is the membership matrix constrained to contain elements in the range [0, 1] so that $\sum_{i=1}^{l} u_{ik} = 1, \forall k = 1,2,\ldots,n$, and u_{ik} is the membership value of $\mathbf{x}_k$ for class i; $\mathbf{C}$ is the set of cluster centroids, $\mathbf{c}_k$ is the centroid of kth class; and $w \in 1,\infty)$ is a weighting exponent determining the fuzziness of each membership. Gaussian radial basis function is usually applied in practice as the kernel function, which is defined as

$$K(\mathbf{x},\mathbf{y}) = \exp(-\|\mathbf{x}-\mathbf{y}\|^2 / 2\sigma^2). \tag{7.9}$$

For minimizing the objective function (7.8), there are two necessary conditions that could be derived as follows

$$u_{ik} = \left(\sum_{j=1}^{l} \left(\frac{K(\mathbf{x}_k,\mathbf{x}_k) - 2K(\mathbf{x}_k,\mathbf{c}_i) + K(\mathbf{c}_i,\mathbf{c}_i)}{K(\mathbf{x}_k,\mathbf{x}_k) - 2K(\mathbf{x}_k,\mathbf{c}_j) + K(\mathbf{c}_j,\mathbf{c}_j)} \right)^{1/(w-1)} \right)^{-1} \tag{7.10}$$

$$\mathbf{c}_i' = \frac{\sum_{k=1}^{n} u_{ik}{}^{w} K(\mathbf{x}_k,\mathbf{c}_i)\mathbf{x}_k}{\sum_{k=1}^{n} u_{ik}{}^{w} K(\mathbf{x}_k,\mathbf{c}_i)}. \tag{7.11}$$

An example of beefsteak is shown in Fig. 7.3a. The beefsteak contained nearly 70 % moisture, which caused heavy reflection during image acquisition and made the segmentation a challenge. Figure 7.3b shows the segmented marbling result using KFCM method. Since the objects in the beef image include fat, lean, and background, the number of cluster classes was set to three for KFCM algorithm. It can be observed that the performance is quite good, where most of the fat particles were segmented successfully.

7.3.3 *Object Measurement*

Once the image is segmented into discrete objects of interest, they can be described and represented for further processing and analysis by measuring their individual features. In general, a segmented object can be represented in features of its external characteristics or internal characteristics. Many features can be used to describe an object. Generally, the features that are the simplest to measure and that contribute substantially towards the classification are the best to use. Measurements that can be performed on features in images for food quality assurance can be grouped into four classes: size, shape, colour, and texture. For each class, a number of different specific measurements can be made, and there are a variety of different ways to perform the operations. Most computer vision systems offer at least a few measures in each class and produce a numeric output suitable for further image analysis.

7.3.3.1 Size

Three commonly used features for size measurement of an object can be found for food quality assurance: area, perimeter, and length and width. The most basic convenient measurement for size is area. For pixel-based representation, this is the number of pixels within an object, which can be straightforwardly determined by counting. The perimeter of an object is particularly useful for discriminating between objects with simple and complex shapes. Perimeter measurements can be easily computed during the extraction of an object from a segmented image. The length and width can also be used to measure the size of an object. It is necessary to locate the major axis of the object and measure its relative length and width.

7.3.3.2 Shape

Shape features are generally invariant to translation, rotation, and scaling. Usually, objects of one class can be distinguished from the others by their shapes, which are physical dimensional measurements that characterise the appearance of an object. Shape features can be measured independently and by combining size measurements.

Table 7.1 Shape features with combinations of size measurements

Shape feature	Formula
Area ratio	$\frac{\text{Area}}{\text{Max diameter} \cdot \text{Min diameter}}$
Aspect ratio	$\frac{\text{Max diameter}}{\text{Min diameter}}$
Circularity	$\frac{\text{Perimeter}^2}{\text{Area}}$
Compactness	$\frac{4\pi \cdot \text{Area}}{\text{Perimeter}^2}$
Diameter range	$\text{Max diameter} - \text{Min diameter}$
Eccentricity	$\sqrt{1 - \frac{\text{Semi-minor}^2}{\text{Semi-major}^2}}$
Roundness	$\frac{4 \cdot \text{Area}}{\pi \cdot \text{Max diameter}^2}$
Shape factor 1	$\frac{4\pi \cdot \text{Area}}{\text{Perimeter}^2}$
Shape factor 2	$\frac{\text{Max diameter}}{\text{Area}}$
Shape factor 3	$\frac{\text{Area}}{\text{Maxdiameter}^3}$
Shape factor 4	$\frac{4 \cdot \text{Area}}{\pi \cdot \text{Maxdiameter} \cdot \text{Mindiameter}}$

Table 7.1 summarizes some of the most widely used shape features with combinations of size measurements for food products.

Various techniques for shape description independent of size measurements have been investigated for food quality assurance. These techniques can be generally classified into three categories: boundary encoding, Fourier descriptor, and invariant moments. The boundary chain code shows the boundary tangent angle as a function of distance around an object, while the differential chain code reflects the curvature of the boundary. Both functions can be further analysed to obtain measurements of shape. The Fourier transform of one cycle of the boundary function is an alternative representation of the associated object's shape. Using Fourier transform, the boundary function spectrum can be low-pass filtered without destroying the characteristic shape of the object. Only the amplitudes and phases of the low-frequency impulses in the spectrum, i.e. the low-order Fourier coefficients, are required to characterise the basic shape of the object. These values are candidates for shape descriptors.

Invariant moments have some of the properties that good shape features must have. The magnitudes of invariant moments reflect the shape of an object and can be used in computer vision to distinguish different objects. Given a boundary function $B(x, y)$, its moments can be computed by

$$M_{uv} = \int_{-\infty}^{\infty}\int_{-\infty}^{\infty} x^u y^v B(x,y)dxdy \qquad u,v \in [0,1,2\ldots). \tag{7.12}$$

The set of moments $\{M_{uv}\}$ is unique for the boundary function. The invariant moments that are insensitive to translation, rotation, and scale changes are

$$\mu_{20} + \mu_{02} \quad \text{and} \quad (\mu_{20} - \mu_{02})^2 + 4\mu_{11}^2, \tag{7.13}$$

where $\mu_{20} = M_{20} - M_{10}^2 / M_{00}$, $\mu_{02} = M_{02} - M_{01}^2 / M_{00}$, and $\mu_{11} = M_{11} - M_{10}M_{01} / M_{00}$. They characterise the object dispersion over the *x*-axis, *y*-axis, and both orientations, respectively.

7.3.3.3 Colour

In image analysis of food products, colour is an influential attribute and powerful descriptor that often simplifies object extraction and identification. Colour vision offers a tremendous amount of spatial resolution that can be used to quantify the colour distribution of ingredients. Colour features of an object can be extracted by examining every pixel within the object boundaries.

The red, green, and blue (RGB) colour space used in computer graphics is device dependent, which is designed for specific devices, e.g. cathode-ray tube display. Therefore, the RGB space has no accurate definition for a human observer, where the proximity of colours in the space does not indicate colour similarity in perception. Colour space transformations are effective means for distinguishing colour images, which is an operation on the original colour space to produce a new transformed space.

Linear transformation is the simplest method for colour conversion from RGB space to others. However, colour space transformations such as HSV (hue, saturation, and value) and L*a*b* are more complex, which are generated by non-linear transformations, and widely used in computer vision systems for food quality assurance.

7.3.3.4 Texture

Image texture is one of the main features measured for food quality assurance using computer vision. It is important to note that in computer vision context, the concept of texture is totally different from the one generally understood and used in the food industry. Food texture refers to the manner in which the food behaves in the mouth and is characterised by parameters such as hardness, cohesiveness, viscosity, elasticity, adhesiveness, brittleness, chewiness, and gumminess. However, image texture is an attribute representing the spatial arrangement of the grey levels of pixels in a region (Anon 1990). As a useful feature for object description, image texture can quantify some characteristics of the grey-level variation within an object, such as fineness, coarseness, smoothness, and graininess.

Initially, image texture analysis is based on the first-order grey-level statistics, which make direct use of the pixel histogram to characterise the relative amount of each intensity value. Four features are normally derived for texture characterisation, including mean, variance, skewness, and kurtosis. Among the other texture analysis methods for food quality assurance, most approaches are statistical, including the run length matrix (RLM) method and the grey-level co-occurrence matrix (GLCM) method. Moreover, several other texture description methods are based on transform, such as Fourier transform and WT. In addition, fractal dimension (FD) has also been employed to numerically describe the image texture characteristics of various food products.

First introduced by Galloway (1975), the RLM approach characterises texture by the grey-level run, which is a set of consecutive pixels with the same grey level. The run length is the number of pixels in a run. Therefore, the run length of coarse textures will be longer than that of fine textures. Define a two-dimensional (2-D) matrix $RLM(g, r)$ as the number of runs with different grey-level g and run length r. Several features can be derived from the $RLM(g, r)$, e.g. short-run emphasis, long-run emphasis, grey-level non-uniformity, run length non-uniformity, run length percentage, and low- and high-grey-level run emphases.

The GLCM method is one of the most frequently cited methods for texture analysis, which is a general procedure presented by Haralick et al. (1973) for extracting image texture information in the spatial domain. To obtain the spatial relationship contained in an image, a spatial-dependence GLCM is first constructed by estimating the second-order joint conditional probability density functions of pixel intensity. Each element (i, j) of GLCM, denoted by CM_{ij}, represents the probability that two pixels with the grey-level i and j co-occur in the image separated by a distance d in direction θ. Theoretically, a variety of GLCMs could be constructed from the image with different values of direction and distance. For the direction θ, the four angles 0°, 45°, 90°, and 135° are commonly used to achieve rotation invariance by averaging the results of each angle. For the distance d, values other than 1 are rarely used in the literature. A set of features can be derived from each GLCM to describe the texture characteristics within the image, such as the homogeneity, contrast, and presence of organized structure.

An image can be viewed as a hilly terrain surface whose height from the normal ground is proportional to the image grey value (Peleg et al. 1984). The image texture features such as roughness, smoothness, and graininess in the image could be described by the FD of this surface, which is invariant under translation, rotation, and certain scale transformation. Several approaches exist to estimate the FD of an image, such as the ε-blanket method that calculates FD by using dilatation and erosion of an image (Peleg et al. 1984), the Frequency domain method that determines FD from the Fourier power spectrum of the image data (Quevedo 2002), and the differential box counting (DBC) method (Sarkar and Chaudhuri 1994). The fractal texture determined by the box counting method varied monotonically with image true values of FD, and could be a useful descriptor in texture recognition. Furthermore, compared with other approaches in terms of computer complexity and accuracy, DBC is a simple, accurate, and computationally efficient method. The image is divided into N_{sr} distinct self-similar pieces, each of which is scaled down by a ratio *sr* in all dimensions, and becomes statistically identical to the original one. The *FD* of image could be computed by

$$FD = \frac{\log(N_{sr})}{\log(1/sr)}. \tag{7.14}$$

Using the DBC method, a number of values of N_{sr} could be estimated for different values of *sr*. Then FD of the image could be estimated by the slope of the least-squares linear fit of the plot of $\log(N_{sr})$ versus $\log(1/sr)$.

The scale of an image is important for texture analysis because there might be several different textures in the same image with different scales. However, the traditional approaches for image texture analysis such as RLM, GLCM methods are limited in that they are restricted to the analysis of an image over a single scale. The development of multi-scale analysis such as WT has been proven to be useful to characterise different scales of textures effectively. WT has not only solid theoretical foundation in formal mathematical theory but also good empirical performance for multi-scale image analysis.

Common WT suffers from a lack of translation-invariant, where a simple shift of the image will result in non-trivial modifications of the values of wavelet coefficients. Since the steerable pyramid transform described by Simoncelli and Freeman (1995) has nice reconstruction properties, in addition to properties of translation-invariance and rotation-invariance, it can be implemented for efficient and accurate linear decomposition of the image into high-pass and low-pass residual bands, and a set of orientated subbands.

Based on the steerable pyramid transform, the texture features can be obtained by computing the energy of each subband, which is widely used for wavelet-based texture characterisation. Assume the *i*th subband image (W_i) has a size of M×N pixels. Then the energy of decomposed subbands in the frequency domain could be calculated as follows:

$$E_i = \frac{1}{M \times N} \sum_{i=1}^{M} \sum_{j=1}^{N} W_i^2(i,j). \tag{7.15}$$

7.3.4 *Classification*

Generally, classification identifies objects by classifying them into one of the finite sets of classes, which involves comparing the measured features of a new object with those of a known object or other known criteria and determining whether the new object belongs to a particular category of objects. Figure 7.4 shows the general classification system configuration used in computer vision for food quality assurance. A wide variety of approaches have been applied for this task. Among

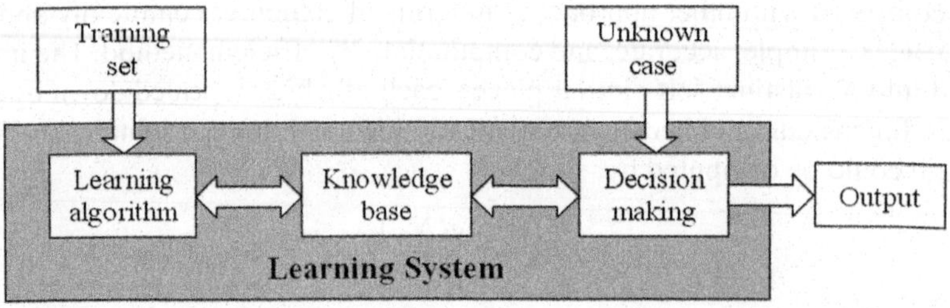

Fig. 7.4 The general configuration of machine learning system

the applications where classification techniques have been employed for building knowledge base, artificial neural network (ANN), and statistical approaches are the two main methods. In the meantime, fuzzy logic, decision tree, and support vector machine (SVM) have been used as well.

Initially inspired by biological nervous systems, ANN approaches combine the complexity of some of the statistical techniques with the objective of imitating human intelligence, which are characterised by their self-learning capability. The complete network represents a very complex set of interdependencies, and may incorporate any degree of non-linearity in theory. For food quality assurance, very general functions can be modelled to transform physical properties into quality factors. ANN technology allows the extension of computer vision technology into the areas of colour, content, shape, and texture inspection at near-human levels of performance.

Statistical classification (SC) utilizes the statistical properties of the observations from the training set. It is generally characterised by having an explicit underlying probability model. There are three kinds of SC techniques mostly used among the applications, i.e. Bayesian classification, discriminant analysis, and nearest neighbour. Bayesian classification is a probabilistic approach to learn and inference, in which probability is used to represent uncertainty about the relationship being learnt. Bayesian learning can produce the probability distributions of the quantities of interest, and make the optimal decisions by reasoning about these probabilities together with observed data. Discriminant analysis takes into account the different variables of an object and works by finding the so-called discriminant functions in such a way that the differences between the predefined groups are maximized. The obtained discriminant rules provide a way to classify each new object into one of the previous defined groups. The nearest neighbour is a non-parametric classification technique by assigning the unknown case as the class most frequently represented among the nearest samples. It involves a training set of both positive and negative cases. A new sample is classified by calculating the distance to the nearest training case.

Fuzzy logic is introduced as a representation scheme and calculus for uncertain or vague notions and could provide a completely different way for the applications like the classification of food products. Compared with traditional techniques, fuzzy logic groups individual samples into classes that do not have sharply defined boundaries. It embodies the nature of human's mind in some sense, as the conception of possibility and probability is truly underlined in this logic. In contrast with the absolute values and categories in the traditional Boolean logic, it mimics more human-like behaviour for decision making and reasoning by extending to handling of the intermediate categories of partial truth or partial false. And thus, it can simulate the human experience of generating complex decisions using approximate and uncertain information.

Decision tree acquires knowledge in the form of tree, which can also be rewritten as a set of discrete rules to make it easy to understand. The main advantage of decision tree classifier is its capability of using different feature subsets and decision rules at different stages of classification. The performance of a decision tree classifier depends on how well the tree is constructed from the training data.

SVM is a state-of-the-art classification algorithm, which has a good theoretical foundation in statistical learning theory (Vapnik 1995). Instead of the minimization of the misclassification on the training set, SVM fixes the decision function based on structural risk minimization to avoid the overfitting problem. It performs classification by finding maximal margin hyperplanes in terms of a subset of the input data between different classes. The subset of vectors defining the hyperplanes is called support vectors. If the input data are not linearly separable, SVM firstly maps the data into a high (possibly infinite) dimensional feature space, and then classifies the data by the maximal margin hyperplanes. Furthermore, SVM is capable of classification in high-dimensional feature space with fewer training data.

To illustrate the performance of SVM classifiers, a 2-D data set with five samples for each class is shown in Fig. 7.5a, where the samples of class + 1 are represented by the dots in yellow colour, while the samples of class − 1 by the dots in blue colour. The performance of a linear SVM is illustrated in Fig. 7.5b. If the input data are not linearly separable, SVM firstly maps the data into a high-dimensional feature space using a kernel function, such as polynomial kernel and Gaussian radial basis function kernel and then classifies the data by the maximal margin hyperplanes as shown in Figs. 7.5c and d, respectively.

7.4 Applications

Computer vision has applicability to various types of food product for quality assurance and process control, including fruits, grains, meats, vegetables, and other foods. Table 7.2 summarizes the reported major applications of computer vision in the food industry from the literature.

For fruit quality assurance, it has been used to classify apples into various grades according to the features of size, shape, colour, surface quality condition, and spectral reflectance of blemishes (Leemans and Kleynen 2008). The size, shape, ripeness, bruise, focal contamination, firmness, soluble-solid content, and anthocyanin distribution of strawberries have been evaluated by computer vision (Nagata and

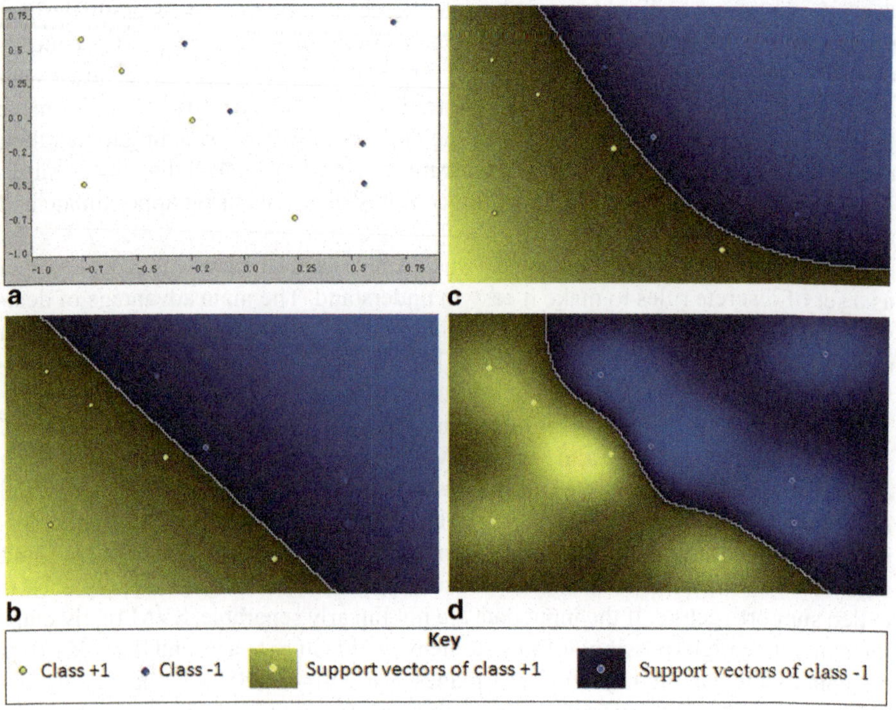

Fig. 7.5 An illustration for the support vector machine classification method. **a** An illustrated data set. **b–d** The performances of a linear SVM classifier (**b**), a polynomial SVM classifier (**c**), a radial basis function SVM classifier (**d**)

Tallada 2008). The mass of mango can be estimated from geometric dimensions such as length, maximum width, and maximum thickness measured by the computer vision method (Spreer and Müller 2011). It is also feasible for computer vision to sort other fruits such as cherries, citrus, pears, pomegranate, raisins, table olives, and tomatoes according to different quality attributes.

The majority of grain applications using computer vision can be found for wheat, rice, and corn quality assurance, such as identification of damaged, impurity kernels, classification of cereal grains and their varieties, and correlating the vitreosity and grain hardness. Substantial work uses morphological (size and shape) features (Igathinathane et al. 2009), while a few investigations are based on colour or textural features (Jayas et al. 2008). It was reported in the work of Majumdar and Jayas (2000) that the highest classification accuracies of CWRS wheat, CWAD wheat, barley, oats, and rye can be achieved when combining morphological, colour, and textural feature sets. To simultaneously discriminate and count filled/unfilled rice spikelets, Duan et al. (2011) use the bimodal imaging system which combines visible light imaging and soft X-ray imaging. Visible light imaging was applied to measure the projected area of the spikelet hull, while soft X-ray imaging yielded the projected area of the inner brown rice kernel.

Table 7.2 Summary of computer vision applications for food quality evaluation

Category	Products	Applications
Fruits	Apple	Size, shape, and colour grading, bruises detection, calyx, and stalk-end recognition
	Cherries	Defect detection
	Citrus fruit	Grading citrus fruits according to their size, shape, and colour features
	Mango	Estimating mass from its geometric dimensions
	Pears	Stem and shape recognition
	Pomegranate	Colour sorting
	Raisin	Classification according to wrinkle edge density, average gradient magnitude, angularity, and elongation
	Strawberry	Evaluating size, shape, ripeness, bruise, focal contamination, firmness, soluble-solid content, and anthocyanin distribution
	Table olive	Grading according to defects on the surface
	Tomato	Determining maturity, shape, and colour classification
Grains	Corn/maize	Classification of different shape and size of kernels, measurement of stress cracking
	Edible bean	Classification according to size, shape, and texture features
	Rice	Monitoring milling quality, assessing breakage and cracks, discrimination and counting rice spikelets
	Wheat, barley, oat, and rye	Classification according to morphological and colour features of healthy and damaged kernels
Fishery and meats	Bivalve	Study of larval growth
	Crassostrea	Detection of hinge lines
	Fish	Fish species recognition, sorting
	Shrimp	Estimating dehydration level
	Beef	Quantifying intramuscular fat, classification, predicting eating quality
	Lamb	Classification, predicting carcass grades, and eating quality
	Pork	Colour evaluation, carcass grading, eating quality prediction
	Poultry	Estimating fat content, classification according to colour and texture features, splenomegaly detection, viscera inspection, wholesomeness characterisation
	Cooked meats	Evaluating the physical changes (shrinkage, pores, and porosity), quality classification
Vegetables	Asparagus	Defect inspection
	Bell pepper	Classification according to colour and damage
	Carrot	Evaluating forking, surface defects, curvature, and brokenness
	Chicory	Evaluating via shape and colour features
	Lentil	Classification according to shape
	Mushroom	Brown blotch or ginger blotch diseases recognition
	Onion	Defect detection
	Potato	Surface defect detection, classification according to size, shape, colour, and texture features
Others	Biscuit	Classification according to size, shape, and colour features
	Cheese	Determination of the shreddability, recognization of cheese shred dimensions, browning, and oiling off properties, inspection of the distribution and amount of ingredients in pasteurized cheese, monitoring curd syneresis

Table 7.2 (continued)

Category	Products	Applications
	Cookies	Quality inspection for size, shape, baked dough colour, and fraction of top surface area
	Muffin	Classification according to shape and colour features
	Oriental noodle	Colour inspection, damage detection
	Pizza	Quality assurance of pizza base, sauce spread, and topping according to size, shape, colour, and texture features
	Potato chip	Classification according to colour and texture features

With advancements in image analysis techniques, computer vision has become a technology of great potential for fish, shrimp, meat, and poultry quality assurance. Using the widths and heights at various locations along fish, the fish species can be recognized (Balaban et al. 2008). By analysing colour during the drying process, shrimp dehydration level can be estimated automatically (Mohebbi et al. 2009). As one of the most important visually assessed attributes that determine the beef quality, the abundance of intramuscular fat (IMF) can be quantified with image processing techniques. By image texture analysis, it is possible to classify tough and tender beef. Using various image features extracted, computer vision has shown practical for pork and lamb quality assurance (Zheng et al. 2008). Substantial progress has been made regarding the inspection of poultry carcasses with computer vision, including splenomegaly detection, viscera inspection, and wholesomeness characterisation (Park 2008). The estimation of fat ingredient content in poultry meat is shown to be possible by computer vision system, even the material obtained in industrial conditions (Chmiel et al. 2011). In addition, for quality assurance of cooked meats as affected by cooking and cooling, computer vision has recently been shown great potential to perform such a task by evaluating the physical changes (shrinkage, pores, and porosity) during the manufacturing procedures and their image features (colour and texture; Du and Sun 2008a).

Furthermore, computer vision has been utilized to classify vegetables, such as bell peppers according to their colour and damage (Shearer and Payne 1990); carrots for forking, surface defects, curvature, and brokenness (Brandon et al. 1990); mushrooms with the brown blotch or ginger blotch diseases (Vízhányó and Felföldi 2000); and onions into good or defective classes (Shahin et al. 2002). In addition, asparagus defects, including spreading tips, broken tips and scarred or cracked spears, can be inspected automatically (Rigney et al. 1992).

It is practical to extend computer vision techniques for quality assurance of other food products. For visual inspection of muffins, Abdullah et al. (2000) developed an automated system incorporating multivariate discriminant algorithms to statistically classify muffins based on surface colour. The biscuits on a moving conveyor belt can be classified in real time into one of four distinct groups: underbaked, moderately baked, overbaked, and substantially overbaked (Nashat et al. 2011). According to the manufacturing procedures of pizza, computer vision has been applied for quality assurance of pizza base, sauce spread, and topping (Du and Sun 2008b). In addition, computer vision has been successfully established as a technique for quality inspection of cheese, which are determination of the shreddability (Apostolopoulos and Marshall 1994), recognition of cheese shred dimensions (Ni and Gunasekaran 1995), measurement of the meltability, browning, and oiling off properties of Cheddar and Mozzarella cheeses under different cooking conditions and sizes of sample (Wang and Sun 2002, 2003, 2004a, b), inspection of the distribution and amount of ingredients in pasteurized cheese (Jeliński et al. 2007), and monitoring curd syneresis in a cheese vat (Everard et al. 2007; Fagan et al. 2008).

7.5 Challenges and Future Perspectives

A number of difficult problems still need to be addressed for the computer vision application in food quality assurance and process control. Mechanically handling and packaging are a remaining challenge in applying computer vision for food quality assurance, especially for those fragile products like apples, pears, and berries that are easily bruised and marked when they are in contact with hard surfaces. To design an effective computer vision system for online applications, mechanical and electrical hardware functionality needs to be considered during the development. For example, in grain-handling facilities, the simple task of identifying a sprocket on the railcar for a robot to open and close the gate requires detection of different types of sprockets, located at different positions on the railcars, under different lighting conditions (Jayas et al. 2005).

The image processing algorithms that are proposed must be put more in touch with image acquisition strategies. In general, the feedback between the output of an image processing algorithm and the setting of image acquisition parameters are largely ignored. The underlying physics of image acquisition are the very basis of the data to be worked with when attempting to quantify image features. It is important to integrate the physics knowledge into image processing strategies, to fully analyse the variability in the acquisition equipment.

The success of a computer vision system highly depends on the segmentation quality of food images. There is no general algorithm available that can robustly segment a variety of relevant structures of food images over a range of data sets. Robust and accurate segmentation of images still remains a challenge in computer vision. Many methods proposed are often sensitive to variation in image acquisition parameters and their own initial settings. Furthermore, most of the segmentation algorithms are still computationally very expensive for online image segmentation. Due to the non-uniform sizes, shapes, surfaces, and colours of food images, it is necessary to incorporate context-based, constraining information into the segmentation algorithm. Two core approaches could be employed, i.e. the expert system-like approach and the mathematical optimization.

A critical issue in terms of all practical and theoretical development is the need to develop appropriate validation and evaluation approaches. There is a shortage of common databases where algorithms can be compared to each other. Not like in the medical image analysis community, almost no test databases in food application are made available to the public, and efforts should be made in setting up such test database. In addition, evaluation methodology should be developed to assess the strengths and weakness of algorithms.

In some application areas for food quality assurance and control, physical models could be developed for simulation of food processing via integrating image-derived quantitative information. A key challenge here is how to choose image features that are valuable, and issues of accuracy, repeatability, and variability must be considered in the long run. Besides, the effective use of these models will require new knowledge from a variety of areas.

Rapid and accurate determination of the internal quality of food products poses technical challenges because of their complex structural, physical, and chemical properties. Researchers are continuing to investigate new, better methods and techniques for assessing internal food quality. The reduction in the price of novel technologies that have been used in medicine, such as MRI and X-rays, opens the door to obtain non-invasive internal images, but these technologies are still far from being applied in real-time systems. The hyperspectral imaging technique described in Chap. 9 provides a new opportunity for determining the optical properties and quality of food products. Compared with other techniques currently available, the hyperspectral imaging technique is faster, simpler, and easier to use for determining the optical properties of turbid and opaque food products.

Food products are 3-D entities, and most activities occur in 3-D space. Thus, 3-D imaging techniques such as MRI and X-rays are needed to improve our ability to study them. An inevitable consequence of the new opportunities offered by these techniques is that the size and complexity of image data are ever increasing. More sophisticated 3-D algorithms that operate on volume data sets are needed to be developed, and richer information could be obtained. Efforts have been directed towards better 3-D visualization and measurement of food structure, but current machines are unable to achieve this objective online and thus results are still at the laboratory level.

It is necessary to integrate the information uncovered at the molecular and cellular level into information uncovered at the macroscopic level. Being a powerful tool, computer vision has an important role in terms of analysing data and integrating information. Currently, computer vision is typically applied for the macroscopic level images of food products. As microstructure is better characterised via advancing microscopy, findings at the molecular and cellular levels may be used to gain new insights.

In conclusion, the ever-improving capabilities of computers, image acquisition, and image processing techniques opens new horizons for the application of computer vision in food quality assurance and control, which will continue to be an active area of research. As data from more research accumulate, hardware becomes faster and more affordable, and image-processing algorithm becomes more intelligent, it is expected that computer vision will find more real-world applications in the food quality assurance and control.

References

Abdullah MZ, Abdul-Aziz S, Dos-Mohamed AM (2000) Quality inspection of bakery products using color-based machine vision system. J Food Qual 23:39–50

Abdullah MZ, Guan LC, Lim KC, Karim AA (2004) The applications of computer vision system and tomographic radar imaging for assessing physical properties of food. J Food Eng 61:125–135

Abdullah MZ, Mohamad-Saleh J, Fathinul-Syahir AS, Mohd-Azemi BMN (2006) Discrimination and classification of fresh-cut starfruits (*Averrhoa carambola* L.) using automated machine vision system. J Food Eng 76:506–523

Anon (1990) IEEE standard 610.4-1990 in IEEE standard glossary of image processing and pattern recognition terminology. IEEE Press, New York

Apostolopoulos C, Marshall RJ (1994) A quantitative method for the determination of shreddability of cheese. J Food Qual 17:115–128

Balaban MO, Odabasi AZ, Damar S, Oliveira ACM (2008) Quality evaluation of seafood. In: Sun D-W (ed) Computer vision technology for food quality evaluation. Elsevier, Amsterdam

Boser BE, Guyon I, Vapnik V (1992) A training algorithm for optimal margin classifiers. In Haussler D (ed) Fifth annual workshop on computational learning theory. ACM Press, Pittsburgh, pp 144–152

Brandon JR, Howarth MS, Searcy SW, Kehtarnavaz N (1990) A neural network for carrot tip classification. ASAE Paper No 90-7549, ASAE, St. Joseph, Michigan, p 13

Chmiel M, Słowiński M, Dasiewicz K (2011) Application of computer vision systems for estimation of fat content in poultry meat. Food Control 22(8):1424–1427

Du C-J, Sun D-W (2004)Shape extraction and classification of pizza base using computer vision. J Food Eng 64:489–496

Du C-J, Sun D-W (2008a) Quality measurement of cooked meats. In: Sun D-W (ed) Computer vision technology for food quality evaluation. Elsevier, Amsterdam

Du C-J, Sun D-W (2008b) Quality evaluation of pizzas. In: Sun D-W (ed) Computer vision technology for food quality evaluation. Elsevier, Amsterdam

Duan L, Yang W, Bi K, Chen S, Luo Q, Liu Q (2011) Fast discrimination and counting of filled/unfilled rice spikelets based on bi-modal imaging. Comput Electron Agric 75(1):196–203

Everard CD, Fagan CC, O'Donnell CP, O'Callaghan DJ, Castillo M, Payne FA (2007) Computer vision and colour measurement techniques for inline monitoring of cheese curd syneresis. J Dairy Sci 90:3162–3170

Fagan CC, Du C-J, O'Donnell CP, Castillo M, Everard CD, O'Callaghan DJ, Payne FA (2008) Application of image texture analysis for online determination of curd moisture and whey solids in a laboratory scale stirred cheese vat. J Food Sci 73:E250–E258

Galloway MM (1975) Texture analysis using gray level run length. Comput Graph Image Process 4:172–179

Gunasekaran S (1996) Computer vision technology for food quality assurance. Trends Food Sci Technol 7:245–256

Haralick RM, Shanmugan K, Dinstein I (1973) Textural features for image classification. IEEE Trans Syst Man Cybern 3:610–621

Igathinathane C, Pordesimo LO, Batchelor WD (2009) Major orthogonal dimensions measurement of food grains by machine vision using Image. J Food Res Int 42:76–84

Jayas DS, Mohan AL, Karunakaran C (2005) Unloading automation implemented in grain industry. Resource September:6–7

Jayas DS, Ghosh PK, Paliwal J, Karunakaran C (2008) Quality evaluation of wheat. In: Sun D-W (ed) Computer vision technology for food quality evaluation. Elsevier, Amsterdam

Jeliński T, Du C-J, Sun D-W, Fornal J (2007) Inspection of the distribution and amount of ingredients in pasteurized cheese by computer vision. J Food Eng 83:3–9

Leemans V, Kleynen O (2008) Quality evaluation of apples. In: Sun D-W (ed) Computer vision technology for food quality evaluation. Elsevier, Amsterdam

Majumdar S, Jayas DS (2000) Classification of cereal grains using machine vision: IV. Combined morphology, color, and texture models. Trans ASABE 43:1689–1694
Mohebbi M, Akbarzadeh-T M-R, Shahidi F, Moussavi M, Ghoddusi H-B (2009) Computer vision systems (CVS) for moisture content estimation in dehydrated shrimp. Comput Electron Agric 69:128–134
Nagata M, Tallada JG (2008) Quality evaluation of strawberries. In: Sun D-W (ed) Computer vision technology for food quality evaluation. Elsevier, Amsterdam
Nashat S, Abdullah A, Aramvith S, Abdullah MZ (2011) Support vector machine approach to real-time inspection of biscuits on moving conveyor belt. Comput Electron Agric 75(1):147–158
Ni H, Gunasekaran S (1995) A computer vision system for determining quality of cheese shreds. In: Food processing automation IV proceedings of the FPAC conference, St. Joseph, Michigan, USA
Novini A (1990) Fundamentals of machine vision component selection. In: Food processing automation II-proceedings of the 1990 conference, ASAE, Hyatt Regency, Lexington, Kentucky, p 60
Otsu N (1979) A threshold selection method from gray level histograms. IEEE Trans Syst Man Cybern 9:62–66
Pandit RB, Tang J, Liu F, Mikhaylenko G (2007) A computer vision method to locate cold spots in foods in microwave sterilization processes. Pattern Recogn 40:3667–3676
Park B (2008) Quality inspection of poultry carcasses. In: Sun D-W (ed) Computer vision technology for food quality evaluation. Elsevier, Amsterdam
Pedreschi F, León J, Mery D, Moyano P (2006) Development of a computer vision system to measure the color of potato chips. Food Res Int 39:1092–1098
Peleg S, Naor J, Hartley R, Avnir D (1984) Multiple resolution texture analysis and classification. IEEE Trans Pattern Anal 6:518–523
Quevedo R, Carlos LG, Aguilera JM, Cadoche L (2002) Description of food surfaces and microstructural changes using fractal image texture analysis. J Food Eng 53:361–371
Ridler TW, Calvard S (1978) Picture thresholding using an iterative selection method. IEEE Trans Syst Man Cybern 8:630–632
Rigney MP, Brusewitz GH, Kranzler GA (1992) Asparagus defect inspection with machine vision. Trans ASABE 35:1873–1878
Sarkar N, Chaudhuri B (1994) An efficient differential box counting approach to compute fractal dimension of images. IEEE Trans Syst Man Cybern 24:115–120
Shahin MA, Tollner EW, Gitaitis RD, Sumner DR, Maw BW (2002) Classification of sweet onions based on internal defects using image processing and neural network techniques. Trans ASABE 45:1613–1618
Shearer SA, Payne FA (1990) Color and defect sorting of bell peppers using machine vision. Trans ASAE 33:2045–2050
Simoncelli EP, Freeman WT (1995) The steerable pyramid: a flexible architecture for multi-scale derivative computation. In: Proceedings of IEEE international conference on image processing, Washington DC, USA, pp 444–447
Spreer W, Müller J (2011) Estimating the mass of mango fruit (Mangifera indica, cv. Chok Anan) from its geometric dimensions by optical measurement. Comput Electron Agric 75(1):125–131
Sun D-W, Brosnan T (2003) Pizza quality evaluation using computer vision–part 1 Pizza base and sauce spread. J Food Eng 57:81–89
Sun D-W, Du C-J (2004) Segmentation of complex food images by stick growing and merging algorithm. J Food Eng 61:17–26
Tillett RD (1990) Image analysis for agricultural processes. Division Note DN 1585, Silsoe Research Institute
Tomasi C, Manduchi R (1998) Bilateral filtering for gray and color images. In: Proceedings of IEEE international conference on computer vision, IEEE, Bombay, India, pp 839–846
Vapnik V (1995) The nature of statistical learning theory. Springer, New York
Vízhányó T, Felföldi J (2000) Enhancing colour differences in images of diseased mushrooms. Comput Electron Agric 26:187–198

Wang H-H, Sun D-W (2002) Correlation between cheese meltability determined with a computer vision method and with Arnott and Schreiber tests. J Food Sci 67:745–749

Wang H-H, Sun D-W (2003) Assessment of cheese browning affected by baking conditions using computer vision. J Food Eng 56:339–345

Wang H-H, Sun D-W (2004a) Evaluation of the oiling off property of cheese with computer vision: correlation with fat ring test. J Food Eng 61:47–55

Wang H-H, Sun D-W (2004b) Evaluation of the oiling off property of cheese with computer vision: influence of cooking conditions and sample dimensions. J Food Eng 61:57–66

Yoo T, Ackerman MJ, Lorensen W, Schroeder W, Chalana V, Aylward S, Metaxas D, Whitaker R (2002) Engineering and algorithm design for an image processing API: a technical report on ITK-The insight toolkit. In: Westwood J et al (eds) Medicine meets virtual reality. IOS Press, Amsterdam, pp 582–592

Zheng L, Sun D-W, Tan J (2008) Quality evaluation of meat cuts. In: Sun D-W (ed) Computer vision technology for food quality evaluation. Elsevier, Amsterdam

Chapter 8
Thermal Imaging

R. Vadivambal and Digvir S. Jayas

8.1 Introduction

All objects above 0 K (−273.15 °C) emit infrared (IR) rays, which are part of the electromagnetic spectrum. Electromagnetic spectrum comprises of radio waves, microwaves, IR rays, visible light, ultraviolet rays, X-rays, and gamma rays. The wavelength of IR rays is in the range of 0.78–1000 μm. The IR region is further divided into different regions: near IR (0.75–3 μm), mid-IR (3–6 μm), far IR (6–15 μm), and extreme IR (15–1000 μm) (Meola and Carlomagno 2004). The intensity of radiation emitted by an object is a function of its surface temperature, i.e., higher the temperature of the body, greater is the intensity of IR radiation emitted by the object. IR thermography was discovered by English physicist Sir William Herschel in the early 1800s when he discovered thermal radiation outside the deep red in the visible spectrum; he termed this as thermometrical spectrum or referred to as dark heat or simply invisible rays, which was later known as infrared. IR radiation can be focused, refracted, reflected, and transmitted, similar to visible light. The emissivity, absorptivity, transmissivity, and reflectivity to IR radiation vary for different materials, but in general, objects which are good absorbers of IR radiation are also good emitters. The list of materials that have good and poor IR radiation properties are provided in Table 8.1 (Anonymous 2002).

Temperature measurement is an important aspect in any industrial process, and IR thermography has revolutionized the concept of temperature measurement. Temperature measurements in agriculture and food industry are mostly performed using conventional methods such as thermometers, thermocouples, thermistors, and resistance temperature detectors (Nott and Hall 1999). These instruments can only determine temperature at specific points and most of these instruments need to establish a contact with the material, whereas infrared thermal imaging (IRTI) is a

D. S. Jayas (✉) · R. Vadivambal
Department of Biosystems Engineering, University of Manitoba,
207 Administration Building, 66 Chancellors Circle, Winnipeg, MB, Canada
e-mail: digvir.jayas@umanitoba.ca

C. P. O'Donnell et al. (eds.), *Process Analytical Technology for the Food Industry*,
Food Engineering Series, DOI 10.1007/978-1-4939-0311-5_8,

Table 8.1 Materials with good and poor infrared radiation properties. (Anonymous 2002)

Infrared property	Good	Poor
Transmissivity	Sodium chloride, germanium, zinc selenide, diamond	Biological materials
Reflectivity	Clean metals, aluminum foil	Paper, rubber
Emissivity/absorptivity	Black electric tape, water, paper, rubber, nonmetallic flat paints	Clean metals, aluminum foil

non-contact, non-destructive technique that can provide temperature mapping of the object of interest. Hence, use of IRTI is widely increasing in many fields.

8.2 Emissivity

The emissivity is a measure of a material's ability to radiate absorbed energy. The emissivity of a material is defined as the ratio of energy radiated by a particular material to energy radiated by a black body at the same temperature. From a black body, at a steady-state temperature, all the energy absorbed is emitted, so the emissivity of black body is equal to 1, i.e., absorptivity = emissivity = 1. But for all objects that are not black bodies: absorptivity + transmissivity + emissivity = 1. Energy radiated from the black body is described by Planck's law as:

$$W_\lambda = C_1/\lambda^5\left(e^{C_2/\lambda T} - 1\right) \tag{8.1}$$

Where

W_λ	Spectral radiant emittance per unit wavelength and unit area (W/m² µm)
λ	Wavelength (µm)
C_1 and C_2	First and second radiation constant
T	Absolute temperature (K)

According to the Stefan–Boltzmann law, the total amount of radiation emitted by an object per unit area is directly related to the emissivity of the object and its temperature:

$$E = \sigma\varepsilon T^4 \tag{8.2}$$

Where

E	Total amount of radiation emitted by an object per unit area (W/m²)
σ	Stefan–Boltzmann constant (5.67×10^{-8} W/m²K⁻⁴)
ε	Emissivity of the object
T	Temperature of the object (K)

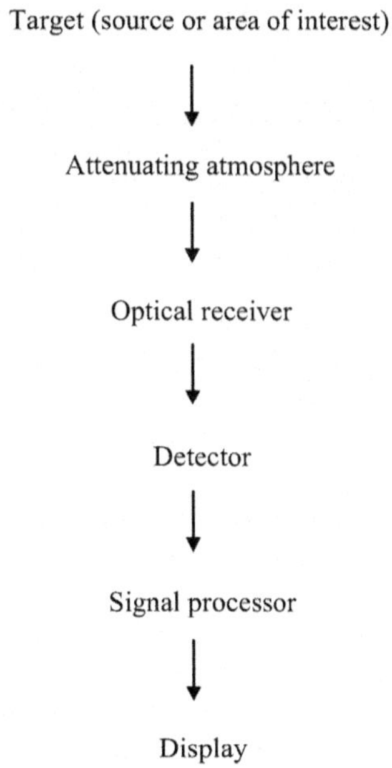

Fig. 8.1 Components of a thermal imaging system

8.3 Instrumentation

8.3.1 Thermal Imaging TI System

IRTI system comprises of a thermal camera equipped with IR detector, a signal processing unit and an image acquisition system, usually a computer. The IR detector absorbs the IR energy emitted by the object and converts it into an electrical impulse. The electrical impulse is sent to the signal processing unit, which translates the information into a thermal image. Most of the thermal imaging (TI) devices scan at a rate of 30 times/s and can sense temperature ranging from −20 to 1500 °C, but the temperature range can be increased further by using filters (Meola and Carlomagno 2004).

Target is the source or the object of interest which emits IR radiation. As the radiation from the source passes through the earth's atmosphere, it is attenuated by scattering and absorption, since atmosphere is not perfectly transparent. The radiation is then received by the optical receiver, which delivers it to the detector. The electrical signal from the detector passes to the signal processor. The signal processor receives the signal from the detector, amplifies it, extracts the information, and delivers the information to the final control device or the display unit (Fig. 8.1).

8.3.2 Detectors

Radiant energy cannot be measured directly, but must be converted to some other form such as electrical, thermal, or chemical energy. These conversion devices are called detectors. IR detectors could be broadly classified into two types: thermal detectors and photon or quantum detectors.

Thermal detectors are made of material with temperature-dependent property. During imaging, when an incident radiation is absorbed, the temperature of the device increases and produces a measurable physical change. In thermal detectors, the heating effect of the incident radiation causes a change in electrical property of the detector, whereas in photon detectors, there is a direct interaction between incident photons and the electrons of the detector material (Hudson 1969). One of the simplest measures for detector performance is responsivity, the detector output per unit input. Thermal detectors are relatively low cost, sensitive over a wide IR range, light, reliable, and convenient to use, but they have slow response and relatively low detectivity (Rogalski 2002). Some of the thermal detectors are thermocouple, thermopile, bolometer, pneumatic detector, and calorimetric detector.

Photon detectors are semiconductors in which the release of electrons is directly associated with photon absorption. In photon detectors, the radiation is absorbed within the material by interaction with electrons either bound to lattice atoms or with free electrons. Depending on the type of interaction, the photon detectors are further subdivided into intrinsic, extrinsic, photoemissive, and quantum well detectors (Rogalski 2002). Their main characteristics are short response time and limited spectral response (Davis and Lettington 1988). The operation of photon detector is based on the measurement of an electrical photocurrent generated by photon absorption in a semiconductor. Photon detectors have high detectivity, fast response, and good material properties, but high cost and difficulty in device processing (Rogalski 2002). Some of the photon detectors are photoelectric detector, photoconductive detector, and photoelectromagnetic detector (Hudson 1969). The principal types of photon detectors used in thermal cameras are mercury cadmium telluride (HgCdTe), indium antimonide (InSb), platinum silicide (PtSi), and quantum well photodetector (QWP) (Williams 2009).

A focal plane array (FPA) is an optical sensor placed at the focal plane of an optical system such as camera or telescope. IR FPA is composed of IR detector array which can be designed and manufactured to be sensitive to small wave IR region to very long wave IR region based on thermal and photon detectors.

TI device can be classified into two types: uncooled and cooled (Sierra Pacific Corporation (2009), Las Vegas, NV). Uncooled TI device is the most common one, and the IR detector elements are contained in a unit that operates at room temperature. Uncooled detectors work by changes in resistance, voltage, or current when exposed to IR radiation. They are less expensive, but their resolution and image quality tend to be lower than the cooled device and they have longer response time.

In the cooled TI device, the sensor elements are contained in a unit which is maintained in a temperature range of − 163 to −265 °C, with − 193 °C being the most common temperature. The different ways to cool the detectors are liquefied gas, cryogenic engine, gas expansion, or thermoelectric effect, and the most common method is cryogenic cooling. The most commonly used cooled IR detectors are mercury cadmium telluride (HgCdTe), indium antimonide (InSb), and indium gallium arsenide (InGaAs). They have a very high resolution and can detect temperature difference as low as 0.1 °C (Abdullah 2008).

8.3.3 Performance of an IR System

An IR imaging system is evaluated based on thermal sensitivity, scan speed, image resolution, and intensity resolution. The thermal sensitivity is expressed as the noise equivalent temperature difference (NETD), which is the difference of temperature at two points of the image which corresponds to a signal equal to the background noise of the camera. The scan speed is the rate at which a complete image is captured. New generation systems are characterized by acquisition speed higher than 1600 Hz. The image resolution is the capability of a system to measure the surface temperature of small objects and is defined as the instantaneous field of view of the detector. The intensity resolution is expressed as the number of gray shades of which the thermal image is composed of, and the latest cameras allow small temperature variation to be determined in a very hot ambience.

8.3.4 Advantages of TI

IR imaging is a noncontact, nondestructive technique which allows online testing during production and gives valuable information without impeding productivity and consuming time.

The IR imaging technique does not need an external source of illumination as required by other imaging techniques. Hence, TI provides a means for seeing at nighttime or under conditions of poor illuminations.

TI could be used under conditions of smoke and mist, because the wavelengths involved are 10–20 times longer than the wavelengths in the visible part of the spectrum; the radiation undergoes less scattering by particles in the atmosphere. Hence, visibility through smoke and mist is increased, which is particularly important in the surveillance used by the firefighters to enable them to see in smoke-filled buildings (Williams 2009).

IR imaging is capable of detecting damage arising in service such as delamination, formation, and propagation of crack, disbands, and impact damage. It is also useful to assess the efficiency of cooling systems, curing processes, plasma treatment, cross-linking processes, material joining (bonding, welding), and simple monitoring of material under thermal or mechanical stimulation (Meola and Carlomagno 2004).

Cost-effective power management is critical to maintain the reliability of electrical and mechanical systems, and IR thermography is the most effective, proven, predictive maintenance technology available to quickly, accurately, and safely locate problems prior to failure (Kouridakis 2007).

Thermography lowers operating costs in the food industry by periodical inspection of equipment with thermal camera, thereby preventing costly failure and subsequent shutdown of the process line.

8.4 Applications of TI in Food Industry

In the food industry, temperature is an important aspect of many processes and products. Basically, temperature is important for two reasons: First, to ensure that sufficiently high temperature has been reached in a certain process or product so that pathogens and microbes are completely killed (a few examples of such processes are cooking, sterilization, pasteurization;products include any meat). Second, to ensure that low temperature has been achieved to preserve the product (examples are freezing, refrigeration, and cooling). In a food industry, the product temperatures may vary significantly due to various factors such as oven temperature, conveyor belt speed, product volume, product composition, and product separation or placement. When product temperature, as they exit the product line, is measured using IRTI, the variation in the temperature could be seen, and if required, suitable modification could be done to the belt speed, or wherever necessary, to maintain safe product temperature, which is otherwise not possible in conventional temperature measurement methods. Other than the temperature measurements, TI is widely useful for many other operations in the food industry such as detecting foreign materials in food and maintaining optimum conditions of storage. The various applications of TI in the food industry are presented in Table 8.2. The details of the thermal camera used in various food industry applications are listed in Table 8.3.

8.4.1 Detection of Foreign Bodies in Food

The presence of foreign bodies in food is a major safety concern and various methods are employed in the food industry. Visual inspection is commonly used but it is affected by several factors. Physical separation methods such as sieving, sedimentation, screening, filtering, and gravity systems are used for the detection of foreign objects as well as are more sophisticated systems such as metal detectors, X-ray machines, optical sensors, and ultrasonic methods. But there is no system capable of determining every contaminant regardless of size and shape.

Warmann and Märgner (2005) studied the detection of foreign bodies in hazelnuts and thermal image analysis of single nuts to inspect the quality of individual nuts using a Thermosensorik CMT 384 thermal camera. The hazelnuts along with foreign bodies were made to pass on a conveyor belt and slightly heated. After a fixed period of cooling time, thermal images were captured. They used image pro-

Table 8.2 Applications of thermal imaging in postharvest and food industry operations. (Source: Vadivambal and Jayas 2010)

Product	Problem	Results from IR thermal imaging studies	References
Apple (*Malus domestica*)	Bruise detection in fruits is a major issue in fruit quality	Possible to determine bruises at an early stage	Baranowski et al. (2009); Varith et al. (2003); Danno et al. (1977)
Apple; Cherry tomato (*Solanum lycopersicum*) Japanese persimmon (*Disopyros kaki* L); Japanese pear (*Pyrus serotina* Rehder); Tomato (*Lycopersicon esculentum* Mill)	Non-destructive method for maturity evaluation is not available	IR thermal imaging makes it possible to determine the maturity of fruits	Hellebrand et al. (2000); Offermann et al. (1998); Danno et al. (1980)
Wheat (*Triticum aestivum*)	Lack of rapid online method to determine varietal purity	IR imaging has a potential to identify wheat classes	Manickavasagan et al. (2008a, 2008b)
Hazel nuts, chocolate chunks	Lack of system to determine all contamination in food, irrespective of shape or size	Possible to determine all sorts of impurities such as leaves, stalks, pedicels, thorns, and foul nuts	Warmann and Märgner (2005); Ginesu et al. (2004)
Potatoe (*Solanum tuberosum*)	Maintaining optimum temperature in a storage facility is a challenge	Optimization of climate control in storage facility is feasible	Geyer et al. (2004)
Wheat	Rapid detection of insect infestation is a challenge	Insect infestation could be determined to certain accuracy in wheat using IR thermal imaging	Manickavasagan et al. (2007)
Ground beef; grain	Temperature mapping not feasible	Temperature mapping enables safe cooking temperature and safe temperature to maintain seed quality	Berry (2000); Manickavasagan et al. (2006)
Citrus (*Citrus sinensis*)	Citrus surface drying results in reduced sensory quality and shelf life	Drying time could be established; fruit quality could be improved	Fito et al. (2004)
Packaging material	Non-destructive technique to detect packaging defect not available	IR imaging has potential to detect cracks, delamination, and voids in packaging material	Liu and Dias (2002)

IR infrared

Table 8.3 Details of infrared thermal cameras used in various food-related studies. (Source: Vadivambal and Jayas 2010)

Camera model	Manufacturer	Spectral range (μm)	Temperature range	Thermal sensitivity	Detector type	Image size	Frame rate	References
AGEMA 880 LW	FLIR Systems, Oregon, USA	8–12	−20 to 1500 °C	0.7 K at 30 °C	HgCdTe, liquid nitrogenor stirling cooled	NA	25 Hz	Offermann et al. (1998)
Infra-Eye 102A	Fujitsu, Tokyo, Japan	8–14	NA	NA	HgCdTe, liquid nitrogencooled	NA	NA	Danno et al. (1977); Danno et al. (1980)
Mikron MCL 160	California, USA	8–14	−40 to 120 °C; 0 to 500 °C	NA	FPA, uncooled	160 × 120	60 Hz	Mikron Infrared Inc. (2009)
Model D500	Raytheon Inc., Waltham, MA	7–14	NA	NA	NA	320 × 240	NA	Catarame et al. (2003)
ThermaCam P25	FLIR Systems, Italy	7.5–13	−40 to 120 °C; 0 °C to 500 °C; optional: 2000 °C	0.08 °C at 30 °C	FPA, uncooled microbolometer	320 × 240	50–60 Hz	Nanni Costa et al. (2007)
ThermaCAM SC500	FLIR Systems, Ontario, Canada	7.5–13	−20 to 500 °C optional: 2000 °C	0.07 K at 30 °C	FPA, uncooled microbolometer	320 × 240	50 Hz	Manickavasagan et al. (2008a; 2008b)
Thermosensorik CMT 384 SM/M	Thermosensorik GmbH, Germany	1.5–5 & 3.4–5.0	NA	NA	HgCdTe, stirling cooled	384 × 288	50–60 Hz	Warmann and Märgner (2005); Ginesu et al. (2004)
Thermovision A40M	FLIR Systems, Danderyd, Sweden	7.5–13	−40 to 120 °C; 0 to 500 °C; optional: 2000 °C	0.08 °C at 30 °C	FPA, uncooled microbolometer	320 × 240	50–60 Hz	Salas-Bringas et al. (2007)
Thermovision 550	AGEMA Infrared Systems, New Jersey, USA	3.6–5	−20 to 250 °C; 1500 °C (with filter)	0.1 °C at 30 °C	FPA, uncooled microbolometer	320 × 240	50–60 Hz	Berry (2000)
VARIOSCAN 2011	Jenoptic Laser, Jena, Germany	10	−40 to 1200 °C	0.1 K at 30 °C	HgCdTe, liquid nitrogen cooled	NA	NA	Hellebrand et al. (2000)
VARIOSCAN 3021 ST	Jenoptic Laser, Jena, Germany	8–12	−40 to 1200 °C	0.03 K at 30 °C	HgCdTe, stirling cooled	360 × 240	NA	Oerke et al. (2006)
VIGOcam v50	Vigo Systems, Warsaw, Poland	8–14	−10 to 100 °C or 0 to 350 °C, optional: 1500 °C	0.08 °C at 30 °C	FPAUncooled microbolometer	384 × 288	30–60 Hz	Baranowski et al. (2009)

InSb indium antimonide, *HgCdTe* mercury cadmium telluride, *FPA* focal plane array, *NA* not available cessing algorithms for thresholding and texture analysis. The study implies that TI could be used to detect foreign materials and determine the quality of individual hazelnuts such as the ones with insect stings or foul nuts. Since their study was tested under laboratory conditions, the authors suggested that extensive test under industrial conditions need to be performed.

Ginesu et al. (2004) studied the potential of TI to detect foreign bodies in food products using a Thermosensorik CMT 384 thermal camera. To distinguish between a food material and a foreign body, either the emissivity or the different heat conductive capacities of the material could be used. Since difference in emissivities may not produce good contrast images, they used the difference in heat capacities of food and other materials to detect undesirable materials. The food materials chosen were almonds and raisin and foreign bodies were wooden stick, stone, metal chip, and cardboard. They used a pulse thermography, and the experimental procedure was that the object was placed (food material and foreign body) on a conveyor belt under the camera and a heat pulse was applied, and then the decrease in surface temperature was observed. Due to difference in heating capacities, different objects cool down at different rates. They recorded a long sequence (500 frames, 80 frames/s) and extracted the thermal images. They applied various image processing techniques such as binarization and statistical and morphological analysis and concluded that results are promising and TI has a potential to detect foreign bodies in food materials.

Meinlschmidt and Märgner (2002) conducted two different studies to detect foreign substances in food using Thermosensorik CMT 384 thermal camera. The first one was to detect the presence of cherries in chocolate chunks by their emissivity coefficient without applying any heat impact. The second study was to detect the presence of leaves, stalks, pedicels, and thorns in a variety of different fruits by difference in the heat conductivity or capacity of different materials by allowing the materials to pass on the conveyor belt with a heat source and a thermal camera captures the image during the state of decreasing temperature. Their results showed that thermography could be used to detect foreign substances in the food material, but they suggested that these methods have to be tested on a larger-scale material in real-time environment.

8.4.2 *Quality of Meat*

Nanni Costa et al. (2007) used thermography for the assessment of pork and ham suitability to be processed as dry-cured ham on the slaughter line. Thermal images were obtained on left and right ham of 40 carcasses of pigs using ThermaCAM P25 thermal camera. Their results showed no difference in the average temperature among the various parameters such as pH, color values, and ham defects, such as veining or red skin. But hams with lower fat cover showed a significantly warmer surface temperature, and it was suggested that lower thermal insulation due to a thinner subcutaneous adipose tissue might be responsible for higher skin temperature. They concluded that IR thermography could be a fast and non-invasive method to estimate the fat content of ham.

Mikron MCL-160 is a highly accurate thermal camera with targeted application in the meat processing industry. This camera could be employed to inspect the meat when it exits the oven to ensure optimum temperature has been attained throughout the product range. It is also possible to remove a particular piece of meat that has not attained the minimum threshold temperature (Foodproduction daily 2008).

8.4.3 *Storage and Postharvest Quality*

To maintain an optimum temperature of 5 °C in a potato storage facility is a challenge, and TI offers a possibility to visualize processes like warming up, cooling, and air flow development in the storage facility. The potential of using TI to optimize the climate control of potato storage was demonstrated by Geyer et al. (2004). An IR thermal camera, ThermaCAM SC500, was used for the experiments. The results of experiments supplied valuable information about temperature distribution in a big box potato store. There was a wide temperature range between the front and the sides of the wooden boxes ranging from 1.5 to 9 °C and wide temperature variation also occurred between the stacks of potatoes in the wooden box. Hence, thermography provided a good view of the temperature differences within a potato storage facility which could be used for designing a temperature control system to provide uniform temperature within the storage.

Varith et al. (2003) employed TI to detect bruises on apples, observing 100 % bruise detection for Fuji and McIntosh apples stored at 3 °C and air-heated at 26 °C within 180 s. Differences in the temperature response between bruised and sound tissue were attributed to thermal property differences. Varith et al. (2003) concluded that the bruise detection was mainly due to the variation in thermal diffusivity, not due to thermal emissivity differences, since they observed no temperature differences between bruised and sound tissue under steady-state conditions. Figure 8.2 shows the temperature variation in bruised and sound tissues for Red Delicious apples. Varith et al. (2003) reported that the bruise damaged tissues were ~ 1–2 °C cooler than the sound tissue, probably due to the fact that bruises warm more slowly than sound tissue, implying that thermal diffusivity α was higher in bruised than in sound tissues. With the higher α, bruises can transfer heat from the apple's exterior into the sound interior tissue, faster than the surrounding sound tissue, resulting in lower surface temperature in bruised than in sound tissue.

8.4.4 *Detection of Insect Infestation in Grain*

In Canadian grain handling facilities, Berlese funnel is the most commonly used method to detect insect infestation (Canadian Grain Commission 2004). It is a time-consuming method and the accuracy is low for developing life stages. The TI could serve as an alternative method to detect insect infestation because the respiration of insects results in heat production higher than that of the grain (Damcevski et al.

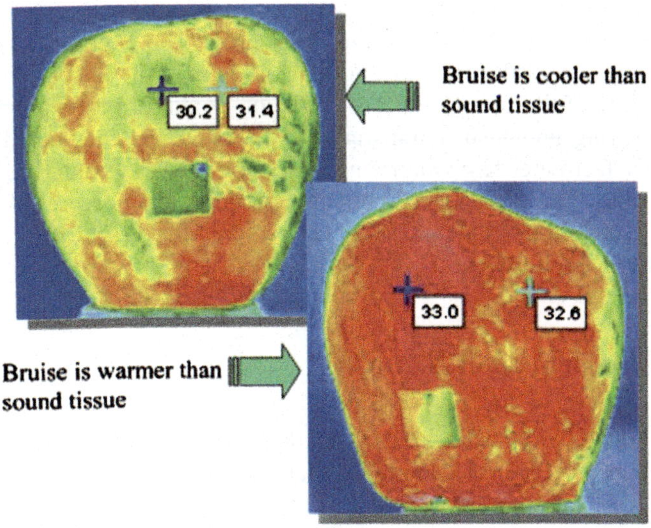

Fig. 8.2 Temperature variation for bruise tissue and sound tissues in red delicious apples (the *top left picture* shows that bruise is cooler than the sound tissues, whereas the *bottom picture* shows the opposite trend). (Varith et al. 2003)

1998; Emekci et al. 2002). An IR thermal camera, ThermaCAM SC500, was used by Manickavasagan et al. (2007) to test whether insect infestation could be determined using TI. The CWRS wheat kernels were artificially infested with eggs of *Cryptolestes ferrugineus* (Stephens) (rusty grain beetle) and thermal images were acquired on days 4, 8, 11, 15, 22, and 27 to represent four larval stages, pupal and adult stages, respectively. The classification accuracy for quadratic function was 83.5 and 77.7% for infested and sound kernels, respectively, and in linear analysis, classification accuracy was 77.6 and 83.0% for infested and sound kernels, respectively. They concluded that insect infestation could be determined by TI to certain accuracy, but it is less effective in identifying the developmental stages of the insect.

8.4.5 *TI in Processing*

Temperature is the most frequently measured variable in any process engineering, but its assessment is always not accurate (Berrie 2001). During pelleting, temperature is a critical parameter, and high temperature in a moist food or feed material results in a wide range of physical and chemical changes. Salas-Bringas et al. (2007) studied the non-contact temperature monitoring of a pelleting process of poultry feed using IR thermography and used ThermoVision A40M thermal camera. The experiment was carried out in a pellet press (Münch Edelstahl, Germany) with two corrugated rollers and assembled with a double conditioner. The IR camera was installed and temperature monitored at various locations such as pellets at the die exit and surface of rotating die/pellets, temperature of meal at the outlet of the conditioner, and pellets at the outlet of the pellet press. The experiment showed that temperature increase in the meal was not only because of the friction in the die hole but also because of the stress, strain, and friction produced in the gap between the rollers and die ring, and the heat transfer from the hotter die. They concluded that IR thermography facilitates temperature measurement of sticky material, moving objects, and temperature distribution within the products in a process. They also suggested that improved instrument design was required for operation in dusty, damp, steamy, and oily environments.

8.4.6 Escherichia coli *Detection*

Traditional methods for isolation and identification of *Escherichia coli* from contaminated food are time consuming and labor intensive (Catarame et al. 2003). Since *E. coli* respiration generates a small but significant amount of heat, this can be detected by thermal camera. Hahn et al. (2006) explored the possibility of determining *E. coli* at their earlier stage using TI. An IR thermal camera, Model D500, was used to image sterile agar and those inoculated with culture of *E. coli*. Detection accuracies ranged between 75 and 100 %, and hence, they concluded that *E. coli* could be detected using TI and the minimum time required for detecting microbial contamination was 5 h. They suggested that *E. coli* should be applied to vegetable and meat surface to determine whether they can be detected at the same rate in the food materials.

8.4.7 Temperature Mapping in Food and Grain

Inadequate cooking of ground beef may result in foodborne illnesses attributable to *E. coli* O157:H7; hence, it is recommended to cook the ground beef patties to at least 71 °C (USDA 1998). Berry (2000) assessed the temperature variability in beef patties cooked from frozen versus thawed state using IR thermography. Cooked beef patties were cut perpendicular to the flat surface within 5 s after removal from the electric griddles and IR images were captured within a second of making the cut using Agema Thermovision 550 thermal camera. The temperatures were observed in the range of 54.4–73.9 °C, but these cannot be assumed as actual temperatures but slightly lower due to rapid evaporative cooling of patty surfaces, on removal from the heating environment. IR imaging showed that internal temperatures were higher and more consistent for patties cooked from thawed state rather than from frozen conditions.

The nonuniformity of surface temperatures of grain after heating in an industrial microwave dryer (2450 MHz) was determined using TI by Manickavasagan et al. (2006). An IR camera, ThermaCAM SC500, was used to determine the surface temperature distribution in barley, wheat, and canola. The average surface temperatures after microwave treatment were between 72.5 and 117.5 °C, 65.9 and 97.5 °C and 73.4 and 108.8 °C for barley, canola, and wheat, respectively. Hence, IRTI could be used to determine the surface temperature mapping of grain which is important in predicting the end use quality of grain.

8.4.8 Drying

Surface drying is an important unit operation in a fresh fruit processing plant. In a citrus surface dryer, using an excessive air temperature or drying the fruit for long time results in loss of sensory quality and reduced shelf life of the fruit. Fito et al. (2004) tested a system to control the surface drying time of citrus using IR imaging. An AGEMA Thermovision 470 camera was installed on the dryer to record the IR emission from the surface of the oranges, var. *Valencia Late (Citrus sinensis)*. Their experiment showed occurrence of two drying steps: the lowest temperature value at the beginning of drying process was considered as the true wet-bulb temperature and the end of first drying step occurred when the entire orange surface was at a higher temperature than the wet-bulb temperature. The next step was the orange peel drying which occurred when no water was present on the orange surface, and it must be avoided because it contributes to the fruit surface damage. Drying time could be established using IR thermal cameras, and an empirical model was developed to correlate drying times with air conditions. Their study revealed that image analysis using IR thermal camera could be used as a non-destructive measure to determine final drying time and hence can improve the fruit quality.

8.4.9 Packaging

One of the key challenges in the packaging industry is to develop a nondestructive technique to detect package defects such as cracking, delamination, and voids. Liu and Dias (2002) studied the potential of TI to identify packaging defects. The principle is that when heat is applied to an object, it diffuses from the source to the surrounding material. Any flaw in the material affects the diffusion rate, which in turn affects the temperature in the vicinity of flaw and, as a result, changes the temperature profile on the surface of the material, which could be detected by surface thermal response of IR imaging. In this study, a thin layer of thermal interface material (TIM) was sandwiched between silicon die and lid, and the four samples with different TIM defect were tested: without TIM defect, TIM delamination, lack of TIM, and no TIM. The testing system consisted of a heating source, an IR camera (3–5 μm) with InSb detector and a data acquisition and image processing system. Their results showed a temperature difference of 0.2–0.3 °C between the normal surface and surface with defects. They concluded that TI is a nondestructive potential tool for detecting packaging defects.

8.5 Conclusions

IRTI technique has been widely used in various fields such as civil, industrial, agriculture, aerospace, and military applications. Use of IRTI in the food industry is gaining popularity, and many researches have been conducted to establish that IR imaging could be a useful tool in improving the efficiency of various operations such as drying, storage, meat processing, and detection of foreign substances in the food material. IR imaging was seen as an expensive technique, but with time and more research, IR imaging has become an affordable technique. TI technique has proven to be a valuable research, educational, and application tool in the food industry, and with TI, temperature testing and control has become more precise and reliable.

References

Abdullah MZ (2008) Image acquisition systems. In: Da-Wen-Sun (ed) Computer vision technology for food quality evaluation. Academic Press, New York

Anonymous (2002) Infrared training: level I course manual. Infrared Training Centre, North Billerica

Baranowski P, Mazurek W, Walczak BW, Sławiński C (2009) Detection of early apple bruises using pulsed-phase thermography. Postharvest Biol Technol 53(3):91–100

Berrie PG (2001) Pressure and temperature measurement in food process controls. In: Kress-Rogers E, Brimelow CJB (eds) Instrumentation and sensors for the food industry. CRC Press, Florida, pp 280–302

Berry BW (2000) Use of infrared thermography to assess temperature variability in beef patties cooked from the frozen and thawed states. Foodserv Res Int 12(4):255–262

Canadian Grain Commission (2004) Managing the quality of stored grain, Manitoba, Canada. http://www.grainscanada.gc.ca/storage-entrepose/monitor-prevent-eng.htm. Accessed 28 Sept 2009

Catarame TMG, O'hanlon KA, Duffy G, Sheridan JJ, Blair IS, McDowell DA (2003) Optimization of enrichment and plating procedures for the recovery of *Escherichia coli* O111 and O26 from minced beef. J Appl Microbiol 95(5):949–957

Damcevski KA, Annis PC, Waterford CJ (1998) Effect of grain on apparent respiration of adult stored product Coleoptera in an air tight system: implications for fumigation testing. J Stored Prod Res 34(4):331–339

Danno A, Miyazato M, Ishiguro E (1977) Quality evaluation of agricultural products by infrared imaging method: grading of fruits for bruise and other surface defects. Memoirs of the Faculty of Agriculture Kagoshima University 14:123–138

Danno A, Miyazato M, Ishiguro E (1980) Quality evaluation of agricultural products by infrared imaging method: Maturity evaluation of fruits and vegetables. Memoirs Faculty Agric Kagoshima Univ 16:157–164

Davis AP, Lettington AH (1988) Principles of thermal imaging. In: Burnay SG, Williams TL, Jones CH (eds) Applications of thermal imaging. Hilger, Bristol, pp 1–34

Emekci M, Navarro S, Donahaye E, Rindner M, Azrieli A (2002) Respiration of *Tribolium castaneum* at reduced oxygen concentrations. J Stored Prod Res 38(5):413–425

Fito PJ, Ortolá MD, De los Reyes R, Fito P, De los Reyes E (2004) Control of citrus surface drying by image analysis of infrared thermography. J Food Eng 61(3):287–290

Foodproduction daily (2008) http://www.foodproductiondaily.com/Processing/New-imaging-tool-detects-thermal-abnormalities-in-meat-says-developer. Accessed 2 Sept 2009

Geyer S, Gottschalk K, Hellebrand HJ, Schlauderer R (2004) Application of a thermal imaging measuring system to optimize the climate control of potato stores. In AgEng 2004 conference, 12–16 Sept 2004, Leuven, Belgium, pp 1066–1067

Ginesu G, Giusto DD, Märgner V, Meinlschmidt P (2004) Detection of foreign bodies in food by thermal image processing. IEEE Trans Ind Electron 51(2):480–490
Hahn F, Hernández G, Echeverría E, Romanchick E (2006) *Escherichia coli* detection using thermal images. Can Biosyst Eng 48:4.7–4.13
Hellebrand HJ, Linke M, Beuche H, Herold B, Geyer M (2000) Horticultural products evaluated by thermography. In AgEng 2000, 2–7 July 2000, Paper No. 00-PH-003, University of Warwick, UK
Hudson RD Jr (1969) Infrared system engineering. Wiley, New York
Kouridakis SJ (2007) Making infrared spectrum visible with infrared cameras: advantages and applications. 4th International conference on NDT. 11–14 Oct 2007, Chania, Greece
Liu Y, Dias R (2002) Evaluation of package defects by thermal imaging. In Proceedings from the 28th international symposium for testing and failure analysis, 3–7 Nov 2002, Phoenix, Arizona
Manickavasagan A, Jayas DS, White NDG (2006) Non-uniformity of surface temperatures of grain after microwave treatment in an industrial microwave dryer. Drying Technol 24(12):1559–1567
Manickavasagan A, Jayas DS, White NDG (2007) Thermal imaging to detect infestation by *Cryptolestes ferrugineus* inside wheat kernels. J Stored Prod Res 44(2):186–192
Manickavasagan A, Jayas DS, White NDG, Paliwal J (2008a) Wheat class identification using thermal imaging: a potential innovative technique. Trans ASABE 51(2):649–651
Manickavasagan A, Jayas DS, White NDG, Paliwal J (2008b) Wheat class identification using thermal imaging. Food Bioprocess Technol. 3:450–460 doi:10.1007/s11947-008-0110-x
Meinlschmidt F, Märgner V (2002) Detection of foreign substances in food using thermography. In: Maldague P, Rozlosnik AE (eds) Proceedings of SPIE thermosense Xxiv, vol 4710. Society of Photo Optical, Bellingham, pp 565–571
Meola C, Carlomagno GM (2004) Recent advances in the use of infrared thermography. Meas Sci Technol 15:27–58
Mikron Infrared Inc (2009) http://www.mikroninfrared.com/content.aspx?id=2192&ekmensel. Accessed 1 Feb 2010
Nanni Costa L, Stelletta C, Cannizzo C, Gianesella M, Pietro Lo Fiego D, Morgante M (2007) The use of thermography on the slaughter-line for the assessment of pork and raw ham quality. Italian J Anim Sci 6(1):704–706
Nott KP, Hall LD (1999) Advances in temperature validation of foods. Trends Food Sci Technol 10(11):366–374
Oerke EC, Steiner U, Dehne HW, Lindenthal M (2006) Thermal imaging of cucumber leaves affected by downy mildew and environmental conditions. J Exp Bot 57(9):2121–2132
Offermann S, Bicanic D, Krapez JC, Balageas D, Gerkema E, Chirtoc M, Egee M, Keijzer K, Jalink H (1998) Infrared transient thermography for non-contact, non-destructive inspection of whole and dissected apples and of cherry tomatoes at different maturity stages. Instrum Sci Technol 26(2/3):145–155
Rogalski A (2002) Comparison of photon and thermal detector performance. In: Henini M, Razeghi M (eds) Handbook of infrared detection technologies. Elsevier, Oxford
Salas-Bringas C, Jeksrud WK, Lekang OI, Schüller RB (2007) Non-contact temperature monitoring of a pelleting process using infrared thermography. J Food Process Eng 30(1):24–37
Sierra Pacific Corporation (2009) http://www.x20.org/thermal/thermal_weapon_sight.htm. Las Vegas, NV. Accessed 15 Oct 2009
USDA (1998) USDA urges consumers to use food thermometer when cooking ground beef patties. FSIS news release food safety inspection service. U.S. Department of Agriculture, Washington, DC
Vadivambal R, Jayas DS (2010) Applications of thermal imaging in agriculture and food industry—a review. Food Bioprocess Technol. 4:186–199 doi:10.1007/s11947-010-0333-5
Varith J, Hyde GM, Baritelle AL, Fellman JK, Sattabongkot T (2003) Non-contact bruise detection in apples by thermal imaging. Innovative Food Sci Emerg Technol 4(2):211–218
Warmann C, Märgner V (2005) Quality control of hazel nuts using thermographic image processing. In IAPR conference on machine vision applications, 16–18 May 2005, Tsukuba Science City, JapanAuthor Query
Williams T (2009) Thermal imaging cameras and their component parts. In Thermal imaging cameras: Characteristics and Performance. CRC Press, Florida, pp 7–34

Chapter 9
Hyperspectral Imaging

A. A. Gowen, E. Gaston and J. Burger

9.1 Introduction

Spectroscopy examines the scattering and absorption of light energy from various regions of the electromagnetic spectrum, including the ultraviolet (UV), visible (Vis) and near-infrared (NIR) wavelength regions. Low-cost sensors have been developed to detect UV-Vis-NIR light reflected from, transmitted through and emitted from various materials. NIR sensing technology is well established as a non-destructive tool in food analysis for raw material testing, quality control and process monitoring, mainly due the advantages it allows over traditional methods, e.g. speed, little/no sample preparation, capacity for remote measurements (using fibre optic probes) and prediction of chemical and physical properties from a single spectrum. Spectrometers integrate spatial information to give an average spectrum for each sample studied; their inability to capture internal component distribution within food products may lead to discrepancies between predicted and measured composition. Furthermore, spectroscopic assessments with relatively small point-source measurements do not contain spatial information, which is important to many food inspection applications. Computer vision systems, which capture spatial information, have been developed for quality control in food processing. Red-green-blue (RGB) colour machine vision systems find widespread use in food quality control for the detection of surface defects and grading operations. However, conventional

A. A. Gowen (✉)
School of Food Science and Environmental Health,
Dublin Institute of Technology, Dublin 1, Ireland
e-mail: aoife.gowen@dit.ie

E. Gaston
IRIS-Innovació i Recerca Industrial i Sostenible,
Avda. Carl Friedrich Gauss n 11., 08860 Castelldefels, Barcelona, Spain

J. Burger
BurgerMetrics SIA, Peldu iela 7, LV-3002 Jelgava, Latvia

C. P. O'Donnell et al. (eds.), *Process Analytical Technology for the Food Industry*,
Food Engineering Series, DOI 10.1007/978-1-4939-0311-5_9,

colour cameras are poor identifiers of surface features sensitive to wavebands other than RGB, such as low but potentially harmful concentrations of contaminants on foods. To overcome this, multispectral imaging (MSI) systems have been developed to combine images acquired at a number (usually <10) of narrow wavebands, sensitive to features of interest on the object.

Hyperspectral imaging (HSI), also known as chemical or spectroscopic imaging, is an emerging technique that integrates conventional imaging and spectroscopy to attain both spatial and spectral information from an object. It was originally developed for remote sensing applications utilizing satellite imaging data of the earth, moon and planets, but has since found application in such diverse fields as astronomy, agriculture, pharmaceuticals and medical diagnostics. HSI, like other spectroscopy techniques, can be carried out in reflectance, transmission or fluorescence modes.

Hyperspectral images are made up of hundreds of contiguous wavebands for each spatial position of a target studied. Consequently, each pixel in a hyperspectral image contains a spectrum representing the light-absorbing and/or scattering properties of the spatial region represented by that pixel (although it should be noted that due to various optical, instrumental and background effects, each pixel spectrum may be influenced by its neighbouring pixels; this becomes a greater problem in high magnification imaging). The resulting spectrum acts like a fingerprint, which can be used to estimate chemical composition of that particular pixel. Hyperspectral images, known as *hypercubes,* can be represented as three-dimensional (3-D) blocks of data, comprising of two spatial and one wavelength dimension, as illustrated in Fig. 9.1. The *hypercube* allows for the visualization of biochemical constituents of a sample, separated into particular areas of the image, since regions of a sample with similar spectral properties tend to have similar chemical composition.

Some advantages of HSI over conventional RGB, NIR spectroscopy (NIRS) and MSI are outlined in Table 9.1. In combining the spectral information provided by spectroscopy and the spatial information provided by imaging, HSI offers improved knowledge on the composition and distribution of components in a product. Moreover, HSI is a rapid method (typical scan time <1 min) compared with traditional quality testing techniques such as high-performance liquid chromatography (HPLC) and gas chromatography–mass spectrometry (GC-MS) which may take hours including sample preparation steps, and since it is a non-destructive and non-contact technique, samples may be further processed or tested as required. The non-destructive, rugged and flexible nature of HSI makes it an attractive PAT tool for identification of critical control parameters that impact on finished product quality.

9.2 Instrumentation

It is currently unfeasible to obtain information in all three dimensions of a *hypercube* simultaneously; one is limited to obtaining two dimensions at a time and a 3-D image is created by stacking the two-dimensional 'slices' in sequence.

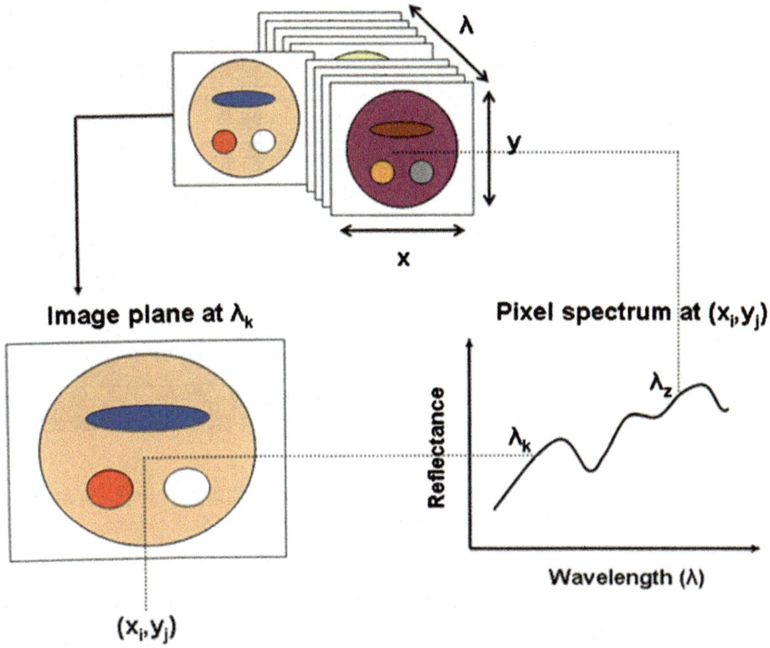

Fig. 9.1 Schematic showing hypercube structure; spatial axes *x*, *y* and wavelength axis (λ)

Table 9.1 Comparison of RGB imaging (RGB), near-infrared spectroscopy (NIRS), multispectral imaging (MSI) and hyperspectral imaging (HSI) techniques

Attribute	RGB	NIRS	MSI	HSI
Spatial information	✓	×	✓	✓
Spectral information	×	✓	Limited	✓
Multiconstituent information	Limited	✓	Limited	✓
Sensitivity to minor components	×	×	Limited	✓

There are three fundamental ways of acquiring a hypercube, commonly known as whiskbroom, pushbroom and staredown. These descriptive names refer to the hardware methodology used to acquire the data stream:

- In a whiskbroom system, a complete spectrum (nW) is acquired at a single spatial location. Once spectral acquisition is completed, the sample is repositioned in both *x* and *y* spatial directions. A back and forth rastering of all *x* and *y* locations (nX × nY) provides acquisition of a complete hypercube. In this mode, data are sequentially saved to file one spectrum at a time. This format is termed band interleaved by pixel, or BIP.
- A pushbroom system utilizes a two-dimensional charge-coupled device (CCD) detector to simultaneously acquire a two-dimensional data matrix representing an image frame 'slice' with spatial x spectral dimensions (nX × nW). The second spatial dimension of the hypercube (nY) is achieved by scanning across the

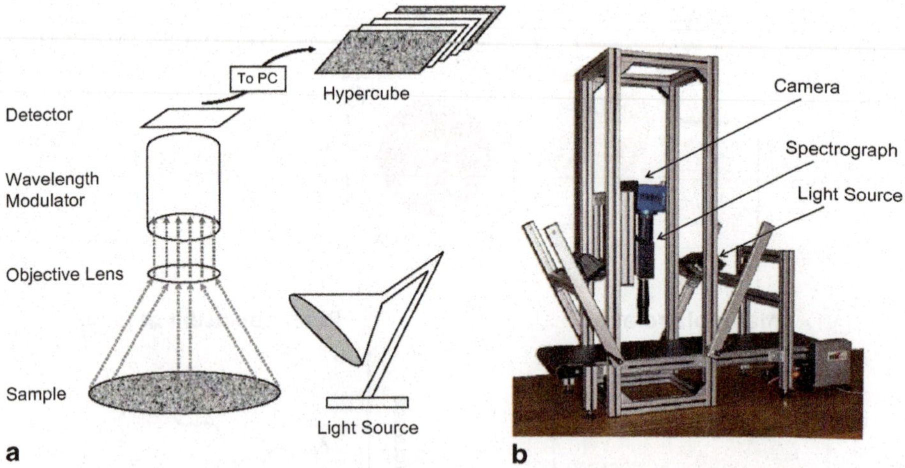

Fig. 9.2 **a** Schematic showing typical components of a hyperspectral imaging system. **b** Example of prototype turnkey pushbroom hyperspectral imaging system

sample surface in a direction perpendicular to the camera imaging line. The individual camera frames are streamed to the data file as sets of spectra, one frame at a time. Each frame is saved such that a complete row or column of the detector representing one complete spectrum is saved, followed by the next spectrum, etc. This is still a BIP format. Some detectors may be rotated 90°, such that the frame is output one complete wavelength channel at a time. This format is line based, and termed band interleaved by line, or BIL.

- A staredown imaging system also acquires two-dimensional data camera frame slices but in this case each frame is a more conventional spatial x spatial image (nX × nY). A complete hypercube is obtained by collecting (and saving) a sequence of these frames (nW) acquired one wavelength band at a time. This file format is known as band sequential, or BSQ.

Typical HSI systems contain the following components: focusing lens, wavelength modulator, detector, illumination and an acquisition system as shown in Fig. 9.2a. In the case of pushbroom line-scanning HSI systems, a spectrograph is used for wavelength modulation; a line of light reflected from or transmitted through the sample under investigation enters the objective lens and is separated into its component wavelengths by diffraction optics contained in the spectrograph; a two-dimensional image (spatial dimension × wavelength dimension) is then formed on the detector; two-dimensional line images acquired sequentially at adjacent positions from the sample target are stacked to form a 3-D *hypercube* which may be processed immediately in real time or stored for further analysis. For such pushbroom systems, relative movement between the object and detector is necessary and this may be achieved either by moving the sample (e.g. via use of a translation stage, see Fig. 9.2b, or a conveyor belt) and keeping the hyperspectral camera in a fixed position or by moving the camera and keeping the sample fixed.

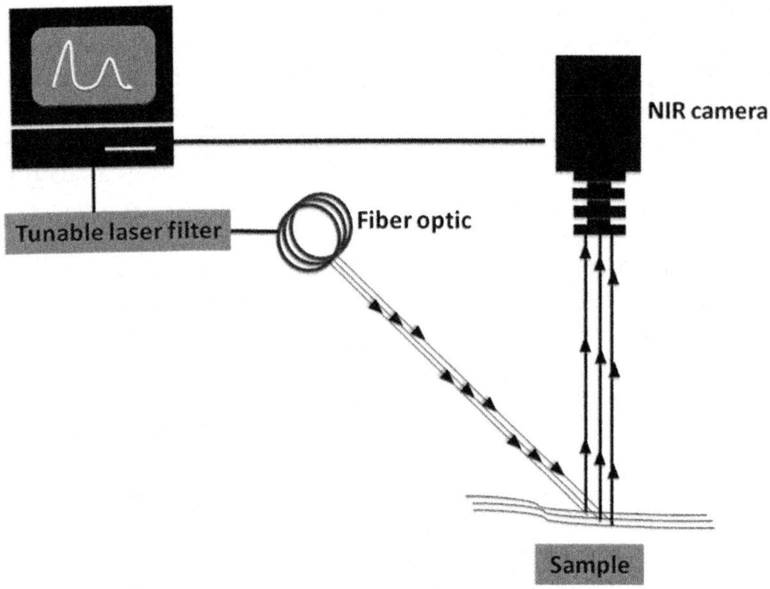

Fig. 9.3 Schematic of the tunable-filter-based hyperspectral imaging system. *NIR* near-infrared

Wavelength of incoming light in the 'staring-imager' configuration is typically modulated using a tuneable filter; acousto-optic tuneable filters (AOTFs) and liquid crystal tuneable filters (LCTFs) are the two most common types employed. More recently, staring-imager systems have been developed that incorporate a tunable laser as the light source, thus removing the need for a wavelength modulator (Margalith 2007). In tunable laser hyperspectral systems, the broadband light source and filters are replaced with a tunable laser system based on optical parametric oscillator (OPO) technology. These systems' main components are the OPO, the camera, and the control software, as presented schematically in Fig. 9.3. Test parameters can be specified in the control software, such as the wavelength range, the spectral resolution and the number of frames at each wavelength. The software then commands the OPO to tune to the first wavelength and fire. The OPO electronics generate a pulse that triggers the data acquisition by the camera. The OPO will then be tuned to the next wavelength and the process repeated until the scan is completed. The OPO beam is collected by optical fibre and directed to illuminate the target. The reflected light from the target is imaged on the focal plane array (FPA) of the camera. The signal collected at each pixel carries information regarding the reflectivity of the target at the position that corresponds to that pixel. The frames are then loaded into the memory of the computer for examination and can be saved in a variety of formats for subsequent analysis by other software packages.

The use of tunable, diffused, pulsed laser is characterised by low average power but very high peak power. This high peak power, coupled with fast camera gating (on the order of microseconds), usually provides signal-to-noise ratio sufficient to collect only one laser pulse per wavelength, allowing entire hyperspectral image stacks to be recorded in well under 1 min. The pulsed laser can also be thought of as a tunable strobe photography light source, permitting motion to be captured without significant blur even at high spatial resolutions. The spectral resolution of this approach is governed ultimately by the linewidth of the laser (Oertel et al. 2009).

The implementation of digital micro-mirror devices (DMDs) is a relatively new development in spectral imaging instrumentation. Kirkhus et al. (2009) recently reported the use of DMDs in NIR spectral imaging. In this setup, a scene is illuminated and imaged onto a camera via two DMDs. A mask is generated by image analysis of the scene in order to locate the region of interest. The mask is then applied to the illumination DMD to ensure that only the region of interest is illuminated. Likewise, the detection DMD receives only the light reflected from the region of interest. The application of this system in reference banking to correct for variation across images, 3-D measurements with structured light and remote interactance measurements is reported by O′Farrell et al. (2010a).

Hyperspectral images can be obtained for reflected, transmitted or emitted light coming from the UV, through the Vis-NIR and up to the short-wave infrared (SWIR) regions of the electromagnetic spectrum. The camera, wavelength modulator and illumination conditions determine the wavelength range of the system. Commercially available Vis-NIR HSI systems typically range between 400 and 1000 nm and utilize cameras with CCD or complementary metal oxide semiconductor (CMOS) sensors; longer wavelength systems require more expensive IR FPA detectors. The sample/target is usually diffusely illuminated by a tungsten-halogen light source. Data acquisition and storage is a major issue in HSI; a typical image of 320×240 pixels in size will contain over 75,000 spectra, each with >100 spectral data points, resulting in a file containing $>7{,}500{,}000$ numbers; if each number is stored in floating point double precision (16-bytes), the resultant image will be >100 MB in size!

9.3 Data Analysis

Numerous techniques exist to analyse HSI data, all of which aim to optimally reduce the immensity of the data while retaining important spatial and spectral information with the power to classify important chemical or physical areas of a scene. Typical steps followed in analysing hyperspectral images are briefly described below. To gain further insights into these and other topics related to hyperspectral image analysis, Grahn and Geladi's comprehensive book is highly recommended (Grahn and Geladi 2007).

Image Calibration Hyperspectral image calibration is required to account for spectral and spatial variations in light source intensity, detector response and system optics. Calibration of spectral response can be achieved using narrow-band light sources (e.g. laser 'pen lights') or calibrated standard reference materials such as National Institute of Standards and Technology (NIST) glasses, and this calibration should be verified periodically. Spatial calibration over the field of view of the HSI instrument should be carried out using a spatially and spectrally homogeneous sample (e.g. flat ceramic tile). Intensity calibration is required to compensate for changes in the detector response and should be carried out using certified reference standards (e.g. Spectralon grayscale standards). Development of suitable reflectance standards and use of correct calibration transformations remains a challenge in HSI. Spatial and intensity calibration should, at the very minimum, be carried out on a daily basis as small changes in electrical power sources, illumination, detector response and system alignment may result in significant changes in the detected response. Inclusion of *internal reference* standards in each hyperspectral image acquired is recommended; this is also a good way to monitor the performance of the system over time.

Preprocessing Preprocessing is usually performed to remove non-chemical biases from the spectral and spatial information contained in a hyperspectral image (e.g. scattering effects due to surface inhomogeneities) and to prepare the data for further processing. Spatial operations usually carried out at the preprocessing stage include (but are by no means limited to): thresholding and masking to remove redundant background information from the *hypercube,* image filtering (e.g. Gaussian filtering) to decrease noise and interpolation (e.g. bilinear interpolation) to decrease image size. A number of spectral preprocessing techniques exist, including polynomial baseline correction, Savitzky-Golay derivative conversion, mean centering and unit variance normalisation.

Classification and Quantification Classification and/or quantification of the material present in a scene are usually the main goals of HSI analysis, and multivariate chemometric methods are often applied to achieve them. There is an abundance of classification and regression algorithms available for hyperspectral image analysis; their performance tends to be goal, data set and to some extent instrument dependent. Classification of hyperspectral images aims to identify objects of similar characteristics using the spectral and spatial information contained in the hypercube. Target or anomaly detection, on the other hand, aims to identify objects with different spectral characteristics as compared to the image background. Various unsupervised methods, including k-nearest neighbours and hierarchical clustering, can be applied in the spectral or spatial domains, or in both simultaneously, to achieve classification. Supervised classification methods, including partial least squares discriminant analysis, linear discriminant analysis and spectral angle mapping require the selection of well-defined and representative calibration and training sets for classifier optimisation.

Hyperspectral image regression enables the prediction of constituent concentrations in a sample at the pixel level, thus enabling the spatial distribution or mapping of a particular component in a sample to be visualised. Numerous approaches are available for the development of regression models (e.g. partial least squares regression, PLSR; principal components regression; step-wise linear regression). One of the advantages of HSI in this respect is the large volume of data available in each hypercube with which to create calibration and training sets. This is also known as the curse of dimensionality, due to the resultant high computational load of high dimensional data. It is thus desirable to decrease the computational burden implied in HSI to manageable levels; this is especially relevant in the development of real-time applications. Using the fact that high dimensional space is mostly empty enables the use of projection methods that can distil only relevant information from data rich hypercubes, e.g. principal components analysis (PCA), independent components analysis (ICA) and multivariate curve resolution (MCR).

Image Processing Images from different planes in a hypercube may be combined using algorithms based on straightforward mathematical operators, e.g. addition, subtraction, multiplication and division. Image processing is also carried out to convert the contrast developed by the classification/regression analysis into a picture depicting component distribution. Greyscale or colour mapping with intensity scaling is commonly used to display compositional contrast between pixels in an image. Image fusion or false colour mapping, in which two or more images at different wavebands are represented as red, green or blue channels and combined to form a new RGB image may be employed to enhance apparent contrast between distinct regions of a sample.

9.4 Applications of HSI in Food Quality Monitoring

9.4.1 Contaminant Detection and Identification

Gómez-Sanchis et al. (2008) studied the feasibility of detecting rot caused by Penicillium digitatum (fungi) in mandarins with a HSI system operating in the 320–1100-nm range. The study concluded that the minimum number of bands to optimise successful classification was 20 and obtained 91 % correct classification. Park et al. (2006) investigated the performance of a Vis-NIR hyperspectral reflectance imaging system for poultry surface faecal contaminant detection. The system allowed for the selection of optimum bandwidths for the construction of an MSI system based on dual-band ratio algorithm to identify ingesta and faeces on poultry carcasses with 96.4 % accuracy. Further investigations (Park et al. 2007) employed the same system to identify the type and source of the faecal contaminants.

Qin et al. (2009) developed a HSI system (450–930 nm) to detect citrus canker in grapefruits, which is a severe disease that can affect the peel of most commercial citrus varieties. The use of the spectral information divergence (SID) classification method, which is based on quantifying the spectral similarities by using a predetermined canker reference spectrum, allowed for correct classification accuracy of 96.2 %.

Hyperspectral reflectance imaging in the NIR region (900–1700 nm) has also been used for the detection of proteins of animal origin (e.g. meat and bone meal) in compound feeds (Fernandez Pierna et al. 2005) and was further demonstrated for the screening of compound feeds (Fernandez Pierna et al. 2006). The proposed method has been tested and validated through studies in line with international standard ISO 17025 (Fernandez Pierna et al. 2010).

The research group of Kim et al. (2001) developed a laboratory-based HSI system with a spectral range of 430–930 nm to conduct food quality and safety research, primarily for the development of MSI systems for food process control, through detection of optimal bands and algorithm development. This system was recently used to conduct hyperspectral reflectance imaging experiments for the detection of apple surface defects/contamination (Kim et al. 2002; Mehl et al. 2004; Liu et al. 2007) and identification of chilling damage on cucumber (Cheng et al. 2004; Liu et al. 2005).

An MSI system was developed by the same research group to detect faecal contamination on apples, based on optimal wavelengths identified by a hyperspectral fluorescence imaging system (Kim et al. 2002). This research team also used hyperspectral fluorescence images to develop a multispectral system for detection of faecal contamination on pork and apple (Kim et al. 2003). Regions of contamination not readily visible to the human eye were easily identified from the multispectral fluorescence images obtained. Vargas et al. (2005) investigated hyperspectral fluorescence imaging for the detection of faecal contamination on cantaloupes, employing PCA to identify dominant wavelengths for the development of a multispectral detection system.

Another research team (Heia et al. 2007) developed a detection method for parasites on codfish, by applying PLSR to transmission hyperspectral images. This method enabled non-destructive identification of parasites 2–3 mm deeper than could be detected by manual inspection of fillets.

The potential of HSI for identification of microorganisms of concern in food has also been reported. Dubois et al. (2005) demonstrated the potential application of NIR HSI as a high throughput technique for the differentiation of bacteria based on their NIR spectra. NIR images of food-specific cards containing both test and calibration bacteria samples were obtained in the spectral region 1200–2350 nm. Some bacteria were identifiable from spectral differences observed at unique wavelengths; however, in situations where particular microorganisms of concern were sought, PLS classification was preferable to separate the genera of bacteria present. The suitability of Raman HSI for the enumeration of waterborne pathogens has also been evaluated (Escoriza et al. 2006). It was shown that while Raman HSI can provide quantitative information for bacterial concentration in water samples, the Raman signal was poor for low bacteria concentration ($\leq 1 \times 10^7$ cells/membrane), necessitating the pre-filtration of dilute water samples prior to examination.

More recently, a hyperspectral fluorescence imaging system has been developed at the US Department of Agriculture (USDA; Jun et al. 2010) to detect microbial biofilms on food contact surfaces such as stainless steel, high-density polyethylene (HDPE), plastic laminate (Formica) and polished granite. Pathogenic strains of *Escherichia coli* and *Salmonella enterica* were used for biofilm formation. High detection rates were reported for the steel, HDPE and granite surfaces. The use of two-wavelength band ratio images made possible the detection of both *E.coli* and *S.*

enterica biofilms. However, a high false-positive rate was reported for the Formica surface; this was attributed to low biofilm growth on this type of surface.

A team of Irish researchers have demonstrated the potential of Vis-NIR HSI reflectance imaging for detection of surface damage on mushrooms caused by bacterial disease (Gaston et al 2010b). Mushrooms were inoculated by *Pseudomonas tolaasii* and PLS discriminant analysis models were developed to classify the mushroom surface into one of the three classes: undamaged, bruise damaged and diseased. High classification rates were reported, indicating the usefulness of this technique for disease detection.

9.4.2 Defect Identification

One research team (Nicolaï et al. 2006) developed an NIR hyperspectral reflectance system with a spectral range of 900–1700 nm to detect the bitter pit defect in apples. The system was capable of identifying bitter pit lesions invisible to the naked eye, but reduced luminosity at the image boundary caused some misclassification errors. Ariana et al. (2006) investigated the application of NIR hyperspectral reflectance imaging in the same spectral region for the detection of bruises on pickling cucumbers. Reflectance for bruised cucumber tissue was generally lower than that for normal tissue, and detection accuracy was dependant on the time after bruising. It was demonstrated that band ratio and difference algorithms were better than PCA for classification of bruised cucumbers.

A Vis-NIR (400–1000 nm) hyperspectral reflectance imaging system was developed to identify bruises on apples (Xing et al. 2005). A PCA analysis based on four wavebands enabled bruise identification with 86 % accuracy. Xing et al. (2006) also developed an MSI system to discriminate between bruises and the stem end/calyx on apples, a well-known problem in computer-vision-based apple sorting. Elmasry et al. (2008) developed a system operating in the same wavelength region but with a different apple variety and found that, by selecting a three-wavelength multispectral system, bruised and sound apples could be successfully distinguished. ElMasry et al. (2009) also investigated the use of a Vis-NIR HSI system (400–1000 nm) and artificial neural network (ANN) for the detection of chilling injury in apples.

Kim et al. (2004) designed a hyperspectral fluorescence system to detect skin tumours on chicken carcasses. UV lamps were used to illuminate samples on a moving stage and hyperspectral images were obtained by acquiring adjacent line scans, as described previously.

Transmission HSI is potentially applicable for the online estimation of internal constituent concentrations and detection of internal defects within foods (Schmilovitch et al. 2004). Qin and Lu (2005) applied hyperspectral transmission imaging to detect pits in tart cherries. Light was transmitted through individual cherries from a light source placed below the sample holder and recorded by an imaging spectrograph placed above the sample. Transmission images for four different sample orientations were tested, and it was shown that sample orientation and colour did not significantly affect classification accuracy. This finding is important for high-throughput operations, where it is difficult to keep sample orientation uniform.

Transmission hyperspectral images may also be obtained from moving samples. Ariana and Lu (2006) employed such an approach to investigate internal damage in cucumbers. Cucumbers were mounted on a rotating stage, illuminated from below and hyperspectral transmission line scans were captured from above the sample. An image thresholding method resulted in higher classification accuracies than PLS analysis, achieving overall classification accuracy up to 94.3 %.

9.4.3 Constituent Analysis

Cogdill et al. (2004) investigated the application of NIR hyperspectral transmission imaging for estimation of oil and moisture content in corn kernels. Stationary samples were illuminated from below via collimating optics through a sample presentation stage: A tuneable filter within the spectrograph removed the need for sample movement. Although this method was capable of predicting moisture content with high accuracy, it was not possible to accurately predict oil concentration.

Ottestad et al. (2009) reported noncontact NIR interactance spectroscopy in combination with multispectral imaging as a rapid non-destructive way to determine the average fraction of water existing as ice, as well as the spatial distribution of ice in super-chilled salmon fillets. Segtnan et al. (2008) achieved noncontact salt and fat distributional analysis in salted and smoked salmon fillets using NIR interactance imaging. It was found that NIR interactance imaging alone was able to predict NaCl contents locally in salted salmon fillets with root mean square error of cross-validation (RMSECV) = 0.56 % and $R = 0.86$. This research team also reported noncontact transflectance NIR imaging for representative on-line sampling of dried salted coalfish. A part of this study compares the principles of reflectance, contact transflectance and noncontact transflectance with regard to water determination in a set of 20 well-defined dried salted cod samples. Transflectance and noncontact transflectance performed equally well and were superior to reflectance measurements for the selected application, since the measured light penetrated deeper into the sample.

ElMasry and Wold (2008) used an online HSI system (460–1040 nm) for quantitative measurements of moisture and fat distribution of fish fillets. Menesatti et al. (2008) used a system working in the 400–970-nm spectral range to provide qualitative evaluation of fish freshness. An objective technique based on a combination of HSI and geometric morphometric tools was presented. This study represented an important methodological evolution in the assessment of fish freshness, as it minimised the error associated with the subjective choice of fish areas by operators.

Burger and Geladi presented the first reported application of HSI to cheese products in 2006 in an article on NIR hyperspectral image regression, where they developed regression models to predict cheese composition from hyperspectral images. The researchers also examined the effects of various spectral preprocessing methods on the prediction ability of the developed regression models. A range of 12 commercial cheese products were tested, specifically selected to span as wide a range as possible in terms of protein, fat and carbohydrate content. The average composition values on the packaging labels were used as standard reference values and a parallel set of reference values for protein and fat content was determined using standard techniques. The challenges of developing accurate calibration models using hyperspectral image data were discussed. One major issue is that reference values were only available for entire bulk samples, not at the individual pixel level! To overcome this limitation, the authors used the mean spectral response from sample images to build calibration models. PLSR models were developed on mean spectra subjected to various spectral pretreatments, and (considering the prediction error of the regression models) results suggested that applying a 1st-derivative Savitsky-Golay smoothing was the most effective spectral pretreatment. Using this approach, a PLSR model with two or four latent variables could be used to satisfactorily predict fat, protein and carbohydrate content. Typical prediction errors of 1–2 % for protein and fat, and 2–3 % for carbohydrate were obtained, which were greater than the errors in the reference measurements (0.14 % protein and 0.41 % fat) but similar to results reported for other NIR spectrometers.

Gaston et al. (2010a) applied Vis-NIR reflectance HSI to predict polyphenol oxidase (PPO) enzyme activity on mushroom caps during the browning process. It was found that HSI can be used for rapid identification of mushrooms with a higher likelihood to develop enzymatic browning, hence aiding produce management decision makers in the industry.

9.4.4 *Quality Evaluation*

Light scattering from a surface is highly dependent on the product density and cell structures, so it follows that scattering profiles may indicate related properties, such as texture. Indeed, the relationship between hyperspectral scattering profiles (in the 500–1000-nm spectral range) and texture has been explored to predict peach firmness (Lu and Peng 2006). In this investigation, a Lorentzian distribution function was fitted to scattering data, and Lorentzian model parameters at each wavelength were used to build an empirical regression model to predict peach firmness.

Polder et al. (2002) showed that a hyperspectral reflectance imaging system in the spectral region of 396–736 nm was more effective than RGB imaging for discriminating ripeness level in tomatoes, regardless of illumination condition tested. ElMasry et al. (2007) used a Vis-NIR HSI system region for non-destructive determination of strawberry quality. A subset of wavelengths was selected and multilinear regression was then used to predict moisture content, total soluble solids content and pH. A similar system was used to evaluate pork quality and marbling level (Qiao et al. 2007), employing a feed-forward neural network to classify samples, with up to 85 % classification accuracy.

Noh and Lu (2007) examined the ability of fluorescence hyperspectral line images to predict apple quality, using a blue-laser diode to produce chlorophyll fluorescence: A hyperspectral line scan located 1.5 mm from the beam centre was analysed using a hybrid PCA–ANN method. No significant differences were observed from fluorescence data obtained after 1–5 min of continuous laser illumination; therefore, fluorescence measurements could be performed within 1 min of illumination. Spectral features were correlated to apple quality characteristics such as firmness and colour. It was noted that the relatively low correlation coefficients obtained in the study could be improved by using multiple line scans rather than single line scans.

Hyperspectral reflectance imaging in the Vis and NIR wavelength ranges has been demonstrated as useful for prediction of a number of quality attributes of mushrooms. Gowen et al. (2008a) developed an HSI system operating in wavelength range of 400–1000 nm to detect bruise damage on white button mushrooms. Quality deterioration of sliced mushrooms was also investigated using the same system (Gowen et al. 2008b). Application of HSI for early detection of freeze damage in *Agaricus bisporus* mushrooms were also investigated by Gowen et al. (2009). A procedure based on PCA and linear discriminant analysis (LDA) was developed with accuracy of higher than 95 % for classification of freeze-damaged mushrooms after only 45 min thawing at which time freeze-thaw damage was not visibly evident. The developed models could be used to identify substandard mushroom batches before surface damage is visibly evident, and developed into a tool for non-destructive grading of post-harvest mushroom quality.

The shelf life of mushrooms packaged using different polymer top films and perforation sizes was also investigated using the same HSI system to extract imaging data through the packaging film (Taghizadeh et al. 2010). Quality indicators such as weight loss, colour, maturity index and in-pack gas composition were measured and PLSR models were built to correlate HSI data with measured quality parameters. Results demonstrated that HSI can be used for rapid evaluation of mushroom quality, facilitating the non-destructive evaluation of the effect of the packaging systems on mushroom shelf life. The results obtained also showed that the polyester (PET) film perforated with holes of 1 mm in diameter was superior in terms of maintaining overall quality. Perforated PET packaging film proved a viable alternative to the conventional PVC film, facilitating an increase in mushroom shelf-life from 10 to 14 days.

9.4.5 PAT Applications of HSI

The growing body of literature on the application of HSI to food quality monitoring (Table 9.2) suggests its suitability as a PAT tool. To date, the majority of work published in this area concerns the small-scale application in a research laboratory setting. However, developments in system components, such as improved cameras, faster hardware and more accurate and efficient algorithms, are resulting in short-

Table 9.2 Summary of measurement mode, product type and wavelength region studied employed in a selection of recent papers published on hyperspectral imaging of food

Mode	Product	Wavelength region (nm)	Author, year
Reflectance	*Apple*	447–951	Liu et al. (2007)
		430–900	Mehl et al. (2004)
		954–1350	Nicolaï et al. (2006)
		500–950	Xing et al. (2005)
		500–950	Xing et al. (2006)
		500–950	Xing et al. (2007)
	Corn	950–1700	Weinstock et al. (2006)
	Cucumber	900–1700	Ariana et al. (2006)
		447–951	Cheng et al. (2004)
		447–951	Liu et al. (2005)
	Citrus fruit	400–970	Menesatti et al. (2005)
	Pasta	400–1700	Menesatti et al. (2004)
	Peach	500–1000	Lu and Peng (2006)
	Pork	430–1000	Qiao et al. (2007)
	Potato	430–1000	Qiao et al. (2005)
	Poultry	430–850	Lawrence et al. (2006)
		430–850	Park et al. (2006)
		430–850	Park et al. (2007)
	Strawberry	400–1000	ElMasry et al. (2007)
	Mushrooms	450–950	Gowen et al. (2008a, b, 2009); Gaston et al. (2010a, b)
	Animal feed	1000–1700	Fernandez Pierna et al. (2005, 2006, 2010)
Fluorescence	*Apple*	500–1040	Noh and Lu (2007)
	Cantaloupe	425–774	Vargas et al. (2005)
	Poultry	425–710	Kim et al. (2004)
	Walnut	425–775	Jiang et al. (2007)
Transmittance	*Cherries*	450–1000	Qin and Lu (2005)
	Codfish	350–950	Heia et al. (2007)
	Cucumbers	450–950	Ariana and Lu (2006)
	Maize	750–1090	Cogdill et al. (2004)
Interactance	*Crab*	760–1040	Wold et al. (2010)
	Pork	760–1040	O'Farrell et al. (2010b)

ening processing and acquisition time, enabling real-time HSI quality monitoring systems. Recently, a number of publications highlight the application of HSI in a high-speed industrial setting.

Wold et al. (2010) reported noncontact NIR interactance imaging spectroscopy can be applied to determine the amount of edible meat in single live crabs on a conveyor belt at high speed. Each crab was scanned individually, 15 NIR images were obtained in wavelength range 760–1040 nm and the total scanning time was about between 0.5–1 s, facilitating scanning of up to 120 crabs per minute. The authors compared PLSR calibration models developed on average spectra from each crab with those based on average spectra from different regions of the samples. The frontal region of the crabs was shown to be optimal for prediction of meat content.

The same system was employed for on-line measurement of fat content in pork trimmings in an industrial setting as reported by O´Farrell et al. (2010a).

Researchers based in the USDA have developed a high-speed pushbroom line-scanning system for inspection of chicken carcasses (Chao et al. 2010). The system is based on hyperspectral reflectance imaging, with the capability to obtain full-range hyperspectral images (55 wavelengths between 389–744 nm) or multispectral images for faster processing. The selection of two wavelengths for the MSI mode, coupled with the short integration time of the detector-facilitated inspection of up to 140 carcasses per minute. The general spectral imaging methodology developed has potential for implementation to other food quality monitoring tasks, such as apple bruise detection and mushroom quality evaluation.

9.5 Summary

HSI is an emerging tool for food quality and safety analysis: The spatial capability of HSI enables characterisation of complex heterogeneous samples, while the spectral capability allows for the identification of a wide range of multi-constituent surface and sub-surface features. Due to the current high cost of HSI systems, most food-related HSI research has been geared towards identification of important wavebands for the development of low-cost MSI systems. However, with recent high-speed applications suitable for industrial settings, it is likely that HSI will be increasingly adopted for safety and quality control in the food industry. Future developments in HSI equipment manufacture, such as lower purchase costs and improvements in processing speed, will encourage more widespread utilisation of this emerging platform technology.

References

Ariana D, Lu R (2006). Visible/near-infrared hyperspectral transmittance imaging for detection of internal mechanical injury in pickling cucumbers. In: ASABE Annual International Meeting, Paper No. 063039, July 2006.

Ariana D, Lu R, Guyer DE (2006) Hyperspectral reflectance imaging for detection of bruises on pickling cucumbers. Comput Electron Agric 53:60–70

Burger J, Geladi P (2006) Hyperspectral NIR image regression part II: Dataset preprocessing diagnostics. J Chemom 20:106–119

Chao K, Yang C, Kim M (2010) Spectral line-scan imaging system for high-speed non-destructive Wholesomeness inspection of broilers. Trends Food Sci Technol 21:129–137

Cheng X, Chen YR, Tao Y, Wang CY, Kim MS, Lefcourt AM (2004) A novel integrated PCA and FLD method on hyperspectral image feature extraction for cucumber chilling damage inspection. Trans ASAE 47:1313–1320

Cogdill R, Hurburgh C, Rippke G (2004) Single-kernel maize analysis by near-infrared hyperspectral imaging. Trans ASAE 47:311–320

Dubois J, Lewis E, Fry F, Calvey E (2005) Bacterial identification by near-infrared chemical imaging of food-specific cards. Food Microbiol 22:577–583
ElMasry G, Wold JP (2008) High-speed assessment of fat and water content distribution in fish fillets using online imaging spectroscopy. J Agric Food Chem 56(17):7672–7677
ElMasry G, Wang N, El Sayed A, Ngadi M (2007) Hyperspectral imaging for nondestructive determination of some quality attributes for strawberry. J Food Eng 81:98–107
ElMasry G, Wang N, Vigneault C, Qiao J, ElSayed A (2008). Early detection of apple bruises on different background colors using hyperspectral imaging. LWT Food Sci Technol 41(2):337–345
ElMasry G, Wang N, Vigneault C (2009). Detecting chilling injury in Red Delicious apple using hyperspectral imaging and neural networks. Postharvest Biol Technol 52(1):1–8
Escoriza M, VanBriesen J, Stewart S, Maier J, Treado P (2006) Raman spectroscopy and chemical imaging for quantification of filtered waterborne bacteria. J Microbiol Methods 66:63–72
Fernandez Pierna JA, Baeten V, Michotte Renier A, Cogdill RP, Dardenne P (2005) Combination of SVM and NIR imaging spectroscopy for the detection of MBM in compound feeds. J Chemom 18(7–8):341–349
Fernandez Pierna JA, Baeten V, Dardenne P (2006) Screening of compound feeds using NIR hyperspectral data. Chemom Intell Lab Syst 84:114–118
Fernandez Pierna JA, Dardenne P, Baeten V (2010) In-house validation of a near infrared hyperspectral imaging method for detecting processed animal proteins (PAP) in compound feed. J Near Infrared Spectrosc 18:121–133
Gaston E, Frías JM, Cullen PJ, O'Donnell CP, Gowen AA (2010a) Prediction of polyphenol oxidase activity using visible and near-infrared hyperspectral imaging on mushroom (*Agaricus bisporus*) caps. J Agric Food Chem 58:6226–6233
Gaston E, Frías JM, Cullen PJ, O'Donnell CP, Gowen AA (2010b) Visible near infrared hyperspectral imaging for the identification and discrimination of brown blotch disease on mushroom (*Agaricus bisporus*) caps. J Near Infrared Spectrosc 18:341–353
Gómez-Sanchis J, Gómez-Chova L, Aleixos N, Camps-Valls G, Montesinos-Herrero C, Moltó E, Blasco J (2008) Hyperspectral system for early detection of rottenness caused by Penicillium digitatum in mandarins. J Food Eng 89(1):80–86
Gowen AA, O'Donnell CP, Taghizadeh M, Cullen PJ, Frias JM, Downey G (2008a) Hyperspectral imaging combined with principal component analysis for bruise damage detection on white mushrooms (*Agaricus bisporus*). J Chemom 22:259–267
Gowen AA, O'Donnell CP, Taghizadeh M, Gaston E, O'Gorman A, Cullen PJ, Frias JM, Esquerre C, Downey G (2008b) Hyperspectral imaging for the investigation of quality deterioration in sliced mushrooms (*Agaricus bisporus*) during storage. Sens Instrum Food Qual 2:133–143
Gowen A, Taghizadeh M, O'Donnell CP (2009) Identification of mushrooms subjected to freeze damage using hyperspectral imaging. J Food Eng 93:7–12
Grahn HF, Geladi P (2007) Techniques and applications of hyperspectral image analysis. Wiley, Chichester
Heia K, Sivertsen A, Stormo S, Elvevoll E, Wold J, Nilsen H (2007) Detection of nematodes in cod (*Gadus morhua*) fillets by imaging spectroscopy. J Food Sci 72:E011–E015
Jiang L, Zhu B, Rao X, Berney G, Tao Y (2007) Discrimination of black walnut shell and pulp in hyperspectral fluorescence imagery using Gaussian kernel function approach. J Food Eng 81:108–117
Jun W, Kim M, Cho B, Millner P, Chao K, Chan D (2010) Microbial biofilm detection on food contact surfaces by macro-scale fluorescence imaging. J Food Eng 99:314–322
Kim MS, Chen YR, Mehl PM (2001) Hyperspectral reflectance and fluorescence imaging system for food quality and safety. Trans ASAE 44:721–729
Kim MS, Lefcourt AM, Chao K, Chen YR, Kim I, Chan DE (2002) Multispectral detection of fecal contamination on apples based on hyperspectral imagery: Part I. Application of visible and near-infrared reflectance imaging. Trans ASAE 45:2027–2037
Kim MS, Lefcourt AM, Chen YR (2003) Multispectral laser-induced fluorescence imaging system for large biological samples. Appl Opt 42:3927–3933

Kim I, Kim MS, Chen YR, Kong SG (2004) Detection of skin tumors on chicken carcasses using hyperspectral fluorescence imaging. Trans ASAE 47:1785–1792

Kirkhus T, Fismen B, Skotheim Ø, Tschudi J (2009) A DMD (Digital Micro-Mirror Device) based multi-object quasi-imaging spectrometer, ProCams at CVPR 2009 Miami FL. http://graphics.cis.udel.edu/2009/poster/posters.pdf

Liu Y, Chen YR, Wang CY, Chan DE, Kim MS (2005) Development of simple algorithm for the detection of chilling injury in cucumbers from visible/near-infrared hyperspectral imaging. Appl Spectrosc 59:78–85

Liu Y, Chen YR, Kim MS, Chan DE, Lefcourt AM (2007) Development of simple algorithms for the detection of fecal contaminants on apples from visible/near infrared hyperspectral reflectance imaging. J Food Eng 81:412–418

Lu RF, Peng YK (2006) Hyperspectral scattering for assessing peach fruit firmness. Biosyst Eng 93:161–171

Margalith E (2007) US patent 7,233,392

Mehl PM, Chen YR, Kim MS, Chan DE (2004) Development of hyperspectral imaging technique for the detection of apple surface defects and contaminations. J Food Eng 61:67–81

Menesatti P, D'Andrea S, Bucarelli A (2004). Non-destructive spectrometric qualification of Italian wheat durum pasta produced by traditional or industrial technology approaches. In: 2004 CIGR International Conference, Beijing, China, 11–14 Oct 2004

Menesatti P, Urbani G, Lanza G (2005). Spectral imaging Vis-NIR system to forecast the chilling injury onset on citrus fruits. In: Mencarelli F, Tonutti P (eds) ISHS Acta Horticulturae 682: V international postharvest symposium, pp 1347–1354

Menesatti P, Zanella A, D'Andrea S, Costa C, Plagia G, Pallottino F (2008). Supervised multivariate analysis of hyper-spectral NIR images to evaluate the starch index of apples. Food Bioprocess Technol 2:308–314

Nicolaï B, Lötze E, Peirs A, Scheerlinck N, Theron K (2006) Non-destructive measurement of bitter pit in apple fruit using NIR hyperspectral imaging. Postharvest Biol Technol 40:1–6

Noh H, Lu R (2007) Hyperspectral laser-induced fluorescence imaging for assessing apple fruit quality. Postharvest Biol Technol 43:193–201

O´Farrell M, Kirkhus T, Fismen B, Skotheim O, Tschudi J (2010a) Quasi-imaging spectrometer with programmable field of view and field of illumination. NIR News 21:8–10

O´Farrell M, Wold JP, Høy M, Tschudi J, Schulerud H (2010b) On-line fat content classification of inhomogeneous pork trimmings using multispectral near infrared interactance imaging. J Near Infrared Spectrosc 18:135–146

Oertel DC, Grothaus JT, Marcott C (2009) Applications of spectral imaging using a tunable laser source, Proceedings of SPIE 7319: 731906-1

Ottestad S, Høy M, Stevik A, Wold JP (2009) Prediction of ice fraction and fat content in superchilled salmon by non-contact interactance near infrared imaging. J Near Infrared Spectrosc 17:77–87

Park B, Lawrence KC, Windham WR, Smith D (2006) Performance of hyperspectral imaging system for poultry surface fecal contaminant detection. J Food Eng 75:340–348

Park B, Windham WR, Lawrence KC, Smith D (2007) Contaminant classification of poultry hyperspectral imagery using a spectral angle mapper algorithm. Biosyst Eng 96:323–333

Polder G, Heijden G, Young I (2002) Spectral image analysis for measuring ripeness of tomatoes. Trans ASAE 45:1155–1161

Qiao J, Wang N, Ngadi M, Baljinder S (2005). Water Content and Weight Estimation for Potatoes Using Hyperspectral Imaging. Published by the American Society of Agricultural and Biological Engineers, St. Joseph, Michigan www.asabe.org. Paper number 053126, 2005 ASAE Annual Meeting

Qiao J, Ngadi M, Wang N, Gariépy C, Prashe S (2007) Pork quality and marbling level assessment using a hyperspectral imaging system. J Food Eng 83:10–16

Qin J, Lu R (2005) Detection of pits in tart cherries by hyperspectral transmission imaging. Trans ASAE 48:1963–1970

Qin J, Burks TF, Ritenour MA, Bonn WG (2009) Detection of citrus canker using hyperspectral reflectance umaging with spectral information divergence. J Food Eng 93(2):183–191
Schmilovitch Z, Shenderey C, Shmulevich I, Alchanatis V, Egozi H, Hoffman A, Ostrovsky V, Lurie S, Arie R (2004). NIRS detection of mouldy core in apples. In 2004 CIGR international conference, Beijing, China, 11–14 Oct 2004
Segtnan VH, Høy M, Sørheim O, Kohler A, Lundby F, Wold JP, Ofstad R (2008) Noncontact salt and fat distributional analysis in salted and smoked salmon fillets using X-ray computed tomography and NIR interactance imaging. J Agric Food Chem 57:1705–1710
Vargas AM, Kim MS, Tao Y, Lefcourt A (2005) Detection of fecal contamination on cantaloupes using hyperspectral fluorescence imagery. J Food Sci 70:E471–E476
Weinstock BA, Janni J, Hagen L, Wright S (2006) Prediction of oil and oleic acid concentrations in individual corn (*Zea mays L.*) kernels using near-infrared reflectance hyperspectral imaging and multivariate analysis. Appl Spectrosc 60:9–16
Wold JP, Kermit M, Woll A (2010) Rapid nondestructive determination of edible meat content in crabs (*Cancer Pagurus*) by near-infrared imaging spectroscopy. Appl Spectrosc 64:691–699
Xing J, Bravo C, Jancsók P, Ramon H, De Baerdemaeker J (2005) Detecting bruises on 'Golden Delicious' apples using hyperspectral imaging with multiple wavebands. Biosyst Eng 90:27–36
Xing J, Jancsók P, De Baerdemaeker J (2006) Stem-end/calyx identification on apples using contour analysis in multispectral images. Biosyst Eng 96:231–237
Xing J, Saeys W, De Baerdemaeker J (2007) Combination of chemometric tools and image processing for bruise detection on apples. Comput Electron Agric 56:1–13

Chapter 10
Diagnostic Ultrasound

Tat Hean Gan

10.1 Introduction

Ultrasound is now widely used for the diagnostics of materials and engineering structures, medical imaging, and is also used in food quality characterization and inspection. In most ultrasonic systems, piezoelectric contact transducers are used when inspecting solid materials and these require a coupling medium to be used between the material and the transducer.

This chapter looks into the application of non-contact ultrasound for food property measurements. The development of various non-contact transducers, e.g. electrostatic, piezoelectric and electromagnetic transducers, will be described. In situations where the signal-to-noise ratio (SNR) is too low, i.e. high acoustic impedance mismatch between the air surface and container wall, additional signal-processing techniques have to be implemented. This chapter will also describe various signal-processing techniques that can be used for air-coupled ultrasonic techniques.

In summary, the advantages and disadvantages of non-contact ultrasonic techniques for food application are given below (Gan et al. 2003; Bhardwaj 2001):

- Does not need special coupling media or gel which can sometimes damage the test sample
- Can be used for complex geometry inspection
- Can be used for high-temperature application
- Less problems of accessibility and complex geometry because the inspection can be carried out from one side
- Can be used to inspect material properties in a food container on a conveyor belt

T. H. Gan (✉)
Brunel University, Middlesex UB8 3PH, UK
e-mail: tat-hean.gan@twi.co.uk

C. P. O'Donnell et al. (eds.), *Process Analytical Technology for the Food Industry*,
Food Engineering Series, DOI 10.1007/978-1-4939-0311-5_10,

10.2 Theory of Diagnostic Ultrasound

Ultrasound (above 20 kHz range) is produced by movement of particles from a vibrating body in a given medium. The mechanical vibrations generated by these elements may travel through solids, liquids and gases. Sound can be described as the variations in pressure, particle displacement or particle velocity that propagate through any medium (Gan 2002). There are two major types of sound waves; longitudinal (compressional) waves and shear (transverse) waves. The longitudinal wave has particle motions that travel in the same direction (parallel) as the direction of propagation of the signal. This type of wave can travel through solid, liquid and gas (Rose 2004). The velocity of the longitudinal signal, C_L can be calculated from the elastic constants of a material using:

$$C_L = \sqrt{\frac{E(1-\sigma)}{\rho(1+\sigma)1(1-\sigma)}} \tag{10.1}$$

where

E	Young's modulus of elasticity
ρ	Density of the medium
σ	Poisson's ratio for the material

In the shear wave mode the wave particles move or vibrate at 90° to the direction of the wave motion. It is typically observed in solids. The velocity of the shear wave can be calculated using:

$$C_S = \sqrt{\frac{E}{2\rho(1+\sigma)}} \tag{10.2}$$

Any change or discontinuity in the medium of travel will affect the ultrasonic properties (Blitz 1967). At any angle other than normal incidence, when a wave passes from one medium to another, which have different velocities, the wave is subjected to a phenomenon known as refraction. This is as shown in Fig. 10.1a. Refraction modifies the mode and direction of sound, and these modifications could be predicted using the Snell's law, given by:

$$\frac{\sin\theta_1}{\sin\theta_2} = \frac{c_1}{c_2} \tag{10.3}$$

where

θ_1	Angle of incidence
θ_2	Angle of refraction

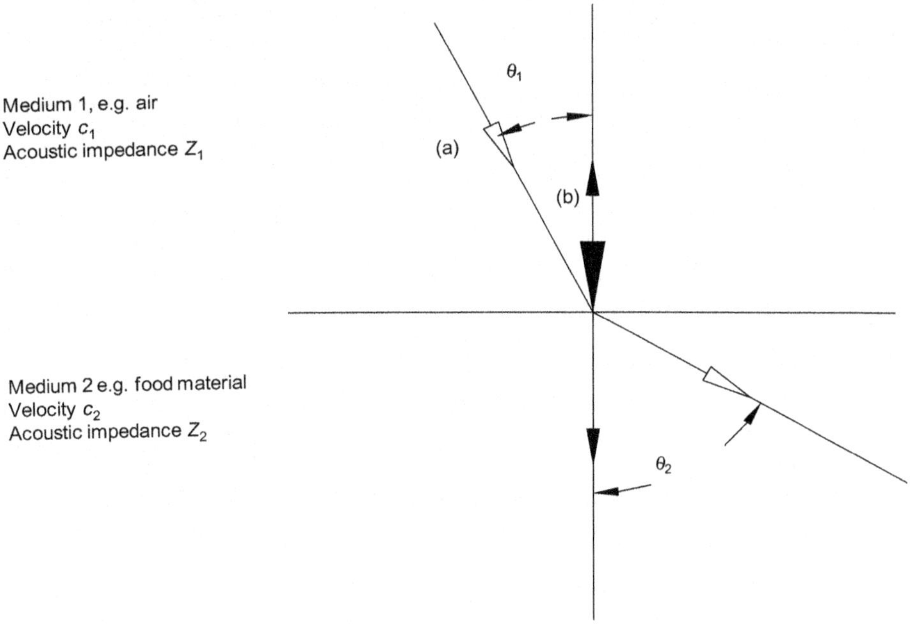

Fig. 10.1 Snell's law for wave travelling across an interface of two media (**a**) and transmission and reflection of incident wave at 0° impinging an interface (**b**)

c_1 Velocity in medium 1
c_2 Velocity in medium 2

When a wave hits an interface at normal incidence, part of the energy will be reflected and some will be transmitted into the second medium, as shown in Fig. 10.1b. The amount of energy that is transmitted across the interface depends on the acoustic impedance Z of both media. The acoustic impedance Z is given by:

$$Z = \rho \cdot c \tag{10.4}$$

where

ρ Density of the medium
c Longitudinal velocity of the medium

Hence, the amount of the transmitted energy across an interface could be estimated by the transmission coefficient T_0 given by:

$$T_0 = \frac{4Z_1Z_2}{(Z_1Z_2)^2} \tag{10.5}$$

and the amount of reflected energy is determined by the reflection coefficient R_0 given by:

$$R_0 = \left(\frac{Z_1 - Z_2}{Z_1 + Z_2}\right)^2 \tag{10.6}$$

As a wave propagates in a medium, the amount of energy from one point to another could be different (e.g. from medium 1 to medium 2). The loss of acoustic energy may be caused by absorption, diffraction, scattering or other interactions with the medium where most wave particles are converted into thermal energy due to internal friction and thermal conductivity (Kocis and Figura 1996).

In most situations, attenuation of sound α increases with frequency (Grandia and Fortunko 1995). When the signal attenuation is purely caused by absorption, the attenuation factor is usually dependent on the square of the frequency.

As a simple example, attenuation of signal α is usually calculated in decibels using the ratio of signal amplitudes A which is given by:

$$\alpha = 20\log_{10}\frac{A_1}{A_0} \tag{10.7}$$

By knowing the attenuation coefficient of a medium, the input signal amplitude can be corrected so that the desired output can be compensated for any loss of energy (Bushong and Archer 1991).

The change in propagation medium (from 1 to 2) will also affect the speed of the ultrasound c. In a simple case (in a contact mode as shown in Fig. 10.2a), the speed of ultrasound through transmission mode in a specific medium c is given by:

$$c = \frac{L}{t} \tag{10.8}$$

where

L Distance of travel by the ultrasonic signal (in the case of pulse-echo approach, L will be doubled)

t Time taken for the ultrasonic signal from the source to receiver

An alternative method to carry out food property characterization or inspection is to use the non-contact ultrasonic. For non-contact ultrasonic method, two approaches can be used: (1) through transmission and (2) single sided, e.g. pulse echo and pitch catch (Castaings et al. 1998). The single-sided inspection is not so common for non-contact food inspection due to the large reflection on the first interface, which masks other signals that are reflected from the internal structure. Due to the large impedance mismatch between the air–material interfaces, the echo from the internal material will be small. For these reasons, the through-transmission method is preferred. The speed of sound for this configuration in the medium of interest c_2 can be determined from (Schindel and Hutchins 1995):

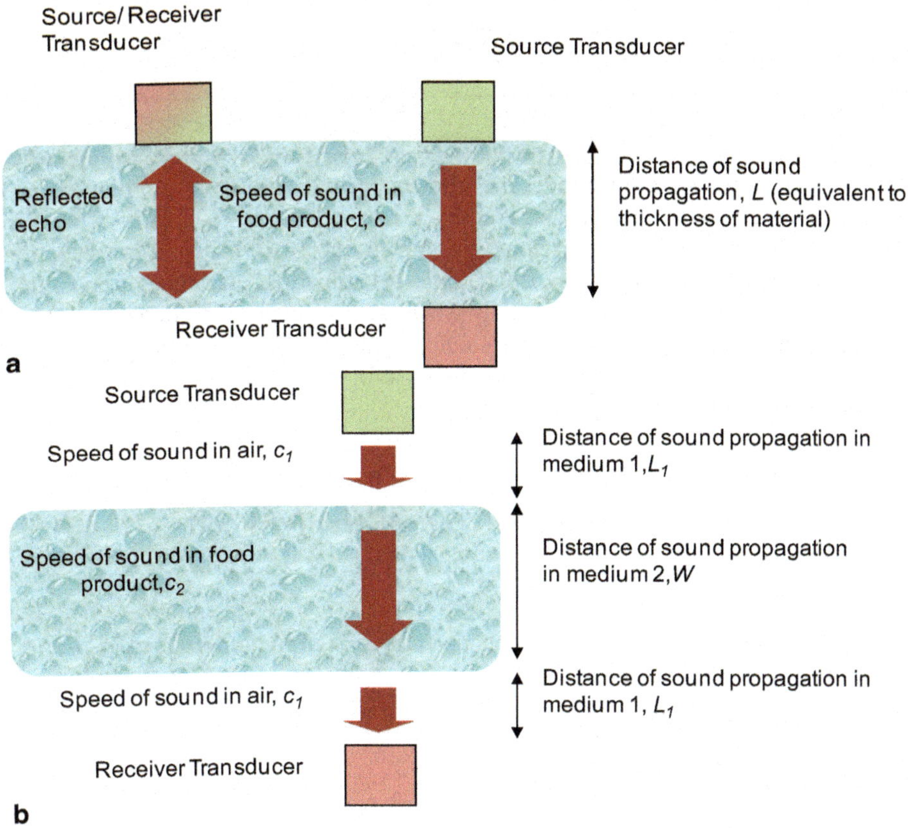

Fig. 10.2 Common methods of ultrasonic inspection of food products and food properties. **a** Contact technique. **b** Noncontact technique

$$t_2 = \frac{(L_1 + L_2 - W)}{c_1} + \frac{W}{c_2}, \tag{10.9}$$

$$t_1 - t_2 = W\left(\frac{1 - \frac{c_1}{c_2}}{c_1}\right) \tag{10.10}$$

$$c_2 = \frac{c_1}{1 - c_1\frac{(t_1 - t_2)}{W}} \tag{10.11}$$

$$\Delta W = \frac{(t_{w-\Delta W} - t_w)c_1}{1 - \frac{c_1}{c_2}} \tag{10.12}$$

In the through-transmission mode, the ultrasound will pass from medium 1 (air) into medium 2 (food material) and then back into medium 1 (air) as illustrated in Fig. 10.2b. The through-transmission technique in Fig. 10.2b is useful for detection of material properties, anomalies, debonding, delamination, etc., using transmitted amplitude *A* and arrival time *t* especially when the test specimen is nonstatic. Other derived parameters such as speed of sound, *c*, distance of sound propagation, *d*, and frequency, *f*, can also produce good and useful information regarding the food sample under evaluation (Vun et al. 2006).

10.3 Non-contact Ultrasonic Transducers for Food Inspection

There are four main types of transducers that can be used for non-contact inspection of food products and these are:

- Piezoelectric transducer
- Electrostatic capacitance transducer
- Electromagnetic transducer
- Lasers

In this section, we will only focus on transducers type1, 2 and 3.

Piezoelectric Transducers The piezoelectric transducer has been widely used for materials testing (Edmonds and Hickman 2000; Garcia and Tanarro 1998). Figure 10.3 shows a configuration of a contact piezoelectric transducer. Different types of wave modes could be generated depending on the geometry and the polarization of the active element. At the back of the element is a high-density backing material that has similar characteristic impedance to the active element. The aim of the backing material is to dampen and absorb any energy from the back of the element, giving a broad bandwidth signal. In order to generate sound waves, a transient voltage is applied across the electrode, causing the element to vibrate and generate ultrasound. The captured signal at the receiver turns the mechanical vibration into electric chargers, which may be sensed using a suitable amplifier.

To transfer the sound generated by the active element into another medium, a suitable form of coupling medium, e.g. water, is required. Ultrasonic immersion testing is a popular technique for the investigation of materials such as metals, fibre-reinforced polymers and many other materials (Xiang et al. 1998). However, not all materials can be immersed into water. For this reason, there has been increased interest in using air as the coupling medium. Piezoelectric air transducers are inherently

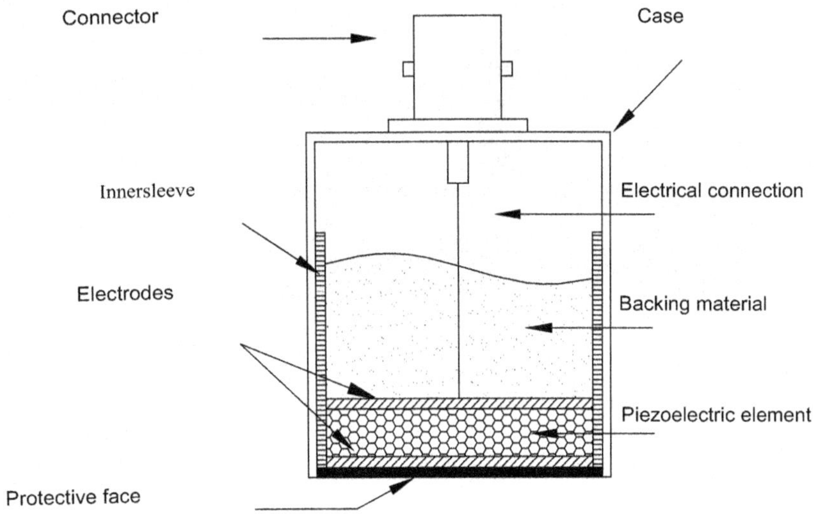

Fig. 10.3 Construction of a typical piezoelectric transducer

resonant devices and require special backing and construction to obtain suitable damping coefficients. The characteristic impedance of the piezoelectric element is very different to that of air. Therefore, a quarter-wavelength-thick matching layer at the frequency of interest is usually introduced at the front surface (Pedersen et al. 1982). Another way of reducing the impedance of the material is to use 1–3-connectivity piezo-polymer composites (Hayward and Gachagan 1996), which contain an array of piezoelectric ceramic rods in a polymer filler matrix.

A 2–2-connectivity composite that consists of alternative layers of ceramic and fillers (Möckl et al. 1990) could also be used. These have a wider bandwidth than traditional piezoelectric materials. Another method, which could be used to improve the impedance mismatch, is to use a piezoelectric polymer such as polyvinylidene difluoride (PVDF) although the material has been found to perform better as a receiver (Manthey et al. 1992).

Electrostatic Capacitance Transducers An alternative transducer design is based on the capacitance or electrostatic principle. This has received much interest recently because of the excellent bandwidths that can be achieved. These devices consist of a thin metalized membrane film and a rigid contoured conducting backplate. These two structures form a capacitor. Applied voltages cause the membrane to vibrate, and hence generate ultrasound, whereas ultrasound impinging on the membrane changes the device's capacitance, allowing it to be used for detection. Metallic backplates can be used (Carr and Wykes 1993), many employing a regular grooved backplate, which was found to improve the acoustic properties of the transducer (Rafiq and Wykes 1991). This type of transducer was found to be very sensitive at high bias voltages when a thin polymer membrane was used (Hietanen et al. 1993). Further investigation showed that the sensitivity of the transducer

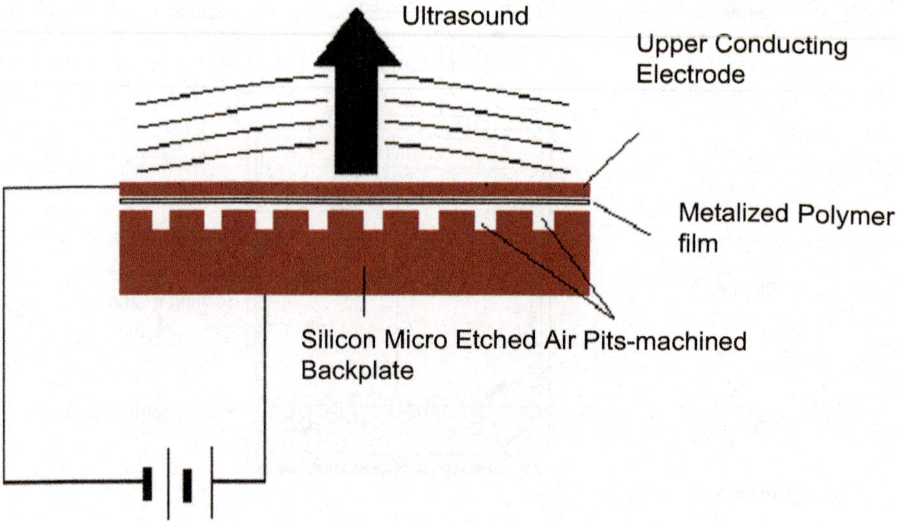

Fig. 10.4 Schematic diagram of an air-coupled micromachined backplate transducer

could be increased by modifying the shape of the groove, e.g. to form a V-grooved backplate device (Pizarro et al. 1999). An additional approach is the fully micromachined device, where the complete structure is fabricated using complementary metal oxide semiconductor (CMOS) technology from a silicon wafer. Typically, such designs use a silicon substrate and a silicon nitride membrane (Ladabaum et al. 1998), although another work has used a hexagonal cell structure with polysilicon membranes (Eccardt et al. 1997).

In this chapter, air-coupled capacitance devices were selected for non-contact food characterization and inspection. The configuration of this device is shown in Fig. 10.4. The design was based on silicon micromachining and had a layer of metalized dielectric membrane where its insulating side is placed against the surface of a rigid conducting backplate (Schindel 1995). The membrane usually has a thickness of 3.5–10 μm. The polished backplate was usually made of (110) silicon wafer that was coated with silicon nitride and silicon dioxide. A photoresist layer was applied to the wafer and part of the photoresist was covered before exposing it, using the photolithographic techniques (Schindel 1995). This produced small patterns of holes with a depth of 40 μm and distance of 80 μm between the centres of the holes. The silicon nitrate and silicon dioxide layers were then removed using phosphoric acid and hydrofluoric acid respectively. The unwanted part of the silicon backplate was etched using potassium hydroxide. Finally 1000 Å of gold was coated on the contoured surface to form a conducting layer.

These small holes help to trap air beneath the membrane and reduce the membrane rigidity, and thus produce a wider bandwidth and enhanced sensitivities (Maxfield et al. 1987). This transducer is affected by change in backplate surface, membrane tension and thickness; however, these features are difficult to control.

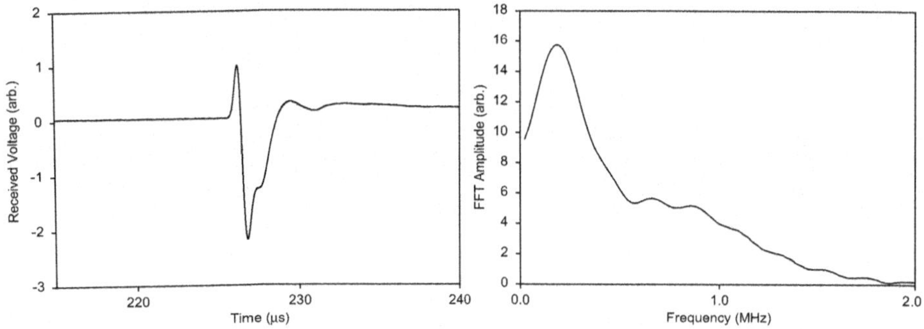

Fig. 10.5 **a** Air-coupled signals transmitted across an air gap driven by a wide band pulse/receiver. **b** Corresponding frequency spectrum

By using a smooth backplate, it is possible to achieve a centre frequency of up to 1 MHz, while a roughened backplate generates lower frequency components (Schindel 1995).

For such a transducer, the bandwidth of the signal is dependent on bias voltage, film thickness and the nature of the transient voltage used for excitation. The bandwidth increases with both bias voltage and reduced film thickness. Ultrasound is generated by applying a transient voltage $V(t)$ across the insulator. The resulting transient electric field then excites vibrations in the membrane. The efficiency and bandwidth are both increased by superimposing a DC bias field upon the transient voltage. As a receiver, the detected sound wave at the membrane varies the capacitance. In the presence of an imposed bias field, a dynamic charge upon the electrodes is generated. The typical response and bandwidth of the transducer is shown in Fig. 10.5.

Electromagnetic Transducers Electromagnetic transducers have been used for the non-contact generation and detection of ultrasonic signals in metals for some time (Maxfield et al. 1987). They consist basically of a coil and a magnetic field, applied to a conducting substrate. The transmitter coil is conventionally driven with a transient current pulse. The characteristics of an electromagnetic acoustic transducer (EMAT) as an ultrasonic source depend on the direction of the applied magnetic field B and eddy current density J induced at a conducting surface. The presence of a Lorentz force F on the substrate causes an elastic wave to propagate into the volume of the material, or along the surface. EMAT detection works via an inverse process, where motion of the surface induces current into the coil.

The exact wave mode to which the EMAT is sensitive depends on the coil and magnetic field configuration. There are many examples, including meander-line, pancake and rectangular coils (Murayama 1996; Hu et al. 1988), and these are shown in Fig. 10.6. Meander-line coils, shown schematically in Fig. 10.6a, can be used for generating Rayleigh (surface) waves, Lamb waves in thinner material, and shear waves at a predetermined angle depending upon frequency of excitation and meander-line geometry. Spiral pancake coils can be used to generate and detect

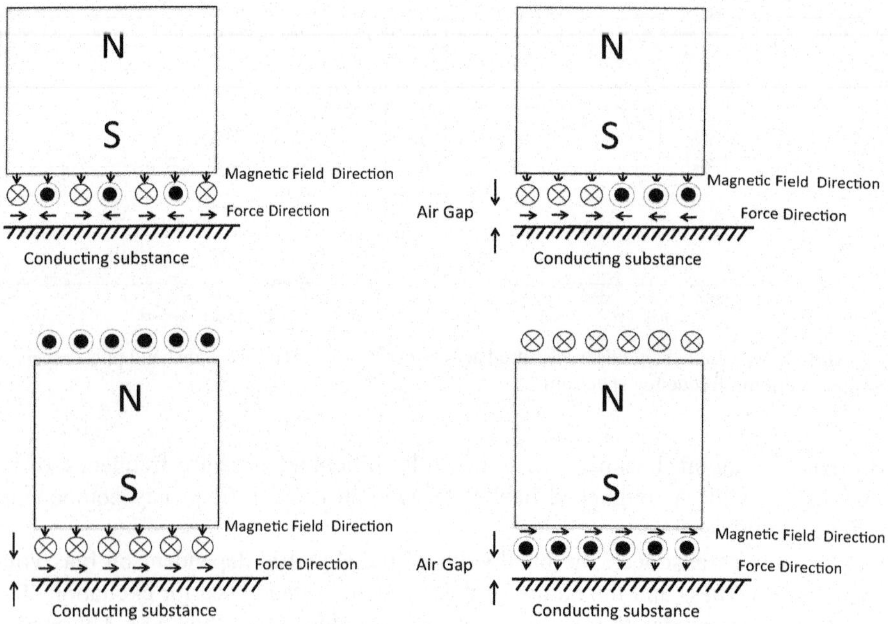

Fig. 10.6 Different types of EMAT configurations. **a** Menderline for Rayleigh and Lamb wave generation. **b** Pancake for radially polarised shear wave generation. **c** Rectangular for shear wave generation. **d** Rectangular for longitudinal wave generation. (Adopted from Ho et al. 2005)

radially polarized bulk shear waves (Fig. 10.6b; Dixon et al. 1994), and rectangular coils are also useful for many types of acoustic modes, as shown in Figs. 10.6c and d. These EMATs have been used for many applications, such as high-temperature measurements (Lee and Ahn 1992; Idris et al. 1994) and canned food inspection (Ho et al. 2007).

10.4 Signal Processing to Improve Low SNR

It has been known that the signal transmitted from an air-coupled transducer is weak due to the high acoustic impedance mismatch between transducer surface and air, and this leads to poor SNR. The SNR problem encountered in air-coupled experiments can be improved by using a high-power tone-burst signal. Such signals are very convenient, in that gated power amplifiers can be used to deliver high powers and, when combined with a broad bandwidth transducer, the frequency of operation can be varied. For piezoelectric transducers in particular, tone-burst excitation leads to considerable advantages for air-coupled testing, in that the frequency can be tuned to the through-thickness resonance of the material. This increases the through-transmission signal levels substantially.

There are, however, disadvantages to using a tone burst. First, the voltage excitation level is limited by the type of transducer used, and in the particular case of a capacitance transducer, the voltage must be restricted to avoid dielectric breakdown of the thin polymer membranes. In addition, the exact frequency of excitation must match the through-thickness resonance of the sample to achieve maximum efficiency, and this might need to be adjusted if the thickness of the material changes (for instance during an imaging experiment involving positional scanning). It may also be the case that the longitudinal velocity is either not known, or might vary. The main disadvantage in the context of defect detection is that a tone burst leads to relatively poor time resolution (Izuka 1998). Defects might be difficult to resolve, because multiple reflections might overlap in time, although cross-correlation can lead to accurate time-of-flight measurements. Because capacitance transducers can operate over a wide frequency range, it is thus better to use a technique that capitalizes on this property. The use of a swept frequency signal, instead of a single transient, allows a high-power, broad bandwidth signal to be used which, when combined with suitable processing, also gives excellent time resolution. It is this property that is used in the pulse compression technique (Rao 1994; Ermolov et al. 1996).

In the context of an air-coupled ultrasound experiment, a tone burst, tuned to the through-thickness resonance, is still likely to give the greatest signal amplitude. However, the advantage of using a wide bandwidth pulse compression approach, using a broadband swept frequency excitation at the source transducer, is that the full air-coupled spectral response of a material (e.g. with multiple resonances) can be obtained instantaneously. This can be achieved without frequency scanning as has been necessary in the past (Folkestad and Mylvaganam 1993). Increased accuracy in time-of-flight measurements can potentially be obtained, and the technique provides the ability to recover small signals from well below the noise floor, although the pulse compression method is only valid for improving the SNR if the noise is random.

10.4.1 Simulation of the Pulse Compression Technique

The pulse compression method can be implemented by driving the ultrasonic source with a so-called chirp or linear frequency-modulated (FM) signal, where the frequency is swept continuously over a predetermined range, and then applying a cross-correlation operation. The chirp is an elongated waveform, with the overall duration of the signal and the rate of frequency sweep defining the chirp characteristics. Note that a chirp signal is not the only suitable source of waveform for pulse compression. Another type is the pseudorandom binary sequence (PRBS) technique (Elias 1980). This is sometimes referred to as bi-phase code modulation (Taseand R. Seller 1998), but is seldom used compared to a linear FM signal. In this technique, the transmitted signal is separated into subpulses, and these subpulses are further modulated into two levels—0° and 180° phase shift in the carrier—according to the code items (Boehmer 1967). This technique is implemented by

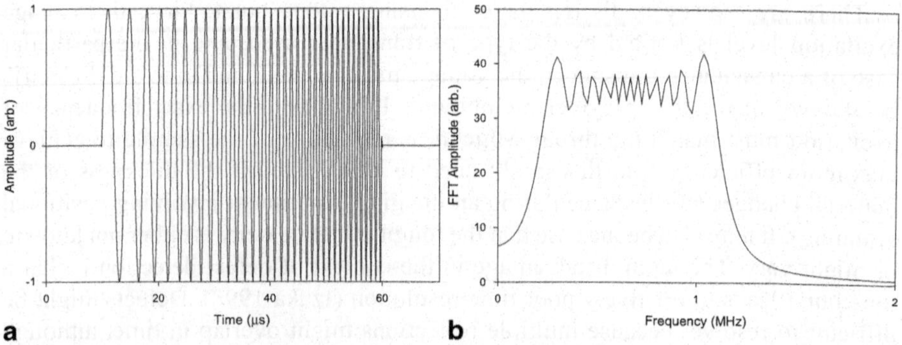

Fig. 10.7 **a** Simulated broadband chirp signal with a duration of 50 μs. **b** Frequency spectrum of the broadband chirp signal. (Adopted from Gan et al. 2001a)

correlating or matching the received PRBS signal with one of the elements of the code. With a single element correlation, the resulting time sidelobe amplitude peaks are much greater compared to the linear FM signal. Finding the proper codes for pulse compression has been the subject of much research, where special Barker (Jia et al. 2000) and Golay (1961) codes have been introduced. However, these codes make the technique more complex, and hence it is rarely used in radar implementation. Even so, these codes do not have sufficiently large time-bandwidth products to make significant improvement of the sidelobes (Arthur 1996). M-sequences (Lee and Furgason 1982) and polyphase modulation (Felhauer 1992) are additional pseudorandom codes. The polyphase is an improved version of the bi-phase codes. In addition, both the bi-phase and polyphase could be treated as an FM code which has been phase quantized. The polyphase modulations require more complex signal processing in the receiver. The drawbacks to the phase-coded technique make the linear FM more efficient and easy to use in various applications.

In the following, it will be shown that the chirp (linear FM) waveform in a pulse compression technique can lead to significant improvements in SNRs when applied to air-coupled through-transmission testing. Note that pulse compression has been used previously for improving the resolution of various measurements for medical applications (Venkatraman and Rao 1996), measuring flow (Gan et al. 2001b), spatial characterisation of scattering microstructure (Rao and Aubry 1994), and food and drinks application (as describe in this chapter).

The pulse compression technique is best described using a simulation, where a chirp signal is buried in noise. This chirp signal can be represented as:

$$C(t) = H(t).sin\left(\omega_s t \frac{\pi B t^2}{T}\right), \tag{10.13}$$

where

$H(t)$ is the Hanning function given by:

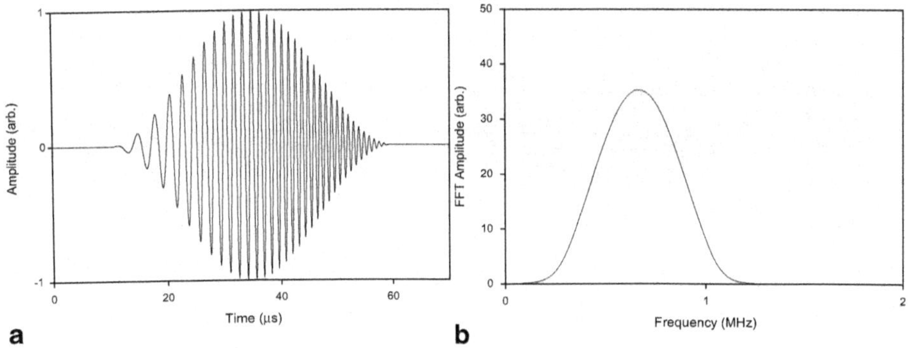

Fig. 10.8 **a** Simulated broadband chirp signal with a duration of 50 μs after a Hanning window was applied. **b** Frequency spectrum of the broadband chirp signal after Hanning window was applied. (Adopted from Gan et al. 2001a)

$$H(t) = \frac{1}{2}\left[1 - \cos\left(\frac{2\pi t}{T}\right)\right] \tag{10.14}$$

The generated signal from Eq. (10.13) is as shown in Fig. 10.8a. The figure shows the bell-shaped envelope resulting from the Hanning filter, and this shape is important to ensure good sensitivity. In addition, the Hanning amplitude weighting also helps to reduce the amount of sidelobes in the signal (Rao et al. 1995). Figure 10.8b shows the frequency spectrum of the Hanning-weighted chirp signal. In the figure, it can be seen that the lower and higher limits of the frequencies are maintained at 200 kHz and 1.2 MHz. The signal is centred at 700 kHz, and the edge ripples have been removed. For this reason, most applications of the pulse compression technique used the Hanning chirp.

It is interesting to demonstrate that a chirp signal is useful for the detection of signals in the presence of high noise levels. The Hanning chirp signal in Fig. 10.8a was thus shifted by 40 μs in time, and mixed into a random noise level of twice the chirp signal amplitude. This simulates the noise levels that could typically be encountered in a real air-coupled material inspection experiment. This is as shown in Fig. 10.9a. In order to produce the compressed pulse signal, $P(t)$, the received signal $CT(t)$ is initially band-pass filtered within the chirp bandwidth. The filtered signal is as shown in Fig. 10.9b, with an SNR of about 6 dB. The band-pass filter removes the noise levels above and below the frequency range of the original chirp-driving signal, but the transmitted chirp signal is still not easily visible. The waveform is now cross-correlated with the reference signal $C(t)$. In the time domain, the cross-correlation function is represented by Eq. (10.15) and this process can be calculated using a matched filter (Millet 1970):

$$P(t) = C(t) \times [CT(t)]. \tag{10.15}$$

The compressed pulse, $P(t)$, is thus produced by the correlation of the received signal $CT(t)$ with the original reference signal $C(t)$. The correlated result is as shown

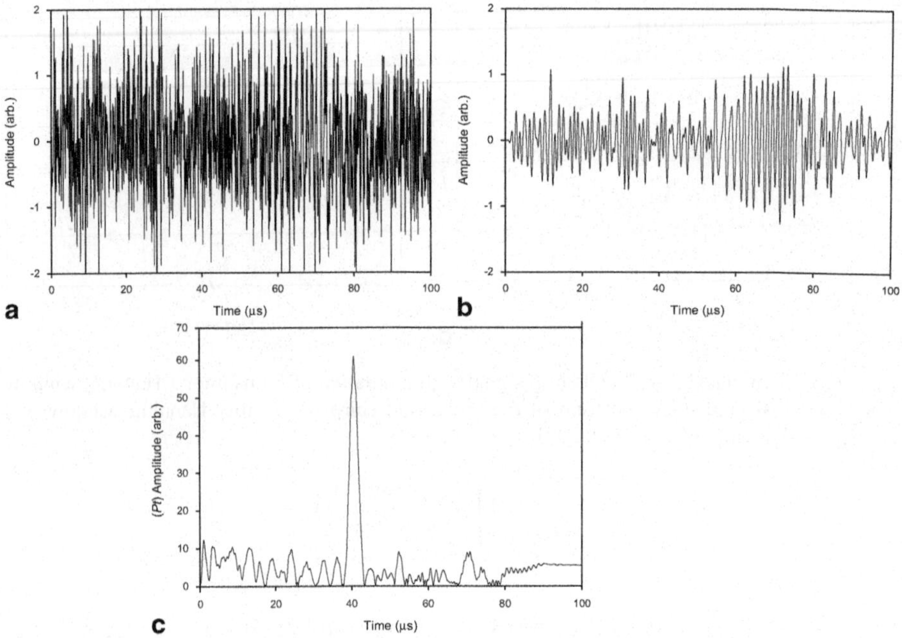

Fig. 10.9 **a** Transmitted broadband Hanning chirp signal which was embedded in noise. **b** Improved noisy waveform using band pass filter, SNR ~6 dB. **c** Compressed pulse signal at 40 μs, SNR ~ 20dB. (Adopted from Gan et al. 2001a)

in Fig. 10.9c and is in the form of a time signal. In the figure, the main peak in the pulse compression output $|P(t)|$ represents the position in time of the transmitted signal, which is at a time delay of 40 μs. The SNR has been greatly improved compared to Fig. 10.9b, as can be seen. The width of the $|P(t)|$ peak can be reduced to give greater time resolution by increasing the bandwidth (B) of the generated chirp signal ($C(t)$), whereas a greater peak amplitude can be obtained by elongating the time duration (T) for the same bandwidth (Fig. 10.9c). It is thus of advantage to use as long a duration T of the chirp as possible, and to also maximize the bandwidth.

However, the pulse compression output can be interpreted much like a conventional ultrasonic waveform, in that the amplitude of the compressed pulse (as a function of time) is related to the amplitude of the received chirp waveforms (as a function of time). Note also that the exact shape of $|P(t)|$ will be modified if the original chirp pulse shape is distorted by the sample. This will almost always happen in the case of a simple plate at normal incidence, where maximum transmission amplitudes occur at well-defined frequencies. In these situations, $|P(t)|$ also contains information concerning the material through which the signal has travelled. In particular, it can be used to determine the different frequencies of resonance that are present. This will be illustrated later in this section.

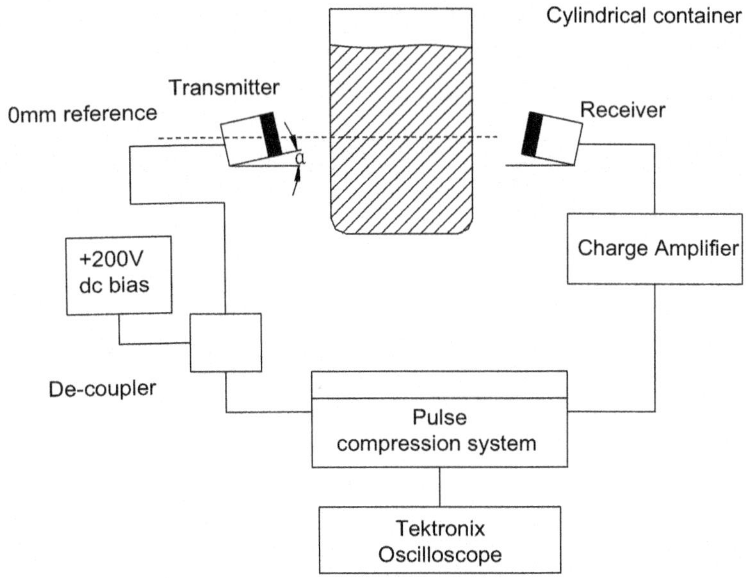

Fig. 10.10 The experimental setup of a liquid level measurement using two broadband capacitance transducers. (Adopted from Gan et al. 2002)

10.5 Applications

This section looks into the application of air-coupled ultrasound for the inspection system for the detection of:

- Internal properties in food containers (Pallav et al. 2008)
- Physiochemical changes in food properties (Meyer et al. 2006; Gan et al. 2006)
- Foreign objects in food products (Cho and Irudayaraj 2006; Pallav et al. 2007; Gan et al. 2002)

10.5.1 Detection of Internal Properties in Food Containers

The non-contact ultrasonic detection of food materials in a container such as liquids and food products is shown in Fig. 10.10. The pulse compression approach was used to detect the signal transmitted across the container. In order to provide initial calibration, the electrostatic transducers were aligned horizontally, with no sample in place, and separated by 170 mm. The applications used a pair of capacitance transducers to generate and detect the chirp signals. The transducers in the present experiments had an active aperture of 10 mm diameter. The transmitter, which had a membrane thickness of 5 μm, was driven by a pulse compression pulser/receiver unit.

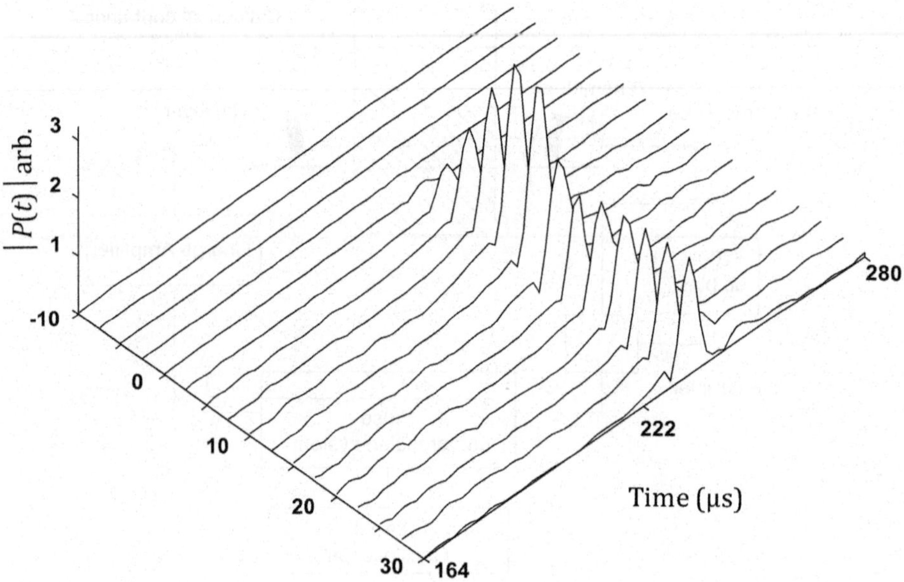

Fig. 10.11 Variation of the pulse compression output as the transducer was scanned vertically down a circular bottle containing liquid. The transducers were aligned axially, with $\alpha = 0°$. (Adopted from Gan et al. 2002)

The transmitted chirp signal across a test sample was captured by another electrostatic transducer, this time with a membrane thickness of 2.5 μm. A thicker membrane was used at the source in order to withstand the high voltage generation pulse and to avoid membrane breakdown. As a receiver, a thinner membrane was used so that the sensitivity can be increased. The output chirp voltage was superimposed upon a +200 V dc bias using a capacitive decoupling circuit, before being applied to a capacitance source of bandwidth 1.5 MHz. The longitudinal waves propagated through the air to the sample. Through-transmitted signals were received by the capacitive receiver, input to a Cooknell CA6/C charge amplifier, and pulse-compressed data was recorded using a digital oscilloscope.

The first example of the non-contact ultrasonic application measured the liquid level in the container. In this application, the transducers were aligned axially ($\alpha = 0°$). The aim was to measure through-transmitted signals as the liquid level changed. A cylindrical bottle with an external diameter of 116 mm and thickness 0.6 mm was then placed between the transducers. Measurements were obtained by varying the volume (and hence the liquid level) of liquid inside the container, with the transducer location fixed. Each variation of 2.6×10^{-5} m^3 of water (corresponding to 2.5 mm in level) was recorded. A second example was performed with the transducers tilted as shown in Fig. 10.10, at angle $\alpha = 12°$. The aim was to try to obtain a signal that reflected from the liquid surface. Timing the arrival of this signal would give an estimate of liquid depth.

Figure 10.11 shows the recorded waveforms when the transducers are at $\alpha = 0°$. The received amplitude remained constant as long as the water level remained

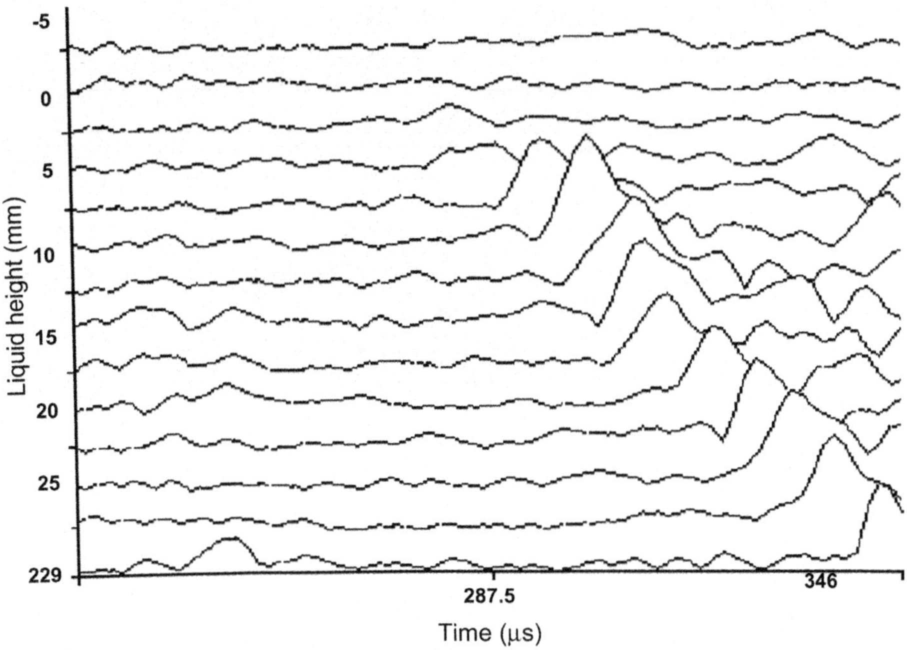

Fig. 10.12 Change in time of arrival of the compressed pulse signal with variation of liquid level using transducers at $\alpha = 12°$. The major peak is a reflection from the liquid surface. (Adopted from Gan et al. 2002)

above 0 m, but as this was approached, the amplitude increased due to signal scattering from the water surface. As the water level reduced further, the peak amplitude reduced to zero as the liquid content was replaced by air. Note that Fig. 10.11 demonstrates an excellent SNR.

A more practical arrangement would be that with tilted transducers, so that a reflection from the underside of the water surface could be obtained. The transducers were thus tilted to $\alpha = 12°$ from the horizontal plane. As before, the water level was varied by 2.5 mm for each measurement. The measured waveforms are shown in Fig. 10.12. As the amount of water in the container reduced by a step size of 2.5 mm, the time of arrival of $|P(t)|$ decreased and the pulse-compressed time peak moved to the left as shown. This was due to the shorter travel time for the signal reflected from the water surface.

The air-coupled setup can also be used for the inspection of canned food products on a production line. Figure 10.13 shows the air-coupled arrangement used to inspect cans in a laboratory-based mock-up of a production line. A transducer–receiver pair was aligned in through-transmission mode across the production line. An infrared proximity sensor was also used to signal the arrival of the can between the ultrasonic transducer pair. The output from the infrared sensor was used to trigger the ultrasonic inspection system.

Fig. 10.13 A scanning system for canned products on a conveyor belt. (Adopted from Gan et al. 2002)

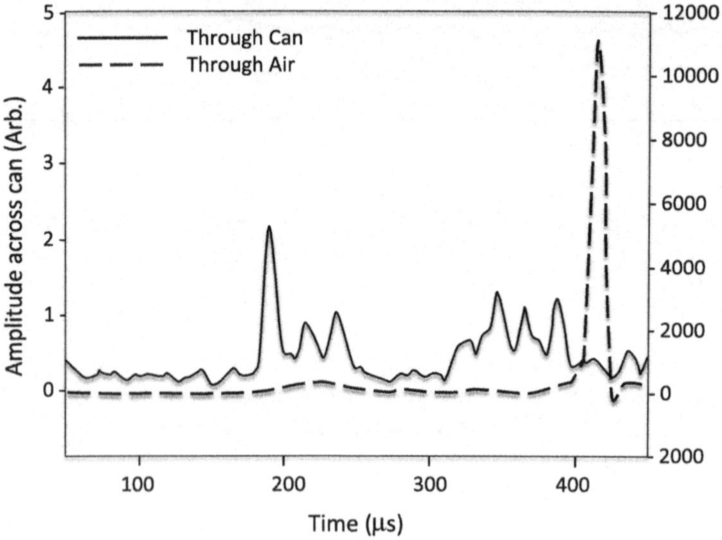

Fig. 10.14 Through-transmitted pulse compression signal (*solid line*) received across a canned meat product (beef chunks in a sauce). The signal through air only is shown as *dotted line*. (Adopted from Gan et al. 2002)

Canned products pose different challenges to ultrasonic propagation. Due to the high acoustic impedance of metal containers, most of the ultrasound is reflected at the air/can boundary. The corrugated surface of some cans leads to further loss of signal. A typical received waveform is shown in Fig. 10.14. It can be seen that the signal through the can arrived sooner than in air only because the canned product has much higher ultrasonic velocity than the air medium. Such signals could be used to determine the properties of the food in the can and whether the product in the can is properly filled.

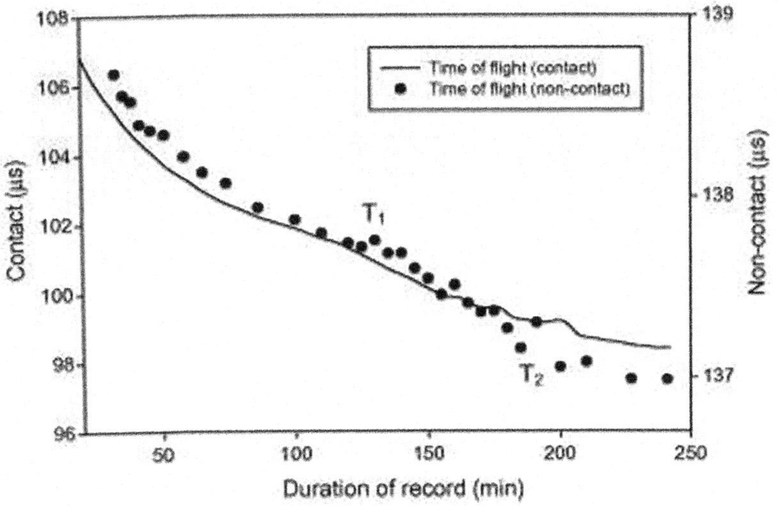

Fig. 10.15 Comparison of signal amplitudes produced by both contact measurements (*continuous line*) and noncontact air-coupled measurements (*dots*) as a function of time, as palm oil was cooled from 60°C to room temperature. (Gan et al. 2006)

10.5.2 Detection of Physiochemical Changes in Food Properties

The same configuration in Fig. 10.10 can also be used to detect physiochemical changes in the food products. For illustration, unrefined palm oil has been used, because it can be easily obtained. The unrefined palm oil used in this experiment was heated to a temperature of 60°C before being placed within a Plexiglas cell with a 3 mm wall thickness and an area of 70 × 70 mm. Time-of-flight data were recorded against temperature, the latter measured with a digital thermocouple device, in order to observe the change in state of the heated palm oil as it cooled down to room temperature (Cho and Irudayaraj 2006).

Changes in the two ultrasonic parameters, i.e. ultrasonic amplitude and time of flight between the two transducers, were recorded in each case, and the results for received amplitude are shown in Fig. 10.15. It can be seen from the results that both contact and non-contact measurements are well correlated. The received amplitudes reduced as the temperature itself decreased, as a function of time. Note the discontinuity in the data. Up to approximately 115 min of elapsed cooling time, the sample was in liquid form (see Fig. 10.16a). As the rate of change of the temperature slows down (i.e. from 115 min onwards), the oil property changed its state and crystallization started to form in the oil. This is shown in Fig. 10.16b. The steep decrease in amplitude between 115 and 175 min in Fig. 10.16 was due to crystallisation, which caused increased scattering of ultrasound. From 175 min onwards, the transition from liquid to solid particles was largely completed. As oil particles merged, attenuation increased due to greater scattering. The amplitude was also affected by the increased density and hence the acoustic impedance (the product of density and

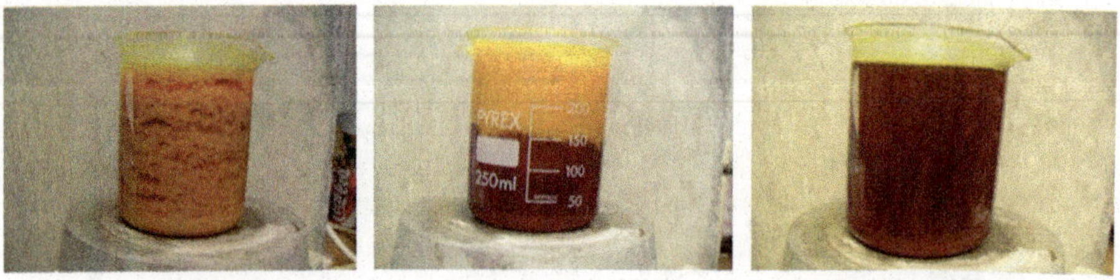

Fig. 10.16 Photograph of palm oil at different times during the experiment of Fig. 10.15, taken at **a** 0 min (60°C), **b** 115 min and **c** 175 min (22 °C). (Gan et al. 2006)

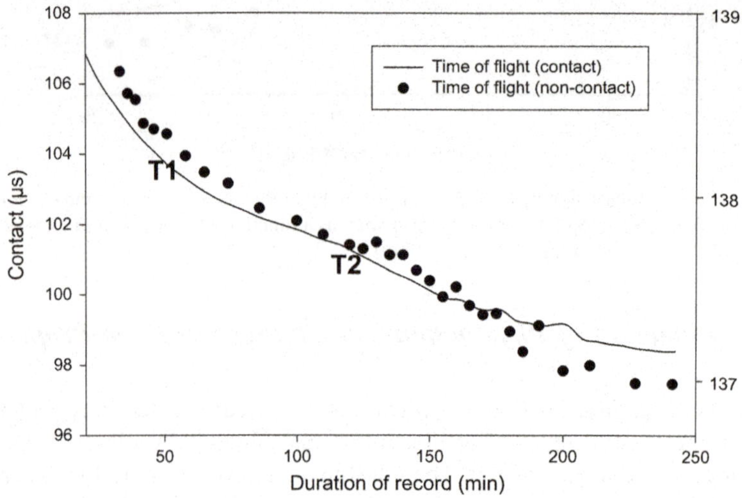

Fig. 10.17 Variation of ultrasonic time of flight across palm oil as temperature changes from T1 of 33 °C to T2 of 22 °C. T1 and T2 denote phase transitions. (Gan et al. 2006)

longitudinal velocity) of the media. The appearance of oil at room temperature is shown in Fig. 10.16c.

The time of flight was also monitored with time, and the results are shown in Fig. 10.17. As will be seen, the time taken to travel across the sample decreased as the temperature reduced (i.e. at greater times). The time-of-arrival data are consistent with the air-coupled through-transmission amplitude data shown in Fig. 10.15 in that there are phase transitions at T1 = 115 min (at 33 °C) and T2 = 175 min (at 22 °C), shown as discontinuities in slope of the graph at these temperatures. It can be seen from the results that there is a slight variation in the results between the contact and non-contact techniques. This is thought to be due to the fact that the temperature of the medium affects the contact transducer, whereas the non-contact technique is not affected. The results show that the noncontact system was measuring the crystallization effect and was less affected by changes in temperature. Experiments were also performed on milk-based products, e.g. full-fat milk at pH ~ 5.5. In order to show that the system can detect changes at different pH levels, various mixtures

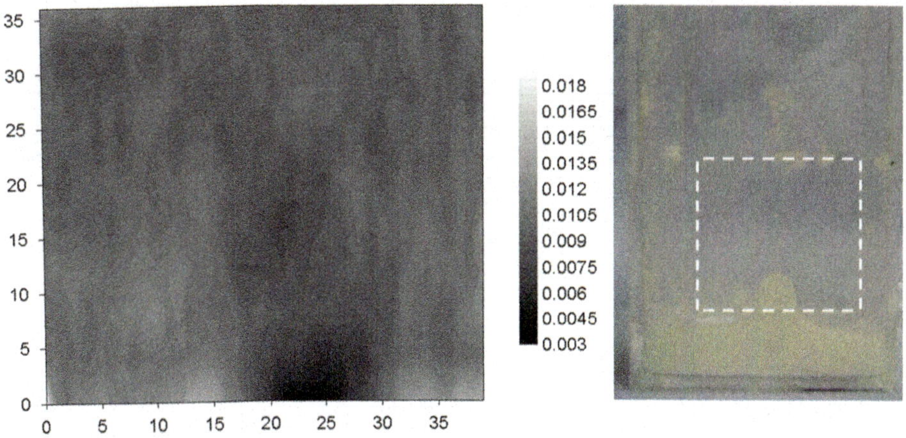

Fig. 10.18 Noncontact through-transmission imaging of the full-fat milk sample mixed with sulphuric acid. **a** Cross-sectional image formed by scanning the air-coupled ultrasonic system. **b** A photograph of the Plexiglas container showing the resultant coagulated milk attached at the bottom of the container. (Gan et al. 2006)

were used. In this case, the product was destabilized by reducing the pH with acid to the 4.5–5.0 range. The pH reading was measured using litmus paper. The amplitude of the transmitted ultrasonic signals were collected at 60-s intervals, with ultrasonic amplitude the principal parameter measured. Imaging experiments were also performed after the pH had been reduced, by scanning the transducer pair in air parallel to the flat surfaces of the container walls. The transducer pair was moved along the x-axis with a step size of 1 mm to a total range of 40 mm (see Fig. 10.15). When this was done, the transducers were then moved in the z direction with a step size of 4 mm. The total area of the side scan was 40×36 mm. Images were then formed of spatial variations in through-transmitted amplitude.

The results showed that the transmitted signal is highly attenuated at certain locations due to significant destabilization of the milk product. An optical photograph of the container was then taken at the end of the experiment and this is shown in Fig. 10.18. The photograph was taken so as to illustrate the x–z plane. Comparison of Figs. 10.18a and b indicate that the air-coupled technique was able to detect structural changes within the sample as variations in the peak amplitude of the transmitted signals. The size of the coagulated areas could also be estimated.

10.5.3 Detection of Foreign Object in Food Products

An experimental arrangement was also established to allow cross-sectional air-coupled imaging, using a combined linear (x) and a rotational (θ) scanning stage, as shown in the block diagram of Fig. 10.19. The PC used for data storage was also used to control the X-rotational stages. The total area of the scan varied depending

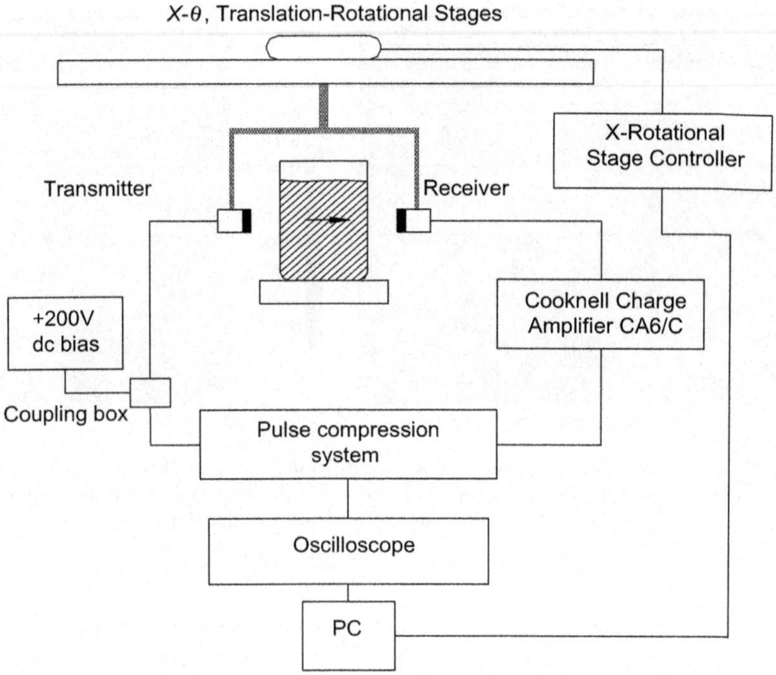

Fig. 10.19 An example of tomography system for food contamination cross-sectional imaging. (Adopted from Gan 2002)

on the size of the test sample involved. The scan had a dimension of 45 mm with spatial intervals of 1 mm. In addition, 180° of data were collected with an angular step of 3°, giving 61 projections. All scanning and data acquisition were controlled by Labview.

A difference technique was used to enhance the resolution of the reconstructed tomographic image, by normalizing the collected data to a reference data set, collected for a uniform liquid of known properties. The image was then that of the difference in properties between the imaged cross-section and that of the reference liquid. The scan dimension was limited to 45 mm (for this application) mainly due to total reflections of the transmitted signal towards the edge of the scan (Gan 2002). Two types of artificial defects or foreign objects present in the liquid container were tested and the images were tomographically reconstructed. Initially, a circular aluminium rod with a diameter of 10 mm was placed inside the container. The container was then scanned across its cross-section. When this was completed, the circular defect was then replaced by a stainless steel plate of dimension 1.5×7 mm.

The image obtained from the peak signal amplitude is as shown in Figs. 10.20 and 10.21. The defect of 10-mm diameter is shown in Fig. 10.20. It is distorted due to diffraction and refraction effects. However, it can be seen that the rod attenuated the through-transmitted signal. Figure 10.21 shows the image of the inserted plate and again, the plate has been detected. Both images indicate that an object with a

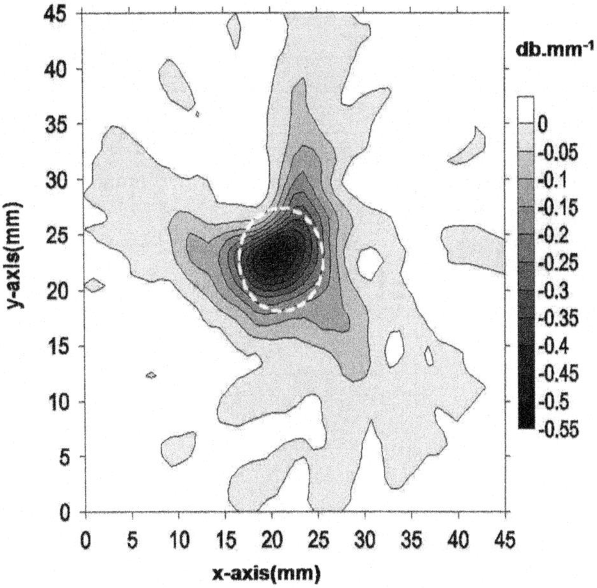

Fig. 10.20 Air-coupled tomographic reconstruction of a polymeric drinks bottle, containing a circular aluminium rod of diameter 10 mm

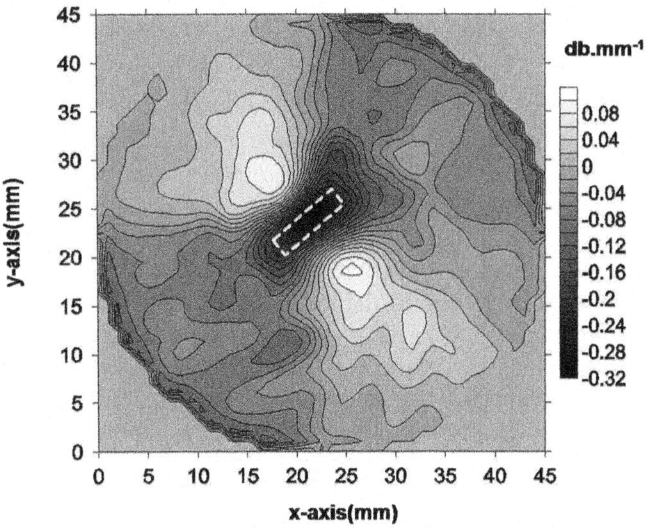

Fig. 10.21 Air-coupled tomographic reconstruction of a polymeric drinks bottle, containing a thin plate of length of 7 mm and width of 1.5 mm

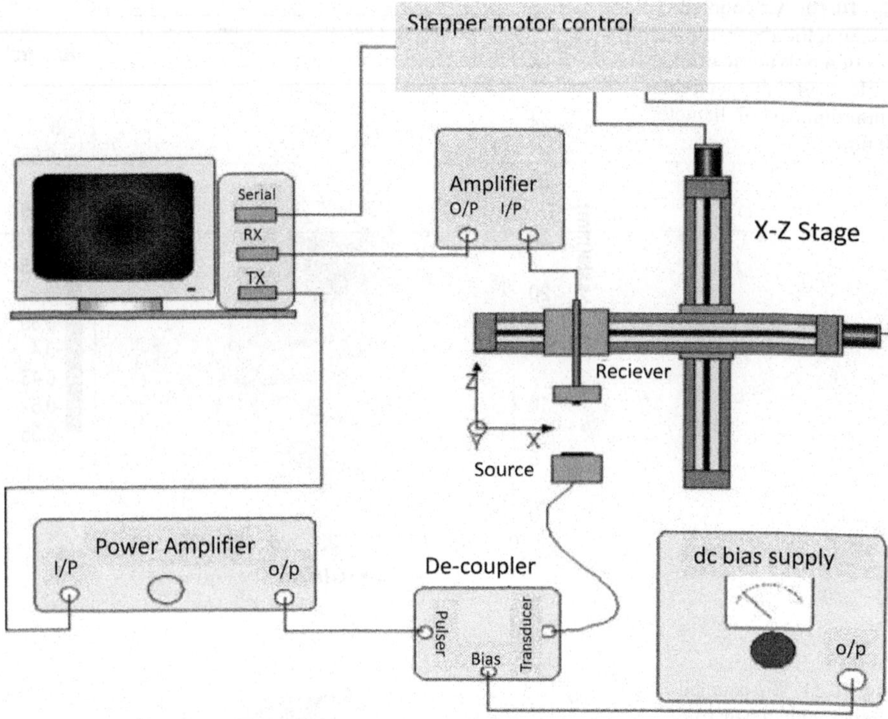

Fig. 10.22 Experimental arrangement for air-coupled ultrasonic inspection of food samples. (Adopted from Pallav et al. 2008)

different acoustic impedance to the water could be detected using transducers located in the air outside the container.

The above setup (Fig. 10.22) shows an illustration of how air-coupled ultrasound could be used to detect inclusions which were not visible externally by eye. Whole chocolate bars with different types of inclusions have been scanned, typical results being shown in Fig. 10.23. Here, Fig. 10.23a shows a conventional milk chocolate bar without inclusions, where the individual chocolate 'squares' are visible in the image. This is then modified when additives such as hazelnuts are added (Fig. 10.23b), with the nuts appearing as the darker areas. The nut inclusion is clearly visible in both amplitude and time of flight images. Such nuts could easily be counted as part of an on-line quality control system.

Another set of experiments was also carried out on a microwaveable polymer food container with dimensions of 180 × 110 × 49 mm. These experiments were conducted to show that contamination or changes in density of the food in the form of liquid sauce, which could not be seen by naked eye, could be detected using the air-coupled scanning technique. The container had a wall thickness of approximately 0.45 mm. The container was placed between the transducers as shown in Fig. 10.24. A linear scan with a dimension of 100 mm and a step size of 1 mm was

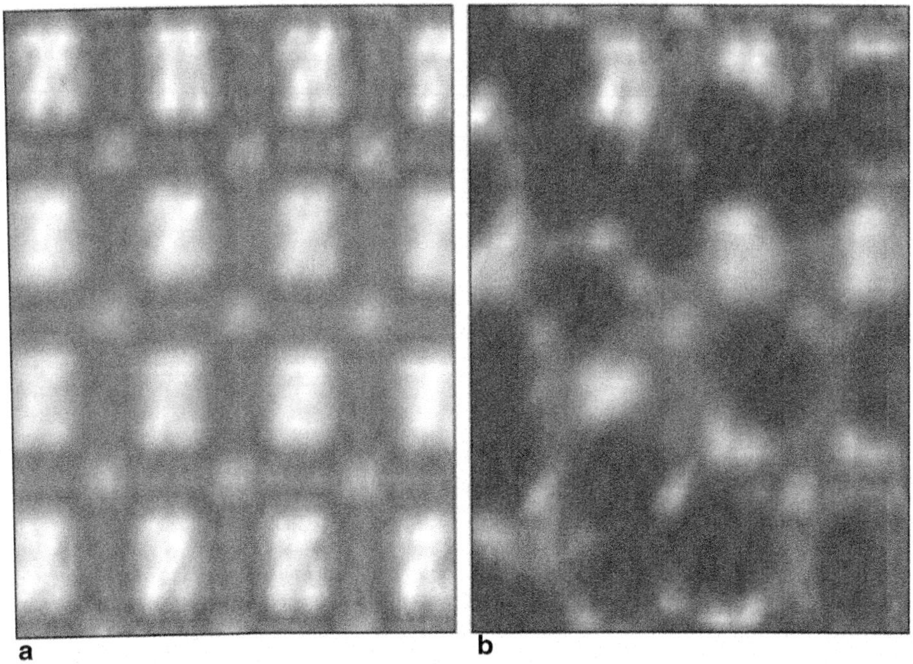

Fig. 10.23 Air-coupled ultrasonic images of **a** a milk chocolate bar and **b** a similar bar containing hazelnuts. (Adopted from Pallav et al. 2008)

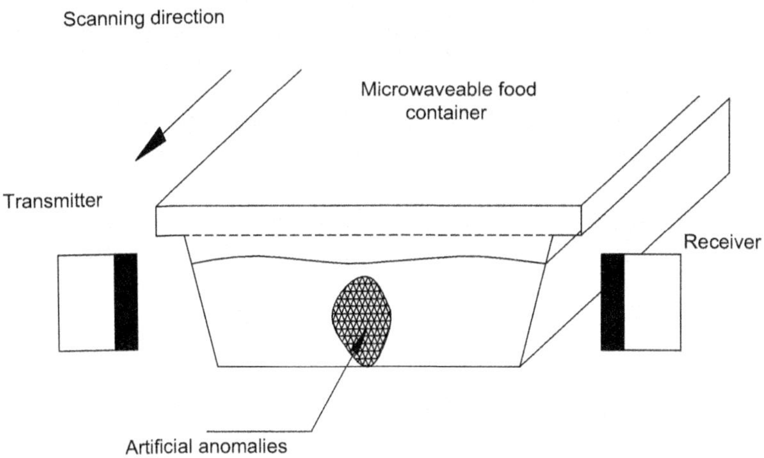

Fig. 10.24 Experiment arrangement to detect the content in a microwaveable polymer food container

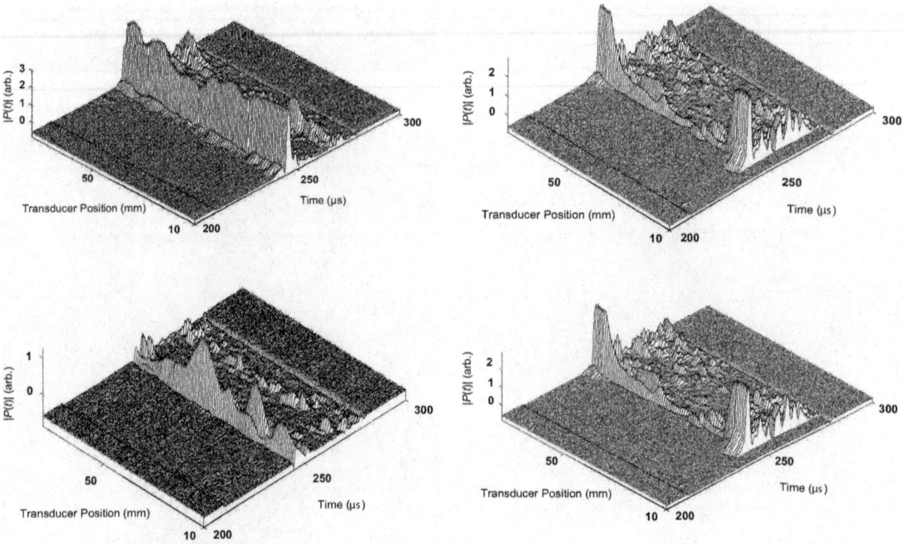

Fig. 10.25 Detected signal through a microwaveable food container, randomly filled with solid starch of **a** 10 %, **b** 20 %, **c** 50 % and **d** 70 % by volume

performed across the cross-section of the container. The transducers were separated by 245 mm. The centre point of the transducers was placed at 25 mm from the bottom of the container. In order to create artificial anomalies, corn flour was mixed into hot water to create starch. The starch that was not properly dissolved was firstly placed at the centre of the container and the results were recorded. The starchy areas were then randomly distributed and the scanned waveforms were recorded.

Figure 10.25 shows various results where the quantity of starch was randomly distributed in the containers. The density of distributed starch particles was gradually increased by 10, 20, 50 and 70 %. Figure 10.25 shows that the waveforms have very good SNR, and that the presence of the cube of starch could be clearly identified. In this figure, the first arrival peak was from the direct transmission of the signal while the second peak was suspected to be from multiple reflection of the signal.

References

Arthur JW (1996) Modern SAW-based pulse compression systems for radar applications, part 2: practical systems. J Elec Comm 8:57–78

Bhardwaj MC (2001) Non-contact air-coupled ultrasonic analysis: introduction to transducers, analytical system, and applications. NDT.net—April, vol 6, No. 04

Blitz J (1967) Fundamentals of Ultrasonics, Butterworths, London

Boehmer AM (1967) Binary pulse compression codes. IEEE Trans Info Theory 13:156–167

Bushong SC, Archer BR (1991) Diagnostic ultrasound: physics, biology and instrumentation. St Louis, Mosby Year Book, pp 1–4

Carr H, Wykes C (1993) Diagnostic measurements in capacitance transducers. Utrasonics 31:13–20
Castaings M, Cawley P, Farlow R, Hayward G (1998) Single sided inspection of composite materials using air coupled ultrasound. J Nondestruct Eval 17(1):37–45
Cho B-K, Irudayaraj JMK (2006) Foreign object and internal disorder detection in food materials using noncontact ultrasound imaging. J Food Sci 68(3):967–974
Dixon S, Edwards C, Palmer SB (1994) The analysis of adhesive bonds using electromagnetic acoustic transducers. Ultrasonics 32:425–430
Eccardt PC, Niederer K, Fischer B (1997) Micromachined transducers for ultrasound applications. Proc IEEE Ultra Symp 2:1609–1618
Edmonds J, Hickman G (2000) Damage detection and identification in composite aircraft components. Proc IEEE Aero Conf 6:263–269
Elias CM (1980) An ultrasonic pseudorandom signal-correlation system. IEEE Trans Sonics Ultra SU-27:1–7
Ermolov V, Stor-Pellinen J, Luukkala M (1996) Analog pulse compression system for real time ultrasonic nondestructive testing. Ultrasonics 34:655–660
Felhauer T (1992) New class of polyphase pulse compression code with unique characteristics. Elec Lett 28:769−771
Folkestad T, Mylvaganam KS (1993) Chirp excitation of ultrasonic probes and algorithm for filtering transit times in high-rangeability gas flow metering. IEEE Trans Ultras Ferr Freq Contr 40:193–215
Frost HM (1979) Electromagnetic-ultrasound transducers: principles, practice and applications. Academic Press, New York (Phys Acous XIV (ed Mason WP, Thurstion RN))
Gan TH (2002) New approaches to ultrasonic imaging using air-coupled and contact techniques. PhD Thesis
Gan TH, Hutchins DA, Billson DR, Schindel DW (2001a) The use of broadband acoustic transducers and pulse compression techniques for air-coupled ultrasonic imaging. Ultrasonics 39:181–194
Gan TH, Hutchins DA, Carpenter PW, Billson DR (2001b) Simultaneous reconstruction of flow and temperature cross-sections in gas jets using air-coupled ultrasonic tomography. Proc IEEE Ultrasonics Symp 1:623–626
Gan TH, Hutchins DA, Billson DR (2002) Preliminary studies of a novel air-coupled ultrasonic inspection system for food containers. J Food Eng 53(4):315–323
Gan TH, Hutchins DA, Billson DR, Schindel DW (2003) High resolution air-coupled imaging of thin materials. IEEE Trans Ultras Ferr Freq Contr 50:1516–1524
Gan TH, Pallav P, Hutchins DA (2006) Non-contact ultrasonic quality measurements of food products. J Food Eng 77:239–247
Garcia A, Tanarro A (1998) Ultrasonic inspection using array transducers. Insight 40:841–844
Golay MJE (1961) Complementary series. IEEE Trans Info Theory 7:82–87
Grandia WA, Fortunko CM (1995) NDE applications of air-coupled ultrasonic transducers. IEEE Ultrasonics Symp 1:697–709
Hayward G, Gachagan A (1996) An evaluation of 1–3 connectivity composite transducers for air-coupled ultrasonic applications. J Acoust Soc Am 99:2148–2157
Hietanen J, Mattila P, Storpellinen J, Tsuzuki F, Vaataja H, Sasaki K, Luukala M (1993) Factors affecting the sensitivity of electrostatic ultrasonic transducers. Meas Sci Technol 4:1138–1142
Ho KS, Gan TH, Billson DR, Hutchins DA (2005) Application of pulse compression signal processing techniques to Electromagnetic Acoustic Transducers for non-contact thickness measurements and imaging. Review of Scientific Instruments 76, paper 054902-1-8
Ho KS, Billson DR, Hutchins DA (2007) Inspection of drinks cans using non-contact electromagnetic acoustic transducers. J Food Eng 80:431–444
Hu JK, Zhang QL, Hutchins DA (1988) Directional characteristics of electromagnetic acoustic transducers. Ultrasonics 26:5–13
Hurrell A, Duck F (2000) A two-dimensional hydrophone array using piezoelectric PVDF. IEEE Trans Ultras Ferr Freq Contr 47:1345–1353

Idris A, Edwards C, Palmer SB (1994) Acoustic wave measurements at elevated temperature using a pulsed laser generator and an electromagnetic acoustic transducer detector. Nondest Test Eval 11:195–213

Izuka Y (1998) High signal to noise ratio ultrasonic testing system using chirp pulse compression. Insight 40:282–285

Jansen DP, Hutchins DA, Mottram JT (1994) Lamb wave tomography of advanced composite laminates containing damage. Ultrasonics 32:83–89

Jia Z-C, Wang Z-G, Mottram X-Q (2000) A conclusion on bi-weighted Barker codes and tri-weighted Barker codes. J Hebei Uni Technol 29:13–15

Kocis S, Figura Z (1996) Ultrasonic measurements and technologies. Chapman & Hall, London

Ladabaum I, Jin XC, Soh HT, Atalar A, Khuri-Yakub BT (1998) Surface micromachined capacitive ultrasonic transducers. IEEE Trans Ultras Ferr Freq Contr 45:678–690

Lee SS, Ahn BY (1992) EMAT application at high temperature. Nondest Test Eval 7:253–261

Lee BB, Furgason ES (1982) An evaluation of ultrasound NDE correlation flaw detection systems. IEEE Trans Sonics Ultra 29:359–369

Manthey W, Kroemer N, Mágori V (1992) Ultrasonic transducers and transducer arrays for applications in air. Meas Sci Technol 3:249–261

Maxfield BW, Kuramoto A, Hulbert JK (1987) Evaluating EMAT designs for selected applications. Mat Eval 45:1166 − 118

Meyer S, Hindle SA, Sandoz J-P, Gan TH, Hutchins DA (2006) Non-contact evaluation of milk-based products using air-coupled ultrasound. Meas Sci Technol 17:1838–1846

Millet RE (1970) A matched filter pulse-compression system using a nonlinear FM waveform. IEEE Trans N Aero Elec Sys 6:73–78

Möckl T, Magori V, Eccardt C (1990) Sandwich layer transducer- a versatile design for ultrasonic transducers operating in air. Sens Actuator A 21–23:687–692

Murayama R (1996) Driving mechanism on magnetostrictive type electromagnetic acoustic transducer for symmetrical vertical-mode Lamb wave and for shear horizontal-mode plate wave. Ultrasonics 34:729–736

Pallav P, Hutchins DA, Gan TH (2007) Air-coupled ultrasonic evaluation of food materials. International Congress on Ultrasound (ICU2007), Vienna, Austria

Pallav P, Hutchins DA, Gan TH (2008) Air-coupled ultrasonic evaluation of food materials. Ultrasonics 49(2):244–253

Pedersen PC, Tretiak O, Ping H (1982) Impedance-matching properties of an inhomogeneous matching layer with continuously changing acoustic impedance. J Acoust Soc Am 72:327–336

Pizarro L, Certon D, Lethiecq M, Hosten B (1999) Airborne ultrasonic electrostatic transducers with conductive grooved backplates: tailoring their centre frequency, sensitivity and bandwidth. Ultrasonics 37:493–504

Rafiq M, Wykes C (1991) The performance of capacitive ultrasonic transducers using v-grooved backplates. Meas Sci Technol 2:168–174

Rao NAHK (1994) Investigation of a pulse compression technique for medical ultrasound: a simulation study. Med Biol Eng Comp 32:181–188

Rao NAHK, Aubry M (1994) Evaluation of a pulse coding technique for spatial structure characterisation. IEEE Trans Ultras Ferr Freq Contr 41:660–663

Rao NAHK, Mehra S, Bridges J, Venkatraman S (1995) Experimental point spread function of FM pulse imaging scheme. Ultra Imag 17:114–141

Rose JL (2004) Ultrasonic waves in solid media. Cambridge University Press, Cambridge

Schindel DW (1995) Progress in non-contact acoustic methods. PhD thesis, Queen's University, Canada

Schindel DW, Hutchins DA (1995) Through-thickness characterization of solids by wideband air-coupled ultrasound. Ultrasonics 33(1):11–17

Schindel DW, Hutchins DA, Zou L, Sayer M (1995) The design and characterization of micromachined air-coupled capacitive transducers. IEEE Trans Ultras Ferr Freq Contr 42:42–50

Török I, Seller R (1998) Pulse compression in search radar. Periodica Polytech Ser El Eng 42:391–408

Tucker B, Diaz A, Eckenrode B (2006) Advanced ultrasonic measurement methodology for non-invasive interrogation and identification of fluids in sealed containers. Proceedings of SPIE vol 6178

Venkatraman S, Rao NAHK (1996) Combining pulse compression and adaptive drive signal design to inverse filter the transducer system response and improve resolution in medical ultrasound. Med Biol Eng Comp 34:318–320

Vun RY, Eischeid T, Bhardwaj MC (2006) Quantitative non-contact ultrasound testing and analysis of materials for process and quality control. 9th European conference on NDT, Berlin

Xiang D, Hsu NN, Blessing GV (1998) A simplified ultrasonic immersion technique for materials evaluation Mat Eval 56:854–859

Chapter 11
Emerging PAT Technologies

Colm P. O'Donnell and P.J. Cullen

11.1 Introduction

The past decade has witnessed significant advancements in process control technologies and strategies. New and adapted process analytical technologies (PAT) are continually being developed, many of which show promise for food process applications. There is also significant potential for technology transfer of proven technologies from complementary process industries including the pharmaceutical and chemical sectors to the food industry. Technology platforms will continue to fuse, as witnessed with spectroscopy and imaging to form hyper-/multispectral imaging systems. This chapter introduces and assesses the potential of emerging PAT for the food industry.

11.1.1 Industrial Process Tomography

The number of applications to which tomographic methods such as electrical resistance tomography (ERT) and optical coherence tomography (OCT) are applied is steadily increasing. Significant improvements in both spatial and temporal measurement limits have recently been reported.

ERT, one of the most common modalities in tomography, is a novel high-speed and relatively low-cost method of process imaging with the ability to perform non-invasive remote internal inspection through volume scanning. ERT measures the distribution of electrical resistance (or conductivity) in a two dimensional (2D) cross-sectional plane of a volume, be it in a pipe or vessel. Measurements can be

C. P. O'Donnell (✉)
School of Biosystems Engineering, University College Dublin, Dublin 4, Ireland
e-mail: colm.odonnell@ucd.ie

P.J. Cullen
School of Food Science and Environmental Health,
Dublin Institute of Technology, Dublin 1, Ireland
e-mail: patrick.j.cullen@dit.ie

C. P. O'Donnell et al. (eds.), *Process Analytical Technology for the Food Industry*,
Food Engineering Series, DOI 10.1007/978-1-4939-0311-5_11,

taken rapidly allowing real-time viewing of data. The information is shown on a tomogram which maps the degree of conductivity. Conductivity measurements are useful as the different phases of a multiphase process have very different values.

ERT has many advantages such as high speed, low cost, no radiation hazard, and non-intrusiveness (Sharifi and Young 2012b). It has the potential to enable for both qualitative analysis of flow by providing three dimensional (3D) conductivity images and quantitative analysis by providing the data required for measurement of some flow parameters such as velocity distribution and flow regime identification. In principle, ERT can be used to investigate and monitor any process where the main continuous phase is at least slightly conducting and the other phases and components have differing values of conductivity. ERT is particularly useful in observing processes such as mixing, flow and separation. It is also useful for monitoring reactions where the reactants or products have different conductivities, or where reactions change overall conductivities such as crystallisation.

Sharifi and Young (2012a) demonstrated that ERT, when accompanied by temperature measurements, has the ability to multidimensionally monitor milk concentration and fat content with high accuracy. The authors concluded that the dynamic, spatially distributed information provided by ERT could be used to facilitate enhanced process control in the milk industry and may be part of the solution to the lack of appropriate sensors to monitor dairy concentration and composition. ERT is particularly suited to monitoring homogeneity of milk, fat content of raw milk, milk standardisation and total solids content of milk concentrate. Further improvement in the resolution of commercially available systems will facilitate adoption of this technique in the food industry.

OCT is a novel optical imaging methodology that uses low-power radiation in the near infrared region of the spectrum to rapidly and non-invasively produce high-resolution images of turbid biological tissue (Ford and Tatam 2013). OCT uses low-coherence interferometry to provide structural images, with a spatial resolution of a few micrometres, to a depth of 1–2 mm below the surface. Advantages of OCT include non-contact, micrometers resolution and high signal acquisition rate up to 300 kHz (Bouma and Tearney 2002). OCT has developed rapidly over the past two decades into a valuable tool mostly for medical applications. However, more recently, the range of OCT applications reported have extended to non-destructive testing (NDT) of materials including the investigation of OCT as a PAT tool for food processing applications.

Leitner et al. (2011) employed spectral-domain and time-domain OCT to investigate food applications. Spectral-domain OCT has advantages in terms of imaging speed and sensitivity, enabling video rate imaging and in-line applications, while time-domain OCT permits the application of dynamic focusing and shows a constant sensitivity over the whole depth range. Leitner et al. (2011) demonstrated the use of OCT for the analysis and the control of wax layer thickness on Braeburn apples. The quality and thickness of the natural wax layer is one important parameter throughout the storage and shelf life of apples. The wax layer, lenticel structures and subsurface structures were clearly visible in the images obtained. Leitner et al. (2011) also reported on the capability of OCT imaging to visualise the structure and effectiveness of coatings during rehydration processes. OCT is well suited to meet future demands of in-process monitoring of coating in food processing applications.

11.2 Dynamic Light-Scattering Technology

Light-scattering techniques have been employed since the 1960s to investigate a wide range of particles. Traditional integrated light scattering and subsequently dynamic light-scattering (DLS) techniques have been employed to study many different types of particles including the study of dynamics and structures of food colloids (Alexander and Dalgleish 2006). In DLS analysis, monochromatic polarised laser light is passed through a dilute single-scattering colloid solution. Information on the size and size distribution of the particles present can be obtained by monitoring the intensity of light at the detector which depends on the interference pattern created by the scattered light from all the particles in the scattering volume. Advantages of the technique include its non-invasive mode of measurement, ease of use, short processing times and relatively low cost. However, a major limitation of DLS is that it is only suitable for use with very diluted systems to avoid multiple scattering effects. Thus, the reported use of DLS in realistic food applications is limited (Alexander and Dalgleish 2006).

DLS has been extended to study optically thick media which exhibit a very high degree of multiple scattering using a technique called diffusing wave spectroscopy (DWS) which was first reported in 1987. DWS exploits the diffusive nature of the transport of light in strongly scattering media to relate the temporal fluctuations of multiply scattered light to the motion of the scatterers (Pine et al. 1990). Advantages of DWS include the capability to study particle motion in concentrated fluids such as colloids, microemulsions and other systems which are characterised by strong multiple scattering (Pine et al. 1990).

There are two different measurement modes in DWS applications, namely the backscattering mode where the laser and detector are placed on the same side of the sample and the transmission mode where light crossing the sample volume is analysed. A backscattering mode DWS system is better suited to process applications (Alexander and Dalgleish 2006).

DWS provides information on average particle size in a sample but not on particle-size distribution (PSD) since all particles within a scattering sample may potentially contribute to the intensity of a scattered path. Mason and Weitz (1995) demonstrated that for concentrated hard-sphere suspensions, the mean-square displacement of the scatters as measured by DWS can be related to the complex viscoelastic function G^* of which G' and G'' are the real and imaginary parts, respectively. A particular advantage of the technique is its ability to measure storage and loss modulus [G', G''] over a wide frequency range.

Since many food systems are suspensions of colloidal particles, DWS is particularly well suited towards monitoring and controlling stability and coagulation processes in the food industry and particularly in the dairy industry. Over the past decade, commercial laboratory and process instruments using DWS systems have been developed. The RheoLight measuring system developed by Optel in cooperation with the NIZO food research institute in the Netherlands has been installed in several cheese factories to monitor milk coagulation and help improve cheese yield and process control. The RheoLight system is also well suited to monitoring product formulation, e.g. mayonnaise manufacture.

In a recent study, Niederquell et al. (2012) demonstrated the potential of DWS to monitor the rheology of self-emulsifying drug delivery systems. The dynamic viscosities obtained from DWS were in accordance with data from capillary viscometry. Also, obtained values of storage and loss modulus were successfully correlated with the weight variability of capsules that were filled on a machine. In conclusion, the DWS technique enabled rheological analysis of self-emulsifying systems over a broad frequency range. It was concluded that DWS has a high potential application in a quality-by-design framework of formulation development and production.

While to date application of DWS in the food industry is very limited, its use is likely to rise in the years ahead with further advances in DWS signal interpretation and more robust equipment design.

11.3 Dielectric and Microwave Sensing

In terms of its use as a PAT tool in food systems, microwave and dielectric sensing have primarily been applied to determination of moisture content, and to a lesser degree salt content. However, other applications have also been investigated such as determination of authenticity, maturity, and other quality parameters. Damez et al. (2008) investigated the potential of dielectric properties to assess beef ageing to facilitate optimisation of storage conditions. They found that parameters such as contact impedance had potential to be used to identify the state of maturations as it was correlated with meat fibre strength. The potential of microwave properties to assess the internal quality of fruit has also been demonstrated.

More recently, Cash (2012) outlined the robust potential of guided microwave spectrometry (GMS) for production measurements in both dairy and meat processing using in-line analysers. Unlike infrared techniques which are limited by sample depth into the process line and are prone to coating effects, GMS systems are based on intrinsic measurement through the sample material cross section.

GMS was validated by Cash (2012) for the calibration and measurement of fat and total solids content in a continuous process stream of fluid milk used in cheese making. For fat content and total solids, the standard errors of prediction with respect to the offline reference method were 0.05 and 0.07%. In ground meat manufacturing, GMS aided in the optimisation of fat to protein ratios and more consistent finished product quality. For fat content, the standard error of prediction of the validation batches with respect to the offline reference method was 0.63 %. GMS was also shown not to be affected by fat build up, inherent in optical measurements that can adversely influence other measurement devices. In future years, GMS is likely to be increasingly exploited in the dairy and meat industries as a robust process analytical tool.

11.4 X-Rays

Since Röntgen first discovered in 1895 that X-rays can identify inner structures, X-ray technology has been continually developed for both medical imaging and NDT applications (Mery et al. 2011). In addition to applications in security, welding quality control, medicine and automotive parts inspection, NDT testing is currently commercially employed in many food product inspection and analysis applications including the poultry, grain, dairy, fruit and fish processing industries.

X-ray inspection has distinct advantages over most other inspection technologies as it enables non-destructive imaging of interior features of a sample for detection of contaminants or compliance with quality parameters. However, X-ray inspection is still a relatively expensive technology which requires high voltage power and radiation shielding (Haff and Toyofuku 2008).

In the past decade, X-ray inspection has been demonstrated through reported research studies to have potential application for bone detection in poultry and fish, identification of insect infection in citrus, detection of codling moth larvae in apples and water content distribution and internal structure in fruit (Mery et al. 2011).

There are three main components in an X-ray inspection system, namely X-ray generator, detector and computer. In most X-ray systems employed in food and pharmaceutical applications, X-rays are funnelled through a collimator from the exit window of an X-ray generator (Mettler-Toledo 2009). A collimator is a mechanical device which narrows down the X-ray stream. In most applications, the X-ray generator is positioned above the processing line and the X-ray beam passes downwards via the collimator through the product and the belt before striking the detector beneath. The beam is about 2 mm wide in the direction of conveyor travel and triangular in shape (Mettler-Toledo 2009). The X-ray detection surface is made from a scintillating material which converts X-rays into visible light. Photosensitive diodes are positioned underneath the scintillator strip and are optically connected to the scintillator and convert the level of visible light into an electrical signal which is sent to the inspection system computer (Mettler-Toledo 2009). The computer compiles a grey-scale image of the inspected product, which is then captured and analysed for acceptance or rejection as appropriate using software tools. The amount of X-ray energy absorbed during the beam's passage through a product is a function of the product thickness, product density

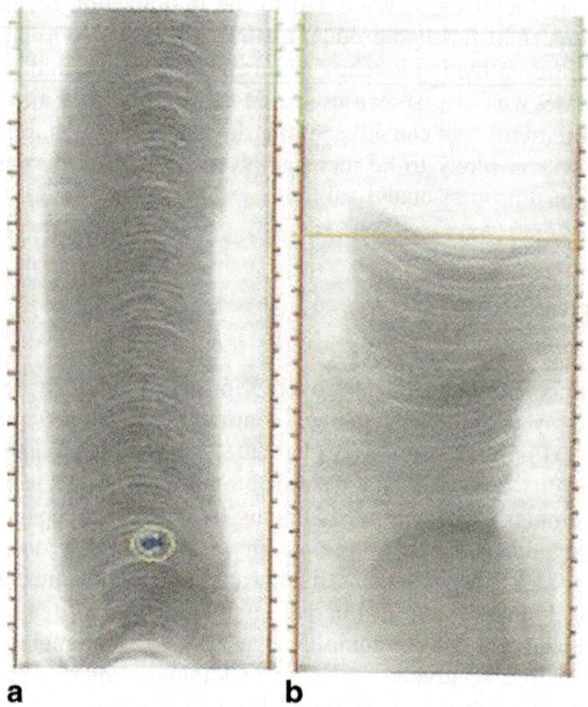

Fig. 11.1 X-ray image of stack of potato chips in a recycled composite can with **a** flavour lumps and **b** collapsed stack of chips. (Mettler-Toledo 2009)

and atomic mass number (Mettler-Toledo 2009). Measurement of the differences in absorption between product and contaminant is the basis of X-ray inspection. Three important parameters which must be considered to improve an X-ray image quality are resolution, signal-to-noise ratio and contrast level. Image quality also depends on the interactions between the different components in the of the X-ray systems including X-ray source, processing conveyor, detector and image analysis system (Haff and Toyofuku 2008).

Modern X-ray systems are capable of simultaneously measuring many food product quality metrics as well as detecting certain contaminants. Quality control checks that can be carried out include measurement of product mass, counting of individual components, checking of fill level, identification of faulty products and inspection of seal quality. Figure 11.1 shows X-ray images of a stack of potato chips in a recycled composite can with a metal base and a foil lid (Mettler-Toledo 2009). Despite the packaging, the machine can still detect a stainless steel contaminant and flavour lumps (hard agglomerates of powder and fat) in the package (Fig. 11.1a). In Fig. 11.1b where the stack of chips has collapsed, X-ray analysis can detect that the fill level has dropped below an acceptable standard.

Despite recent significant research advances, detection of organic contaminants in food products is a particular challenge in quality assurance of food products, as conventional X-ray systems are unsuitable for the their detection due to the low

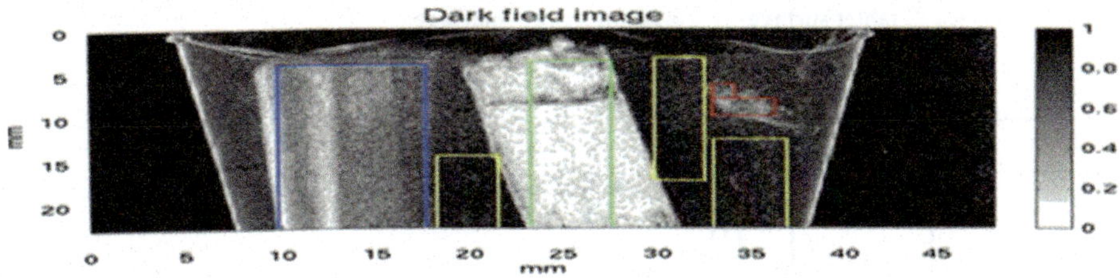

Fig. 11.2 Sour cream: X-ray images of sour cream with three foreign bodies; eight layers of paper (*left*), a cigarette butt (*middle*) and a fly (*right*). (Nielsen et al. 2013)

contrast observed with X-ray absorption. While inorganic contaminants give rise to a high-contrast image in X-ray detectors, organic contaminants are particularly difficult to detect using automated systems. While the absorption contrast can be increased through the application of X-ray energies in the 10–25-keV range to benefit from the higher attenuation, it results in a lower scanning speed or requires high X-ray power levels (Nielsen et al. 2013). Recently, emerging technologies based on techniques employing new X-ray modalities such as phase contrast and dark-field imaging which have higher sensitivity to organic materials have been developed. In a recent study, Nielsen et al. (2013) demonstrated the potential of dark field X-ray radiography with a grating-based interferometer for the identification of organic foreign bodies (Fig. 11.2). However, further development is required before this technology can be exploited by industry for online process inspection applications.

While recent developments in X-ray technology have enabled detection of defects that were not previously possible, a key remaining challenge is to overcome the loss in the signal-to-noise ratio in X-ray images as the speed of the systems increase. Further developments in this field will open new possibilities for automated inspection and quality monitoring in a wider range of food processing applications.

11.5 Terahertz Imaging

The terahertz (THz) region of the electromagnetic spectrum spans between 100 GHz and 30 THz. Recently, there is renewed interest in THz time-domain spectroscopy and imaging as a potential quality control and process monitoring tool due to technological developments in source and detector components.

THz systems are designed to operate in transmission or reflection modes. THz radiation has sufficient energy (1–10 meV) to promote molecular rotations and vibrations; and can thus be used for their identification (Gowen et al. 2012). THz radiation is non-ionising and sensitive to polar molecules such as water but interacts very weakly with materials composed of non-polar molecules, e.g. plastics. Thus, THz systems are particularly suited for analysis of packaged food products. THz

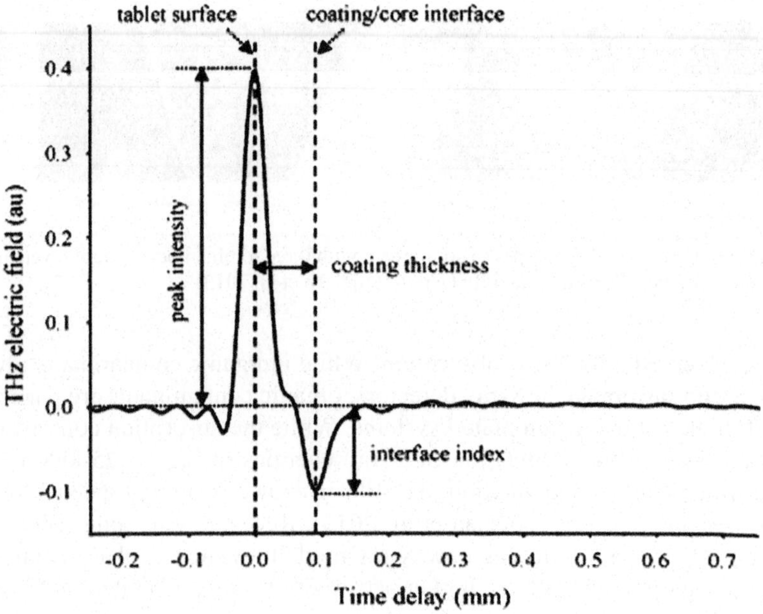

Fig. 11.3 Typical terahertz waveform. Reflection from the tablet surface (i.e. air/coating interface) and from the coating/core interface are indicated by *dashed lines; drawn through arrows* indicate how TPI (peak intensity, interface index, layer thickness) are related to the waveform. (Maurer and Leuenberger 2009)

imaging also enables direct measurement of sample thickness, refractive index and absorption coefficient. A typical THz waveform is shown in Fig. 11.3.

There are many studies of THz imaging applied to pharmaceutical quality control including tablet coating characterisation, evaluation of dissolution properties and characterisation of polymorphisms (Shen 2011).

From a food safety perspective, compounds such as pesticides or antibiotics exhibit characteristic absorption peaks at specific THz frequencies. Many recent studies have been published on THz applications for antibiotic and pesticide detection in foods. Chemometric analysis of the absorption coefficient or refractive index is frequently applied for quantitative analysis of samples (Maurer and Leuenberger 2009).

THz techniques are well suited to moisture content measurement in dried foods. A linear relationship between moisture content and THz time-domain data (peak-to-peak amplitude) or frequency-domain data in the range 0.2–0.6 THz was reported by Parasoglou et al. (2009) who examined the moisture content of confectionary wafers using THz TDS. Moreover, the possibility of characterising the molecular nature of water in low moisture content products (e.g. free vs. bound water) could be used for the determination of shelf life (Gowen et al. 2012).

Although the potential of THz technologies has been demonstrated to be suited to the measurement of quality parameters of selected food and pharmaceutical products, much research and development is required before these technologies are suitable for mainstream adoption by industry. The development of THz spectral library databases is something that is still in its infancy (Gowen et al. 2012). Other significant areas to be addressed include the high cost of THz technologies relative to other sensing technologies, slow speed of THz image acquisition, further research on chemometric modelling techniques for improved analysis and understanding of THz spectra.

11.6 Particle Imaging PATs

PSD of suspensions and emulsions is an important parameter influencing the structural and sensory characteristics of foods. PDS is also an important factor of powder performance, governing flow and reconstitution behaviour. Crystal size distribution (CSD) is also found to influence the quality of many foods. Quantifying particle size is a difficult task due to the nature of particles (vast numbers, small size ranges, nonspherical shapes, cohesive and agglomerated particles, moving nature in liquids or air, particle growth or dissolution). Monitoring PSD has been an identified critical control parameter for the small-molecule pharmaceutical and fine chemical industries for decades, with considerable innovation in in-line particle monitoring in recent times. There are a number of emerging approaches that can provide information on particle size and morphology, which could be employed as PAT tools to control food process operations. Outlined below are a number of emerging particle monitoring approaches, which offer in-line monitoring of particles carried in a liquid suspension or air.

11.6.1 Focused-Beam Reflectance Measurement

This approach focuses a laser beam through a window and collects the light that is scattered back to the probe to provide, in real time, a chord length distribution (CLD). The focused-beam reflectance measurement (FBRM) probe emits its laser beam rotating at a high velocity (9000 rpm), which propagates into the suspension (Amamou et al. 2010), overcoming problems due to particle motion. As the beam intersects suspended particles it is backscattered, this backscatter is collected by the FBRM optics and is converted into chord length as the product of the measured crossing time and the beam velocity (Worlitschek and Mazzotti 2003). The duration of each reaction is multiplied by the velocity of the scanning beam, resulting in a chord length. The measurement range is 0.8 ± 1000 mm, with the distribution sorted by chord length into a 38-channel distribution. Typically, many thousands of chord lengths are measured per second, with the numbers of counts dependent on the concentration of solids present in the suspension (Barrett and Glennon 1999). Both

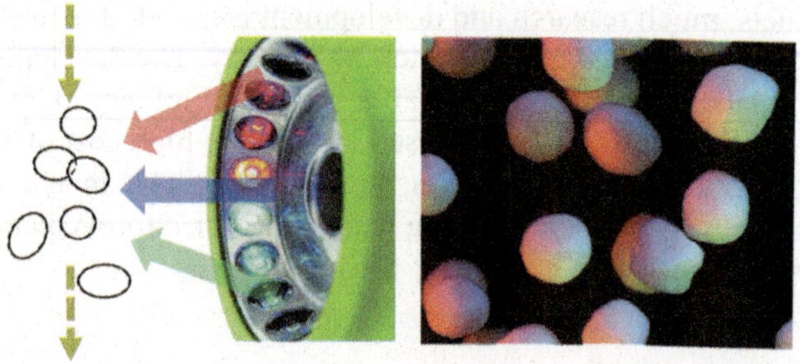

Fig. 11.4 3D particle imaging in real time using stereoscopic technique. (Reproduced by permission of Innopharmalabs)

modelling and empirical approaches have been used to establish the correlation between the chord length measured with FBRM and particle size. Due to its principles of measurement, the FBRM probe can be readily inserted online or into a reaction vessel without the need of installing a pre-dilution side stream as is required for other online previous particle sizing tools. A variety of applications using FBRM were developed and reported in the literature, including crystallization, flocculation, slurries, polymorphic transitions, particle disruption and solubility measurements (Yu and Erickson 2008) and references therein.

The most attractive advantage of this technique is its utility both for in situ measurement of high solid-concentration suspensions and for following rapid crystallization kinetics. However, the CLD is only an indication of the real population. This probe does not detect particle shape. In fact, CSD is more meaningful for the characterisation of a crystal product. Thus, many papers have discussed methods to recover the CSD value from a measured CLD. If the crystal shape is known, particularly when the crystals can be considered as spherical, this restoration is possible. Furthermore, if crystal shape does not change during freezing, the mean chord length can be considered proportional to the mean diameter.

11.6.2 Stereoscopic Particle Imaging

A novel particle imager (Eyecon^{TM}, Innopharmalabs) has been commercialised, which can determine both PSD and shape factors of powder streams in real time (Fig. 11.4). The system's approach uses structural lighting to obtain particle tomography information from images in either stationary or moving environments. The technology is based on high-speed 3D machine vision, enabling both particle size and shape quantification in the range of 50–3000 microns. A continuous image sequence of the particles is captured using illumination pulses with a length of 1 μs, freezing the movement of particles with particle velocities of up to several meters per second. The illumination is arranged according to the principle of photometric stereo for capturing the 3D features of the particles in addition to a regular 2D image. The particle size is estimated from the images using both the 2D and 3D information, applying novel image analysis methods and direct geometrical measurement. A series of algorithms facilitates particle detection, ellipse model fitting and size estimation. As the approach is based on direct measurement instead of indirect, such as laser diffraction, there is no need for material based calibration. In addition, the method it is non-contact and can be applied, e.g. through a view glass on a granulator or dryer, without physical modification of the process equipment. PAT applications include granulation, drying, milling and spheroidisation processes. The novelty of the system includes high-speed measurements, no additional calibrations, ability to deal with complex particle morphology and no sampling requirements.

11.7 Electronic Tongue and Nose

Taste and smell testing are critical quality control tests within the food industry (Herbert et al. 2000). However, simultaneous determination of different compounds has traditionally been challenging to obtain in real time, and normally, classification tools are not always well developed (Gutes et al. 2007). The electronic tongue (e-tongue) and nose (e-nose) are emerging bioinspired analytical systems that provide fast determination of several compounds or sample classification. These technologies are typically an array of sensors combined with chemometric tools to characterise complex samples, and can be considered as analytical instruments that artificially reproduce the taste or smell sensation. Arrays of gas sensors are termed 'electronic noses' (Peris and Escuder-Gilabert 2009) while arrays of liquid sensors are referred to as 'electronic tongues' (Stetter and Penrose 2002). They have been employed for the recognition, classification and quantitative determination of multiple component concentrations. The analytical strategy is based on the measurement of a great number of samples covering the expected variability and afterwards, a visualization tool, such as principal component analysis (PCA), is used for samples grouping, after which pattern recognition variants are used for classification (Gutes et al. 2007).

11.7.1 E-Nose

Despite the importance of aroma as an indicator of quality and product conformity, analytical monitoring by the food industry has traditionally been limited. This is principally due to the lack of reliable odour assessing instruments and the practical impossibility of employing sensory panels to the continuous monitoring of aroma (Ampuero and Bosset 2003). Typically, the instruments consist of headspace sampling, sensor array and pattern recognition modules, to generate a signal pattern to characterise odours. As they are easy to build, cost-effective and provide a short time of analysis, e-noses are becoming increasingly popular as objective automated non-destructive techniques to characterise food flavours (Peris and Escuder-Gilabert 2009). Compared to sensory panels, the main advantage of the e-nose is that once calibrated they can perform odour assessment on a continuous basis at a minimal cost. Furthermore, once established, this technique does not require trained personnel like a sensory panel does, is not subject to individual breakdown or variation of sensitivity (Sarig 2000), is not overloaded under normal operation and takes comparatively very little time. Just like the human olfactory system, e-noses do not need to be specially designed to detect a particular volatile. In fact, they can learn new patterns and associate them with new odours via training and data storage functions as humans do. However, training of e-noses based on sensory panel classifications is required in order to obtain odour-meaningful classifications.

Although still developing, e-noses can potentially be applied to process control and monitoring, acceptance or rejection of raw material, intermediate and final products, assistance in the development of new products, as well as to the assessment of synergistic effects of individual odorants (Ampuero and Bosset 2003). Reported PAT applications are listed in Table 11.1. García et al. (2005) have made use of an e-nose to identify spoiled Iberian hams during the curing process, with a 100 % discrimination of two types of Iberian hams (spoiled and unspoiled) found. E-noses have been also applied to bioprocess monitoring where microbiological processes are involved in food production, i.e. to screen the aroma generation of lactic acid bacteria strains in the production of cheese and other fermented dairy products. Marilley et al. (2004) employed an e-nose to discriminate between seven different genotype strains of *Lactobacillus casei* isolated from Gruyère cheeses. Pani et al. (2008) used an e-nose to monitor changes in the aroma profile of tomato slices during air dehydration processes. The authors reported that the e-nose was able to characterise the process aromatic fingerprint, which could be used to parameterise the degradative events caused by dehydration. Bhattacharya et al. (2008) reported real-time smell monitoring of black tea during the fermentation process using an e-nose as well as prediction of the correct fermentation time. The approach showed promise to be used for the online prediction of optimum fermentation time by the industry.

11.7.2 E-Tongue

Reported applications of the e-tongue (Table 11.2) include freshness evaluation and shelf-life investigation, authenticity assessment, foodstuff recognition, quantitative analysis and process monitoring (Escuder-Gilabert and Peris 2010). These devices are designed to evaluate and compare tastes, imitating the human tongue. The compounds responsible for taste may be identified by the sensory cells of taste, which transforms the information into an electrical signal allowing the brain to identify the basic standards of taste, which are bitter, acid, salty, sweet and, finally, 'umami' (pleasant savoury taste). The e-tongue simulates this approach, where once immersed in a liquid, a pattern of signals that depends on the composition of the matrix solution

Table 11.1 Reported applications of e-noses in food process monitoring applications. (Peris and Escuder-Gilabert 2009)

Sample	Type of study	Sample handling system	Detection system	Data processing algorithm	Ref.
Wine-must	Discrimination between fermentation stages	SHS—pervaporation	A32S AromaScan: 32 CP	PCA	Pinheiro et al. (2002)
Iberian hams ('Montanera')	Spoiling during the curing process	SHS	16 Tin-oxide thin films	PCA, PNN	García et al. (2005)
Milk fermented with *Lactobacillus casei* strains used in Gruyère cheese	Discrimination between genotype strains	INDEXSmart	Nose MS	PCA	Esbensen et al. (2004)
Milk fermented with *Lactococcus lactis* strains	Discrimination between odour intensity scores	SHS	FOX 3000: 12 MOS	PCA	Gutiérrez-Méndez et al. (2008)
Australian red wines	Spoilage caused by Brettanomyces yeast	SHS	HP4440: MS	PCA, PLS, SLDA	Cynkar et al. (2007)
Australian red wines	Spoilage caused by Brettanomyces yeast	SPME (for MOS) SHS (for MS)	FOX 3000: 12 MOS HP4440: MS	PLS	Berna et al. (2008)
Tomato *cv.* Cencara	Dehydration processes of tomato slices	SHS	Air Sense: 10 MOS	PCA	Pani et al. (2008)
Mangoes (*Mangifera indica* L.)	Discrimination between harvest maturities within a ripening stage Discrimination between ripening stages within a maturity stage Discrimination between fruit varieties	SHS	FOX 4000: 18 MOS	DFA	Lebrun et al. (2008)
Black tea	Estimation of optimum fermentation time	SHS	8 MOS	TDNN, SOM	Bhattacharya et al. (2008)

PLS partial least squares, *SPME* solid phase microextraction, *MOS* metal oxide sensors, *DFA* deterministic finite automaton, *TDNN* time delay neural network

Table 11.2 Applications of e-tongues in food process monitoring. (Escuder-Gilabert and Peris 2010)

Sample	Type of study	Chemical sensors	Data processing algorithm	Ref.
Starting culture for light cheese production	Fermentation monitoring	Potentiometric sensors (30 chalcogenide glass and solvent polymeric working electrodes; Ag/AgCl reference electrode)	PLS	Esbensen et al. (2004)
	Discrimination of samples from fermentation batches			
Milk	Online monitoring of sources of raw milk	Voltammetric working electrodes (Au, Pt and Rh) embedded in a dental material	PCA	Winquist et al. (2005)
	Online monitoring of the cleaning process of the pasteurization unit	Voltammetric working electrodes (Au, Pt, Rh and stainless steel) embedded in PEEK™		
Red wine	Monitoring of the ageing process	Voltammetric sensors (7 polypirrole-based sensors, 4 CPEs based on phthalocyanines and 2 CPEs based on perylenes; Ag|AgCl/KCl saturated reference electrode; Pt wire counter electrode)	PCA, SIMCA	Parra et al. (2006)
	Discrimination of samples aged in oak barrels of different wood origin and toasting level			
Wine	Analysis of the effects of several treatments on wine phenolic compounds composition	Potentiometric sensors (14 plasticized PVC sensors, 11 chalcogenide glass sensors,1 glass pH electrode; Ag/AgCl reference electrode)	PCA, ANOVA, ASCA, PLS	Rudnitskaya et al. (2009)
	Prediction of chemical parameters			

PLS partial least squares, *PCA* principal component analysis, *CPE* carbon paste electrode, *SIMCA* soft independent modelling of class analogy, *ANOVA* analysis of variance, *ASCA* ANOVA-simultaneous component analysis is obtained. The signal is subsequently decomposed into qualitative and quantitative information by chemometric methods, with the useful information selected as final result (Dias et al. 2011). The sensor features of the e-tongue are different from those of the traditional chemical sensors, where instead of high selectivity for detecting substances, e-tongues have an ability to obtain global information about the solution, which is called the overall selectivity (Vlasov et al. 2002; Toko 2000; Legin et al. 2005). For the purpose of determining the flavour quality and intensity, the taste evaluation should not be based on specific substances but in all the matrix substances. Therefore, the main analytical properties of an e-tongue are the sensitivity and stability, being selectivity a less important characteristic (Dias et al. 2011).

With respect to process monitoring, Escuder-Gilabert and Peris (2010) reviewed e-tongue applications. Fermentation process monitoring of a starting culture for cheese production (Esbensen et al. 2004) allowed detection of abnormal operating conditions at an early stage. Moreover, the capability of the e-tongue to quantify organic acids (such as citric, lactic and orotic) in the fermentation media was demonstrated with average prediction errors in the 5–13 % range. The authors concluded that the e-tongue was a promising tool for fermentation process monitoring. Winquist et al. (2005) employed an e-tongue as a PAT tool within a dairy process line for direct in-line measurements of different sources of raw milk and to monitor the cleaning process of the pasteurization unit. Parra et al. (2006) designed an e-tongue to monitor the ageing of red wines and to discriminate wine samples aged in oak barrels of different characteristics. Multivariate inspection of voltammetric data showed the high capability of discrimination and classification of this e-tongue.

The main drawback of these systems is the large amount of previous measurements needed for the modelling, calibration or learning stage. However, flow analysis, either flow injection and sequential injection, offer advantages in the development of e-tongues (Gutes et al. 2007) and facilitation of in-line measurement.

11.8 Laser-Induced Fluorescence

Laser-induced fluorescence (LIF) is a spectroscopic method which utilizes the optical emission from molecules that have been excited to higher energy levels by absorption of electromagnetic radiation. It is primarily used for flow visualization, and non-destructive evaluation of materials. In the food industry, it has the potential to be employed as a PAT tool for quality determination of products as well as monitoring of various unit operations. Fluorescence-based measurements also have the advantage of offering high specificity and sensitivity and it is viable to miniaturize and automate the technology (Widengren 2010). The fluorescence of a molecule is the light emitted spontaneously due to transitions from excited singlet states (S1) to various vibrational levels of the electronic ground state ($S_{0,1} \rightarrow S_{0,v}$; Hof et al. 2005). The emitted light can be characterised by a number of parameters, for example, the fluorescence intensity at a given wavelength.

Planar laser-induced fluorescence (planer-LIF) imaging employs the fluorescence of an organic medium induced by a laser sheet combined with image analysis for flow visualization (Fagan et al. 2009). Arratia and Muzzio (2004) stated that it can be used to unveil flow patterns and structures that serve as the starting point to analyse fluid mixing in stirred tanks. The laser, which is usually a neodymium-doped yttrium aluminium garnet or Argon-ion laser, is used to form a thin sheet of light which excites a fluorescent species within a flow. The fluorescent species is typically a tracer compound such as rhodamine B (Fagan et al. 2009). The selection of a tracer is dependent upon its absorption wavelength being compatible with the laser excitation wavelength, a large separation between emission and excitation absorption spectra, and a high quantum efficiency to maximize signal strength (Crimaldi 2008). The emitted fluorescence is optically captured. However, planer-LIF is limited to optically clean systems with a constant refractive index (Wadley and Dawson 2005). Guillard et al. (2000a) developed advanced methods in planer-LIF image analysis in order to study the large-scale mixing structures obtained in a Rushton-turbine-agitated reactor. They stated that while a standard statistical approach allowed the observation of tracer dispersal, a dynamic structural approach, as detailed by (Guillard et al. 2000b), was required for deeper understanding of the 3D mixing process. Planer-LIF has been used to study fast mixing of two liquid streams in flow channels at millimetre size (Luo et al. 2007) while Wadley and Dawson (2005) used the technique to study mixing within static mixers for both turbulent and transitional flow regimes.

The use of naturally occurring fluorophores as indicators of product quality has also been examined. Fluorescence spectroscopy has been widely employed to characterise a range of food products including dairy (Andersen and Mortensen 2008; Christensen et al. 2003; Fagan et al. 2011), egg (Karoui et al. 2008, 2007) and meat (Allais et al. 2004; Frencia et al. 2003; Schneider et al. 2008; Wold et al. 2002). The majority of such studies utilize front-face fluorescence spectroscopy (FFFS) as it overcomes the limitations associated with right-angle fluorescence spectroscopic techniques which cannot be applied to thick substances due to the large absorbance and scattering of light (Genot et al. 1992). Herbert et al. (2000) used tryptophan and vitamin A fluorescence spectra in conjunction with chemometric techniques to discriminate between soft cheeses, resulting in classification rates greater than 90%. Fluorescence spectroscopy has also been utilized to monitor changes in milk powder during manufacture and storage (Liu and Metzger 2007). They found that it was possible to detect changes due to Maillard reaction, modification of the tryptophan environment and degradation of riboflavin. Herbert et al. (1999) monitored the evolution of tryptophan fluorescence resulting from milk coagulation in a 1×1 cm quartz cuvette. They stated that it was possible to detect stages of coagulation and differences between gels with varying rheological characteristics. Fagan et al. (2011) also demonstrated that there is a basis for the development of a syneresis control technology based on the utilization of naturally

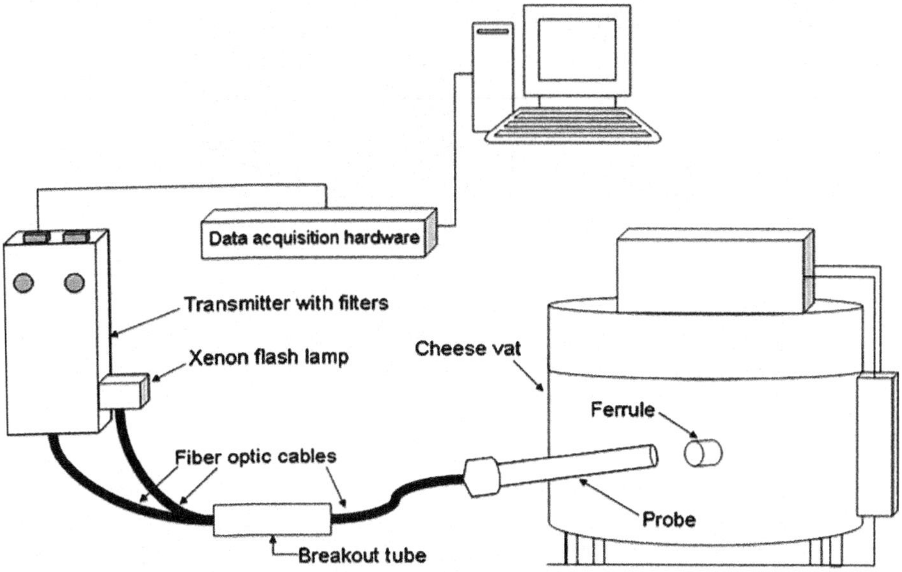

Fig. 11.5 Schematic of the fluorescent measuring equipment used for monitoring milk coagulation and curd syneresis. (Fagan et al. 2011)

present milk fluorophores. Using a FFFS probe installed in the wall of a laboratory scale cheese vat (Fig. 11.5), they were able to monitor changes in tryptophan, and riboflavin fluorescence during milk coagulation and syneresis.

Karoui et al. (2006) evaluated FFFS for its potential to rapidly monitor egg freshness. Utilizing tryptophan emission spectra of tryptophan, it was possible to classify samples with a correct classification rate of 54.3 %, while utilizing the fluorescence spectra due to Maillard reaction products the percentage of samples correctly classified was 91.4 %.

Fluorescence imaging has also been investigated for application in food safety applications. The potential of fluorescence imaging, for example, to detect diluted faecal matters from various parts of the digestive tract, including colon, caeca, small intestine and duodenum, on poultry carcasses has been studied (Cho et al. 2009). Cho et al. (2009) also stated that one of the challenges in using fluorescence imaging to inspect agricultural material is the low fluorescence yield as fluorescence can be masked by ambient light. Using the equipment they developed (Fig. 11.6), the authors found that faeces spots on the carcasses, without dilution and up to 1:5 dilutions, could be detected with 100 % accuracy regardless of faeces type. Detection accuracy for faecal matters diluted up to 1:10 was 96.6 % (Cho et al. 2009).

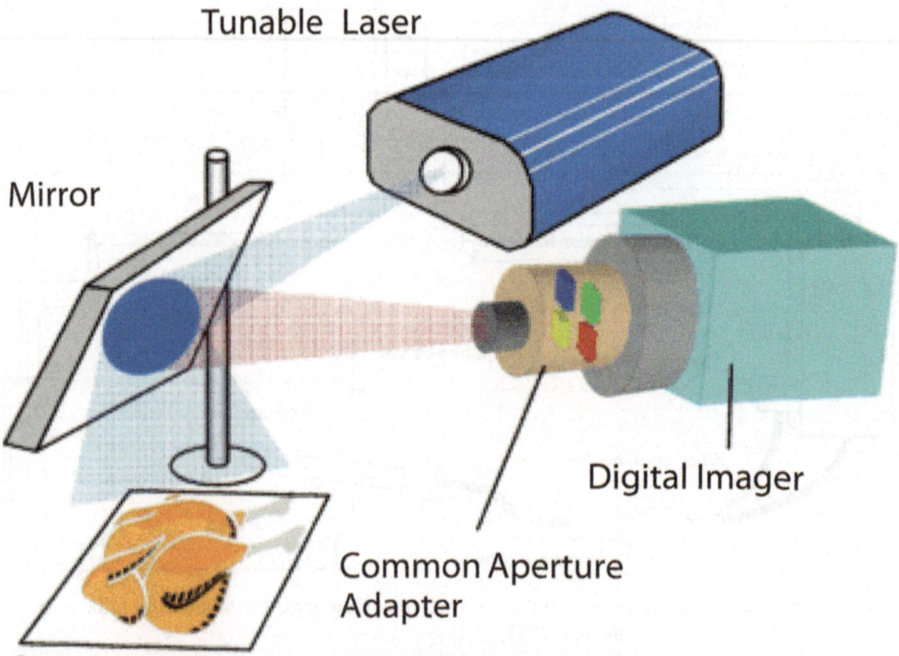

a

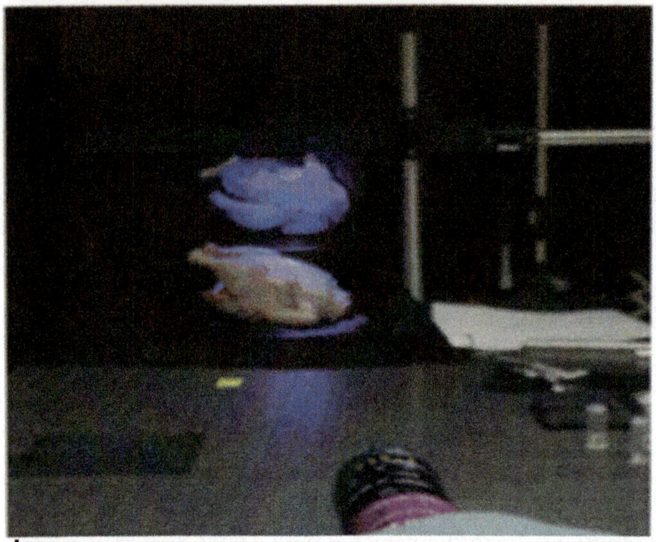

b

Fig. 11.6 A schematic (**a**) and photo (**b**) of the LIF imaging system and a photo (**c**) and 5×5 binned image at 22-ns gate-delay time (**d**) of contamination spots on poultry skin. (Adapted from Cho et al. 2009)

References

Alexander M, Dalgleish DG (2006) Dynamic light scattering techniques and their applications in food science. Food Biophys 1(1):2–13

Allais I, Viaud C, Pierre A, Dufour E (2004) A rapid method based on front-face fluorescence spectroscopy for the monitoring of the texture of meat emulsions and frankfurters. Meat Sci 67(2):219–229

Amamou AH, Benkhelifa H, Alvarez G, Flick D (2010) Study of crystal size evolution by focused-beam reflectance measurement during the freezing of sucrose/water solutions in a scraped-surface heat exchanger. Process Biochem 45(11):1821–1825

Ampuero S, Bosset JO (2003) The electronic nose applied to dairy products: a review. Sens Actuators B Chem 94(1):1–12

Andersen CM, Mortensen G (2008) Fluorescence spectroscopy: a rapid tool for analyzing dairy products. J Agric Food Chem 56(3):720–729

Arratia PE, Muzzio FJ (2004) Planar laser-induced fluorescence method for analysis of mixing in laminar flows. Ind Eng Chem Res 43(20):6557–6568

Barrett P, Glennon B (1999) In-line FBRM monitoring of particle size in dilute agitated suspensions. Part Part Syst Charact 16(5):207–211

Berna AZ, Trowell S, Cynkar W, Cozzolino D (2008) Comparison of metal oxide-based electronic nose end mess spectrometry-based electronic nose for the prediction of red wine spoilage. J Agric Food Chem 56(9):3238–3244

Bhattacharya N, Tudu B, Jana A, Ghosh D, Bandhopadhyaya R, Saha AB (2008) Illumination heating and physical raking for increasing sensitivity of electronic nose measurements with black tea. Sens Actuators B Chem 131(1):37–42

Bouma E, Tearney G (2002) Handbook of optical coherence tomography. Marcel Dekker, New York

Cash R (2012). Guided microwave spectrometry system- production measurements using inline analysers. In: IFPAC Annual Meeting, Jan 17–21, 2012 Baltimore, MD, USA

Cho B, Kim MS, Chao K, Lawrence K, Park B, Kim K (2009) Detection of fecal residue on poultry carcasses by laser-induced fluorescence imaging. J Food Sci 74(3):E154–E159

Christensen J, Povlsen VT, Sorensen J (2003) Application of fluorescence spectroscopy and chemometrics in the evaluation of processed cheese during storage. J Dairy Sci 86(4):1101–1107

Crimaldi J (2008) Planar laser induced fluorescence in aqueous flows. Exp Fluids 44(6):851–863

Cynkar W, Cozzolino D, Dambergs B, Janik L, Gishen M (2007) Feasibility study on the use of a head space mass spectrometry electronic nose (MS e_nose) to monitor red wine spoilage induced by Brettanomyces yeast. Sens Actuators B Chem 124(1):167–171

Damez JL, Clerjon S, Abouelkaram S, Lepetit J (2008) Electrical impedance probing of the muscle food anisotropy for meat ageing control. Food Control 19(10):931–939

Dias LG, Peres AM, Barcelos TP, Morais JS, Machado A (2011) Semi-quantitative and quantitative analysis of soft drinks using an electronic tongue. Sens Actuators B Chem 154(2):111–118

Esbensen K, Kirsanov D, Legin A, Rudnitskaya A, Mortensen J, Pedersen J, Vognsen L, Makarychev-Mikhailov S, Vlasov Y (2004) Fermentation monitoring using multisensor systems: feasibility study of the electronic tongue. Anal Bioanal Chem 378(2):391–395

Escuder-Gilabert L, Peris M (2010) Review: highlights in recent applications of electronic tongues in food analysis. Anal Chim Acta 665(1):15–25

Fagan CC, O'Donnell CP, Cullen PJ (2009) Monitoring and control of mixing operations. In: PJ Cullen (ed) Food mixing: principles and applications. Blackwell, England, pp 107–124

Fagan CC, Ferreira TG, Castillo M, Payne FA, O'Callaghan DJ, O'Donnell CP (2011) Evaluation of naturally occurring milk fluorophores as tracers molecules for curd syneresis. J Dairy Sci (in press)

Ford HD, Tatam RP (2013) Spatially-resolved volume monitoring of adhesive cure using correlated-image optical coherence tomography. Int J Adhes Adhes 42:21–29

Frencia JP, Thomas E, Dufour E (2003) Measure of meat tenderness using front-face fluorescence spectroscopy. Sci Aliment 23(1):142–145

García M, Aleixandre M, Horrillo MC (2005) Electronic nose for the identification of spoiled Iberian hams. In: Spanish on conference electron devices, pp 537–540

Genot C, Tonetti F, Montenaygarestier T, Marion D, Drapron R (1992) Front face fluorescence applied to structural studies of proteins and lipid-protein interactions of viscoelastic food-products.2. Application to wheat gluten. Sci Aliment 12(4):687–704

Gowen AA, O'Sullivan C, O'Donnell CP (2012) Terahertz time domain spectroscopy and imaging: Emerging techniques for food process monitoring and quality control. Trends Food Sci Technol 25(1):40–46

Guillard F, Trägårdh C, Fuchs L (2000a) New image analysis methods for the study of mixing patterns in stirred tanks. Can J Chem Eng 78:273–285

Guillard F, Trägårdh C, Fuchs L (2000b) A study on the instability of coherent mixing structures in a continuously stirred tank. Chem Eng Sci 55(23):5657–5670

Gutes A, Cespedes F, del Valle M (2007) Electronic tongues in flow analysis. Anal Chim Acta 600(1–2):90–96

Haff RP, Toyofuku N (2008) X-ray detection of defects and contaminants in the food industry. Sens Instrum Food Qual Saf 2(4):262–273

Herbert S, Riaublanc A, Bouchet B, Gallant DJ, Dufour E (1999) Fluorescence spectroscopy investigation of acid-or rennet-induced coagulation of milk. J Dairy Sci 82(10):2056–2062

Herbert S, Riou NM, Devaux MF, Riaublanc A, Bouchet B, Gallant DJ, Dufour E (2000) Monitoring the identity and the structure of soft cheeses by fluorescence spectroscopy. Lait 80(6):621–634

Hof M, Fidler V, Hutterer R (2005) Basics of fluorescence spectroscopy in biosciences. In: M Hof, V Fidler, R Hutterer (eds) Fluorescence spectroscopy in biology: advanced methods and their applications to membranes, proteins, DNA, and cells. Springer, Berlin

Karoui R, Kemps B, Bamelis F, De Ketelaere B, Merten K, Schoonheydt R, Decuypere E, De Baerdemaeker J (2006) Development of a rapid method based on front face fluorescence spectroscopy for the monitoring of egg freshness: 1-evolution of thick and thin egg albumens. Eur Food Res Technol 223(3):303–312

Karoui R, Schoonheydt R, Decuypere E, Nicolai B, De Baerdemaeker J (2007) Front face fluorescence spectroscopy as a tool for the assessment of egg freshness during storage at a temperature of 12.2 degrees C and 87 % relative humidity. Anal Chim Acta 582(1):83–91

Karoui R, Nicolai B, De Baerdemaeker J (2008) Monitoring the egg freshness during storage under modified atmosphere by fluorescence spectroscopy. Food Bioprocess Technol 1(4):346–356

Lebrun M, Plotto A, Goodner K, Ducamp MN, Baldwin E (2008) Discrimination of mango fruit maturity by volatiles using the electronic nose and gas chromatography. Postharvest Biol Technol 48(1):122–131

Legin A, Rudnitskaya A, Seleznev B, Vlasov Y (2005) Electronic tongue for quality assessment of ethanol, vodka and eau-de-vie. Anal Chim Acta 534(1):129–135

Leitner M, Hannesschläger G, Saghy A, Chassagne-Berces S, Chanvrier H, Verlinden BE (2011) Optical coherence tomography (OCT) for quality control and microstructure analysis in food. In: InsideFood FP7 Workshop, 26 May. Athens, Greece

Liu X, Metzger LE (2007) Application of fluorescence spectroscopy for monitoring changes in nonfat dry milk during storage. J Dairy Sci 90(1):24–37

Luo P, Cheng Y, Zhao Y, Jin Y, Yang W (2007) Millisecond mixing of two liquid streams in a mixer model. Chem Eng Sci 62(18–20):5688–5695

Marilley L, Ampuero S, Zesiger T, Casey MG (2004) Screening of aroma-producing electronic lactic acid bacteria with an nose. Int Dairy J 14(10):849–856

Mason TG, Weitz DA (1995) Optical measurements of frequency-dependent linear viscoelastic moduli of complex fluids. Phys Rev Lett 74(7):1250–1253

Maurer L, Leuenberger H (2009) Terahertz pulsed imaging and near infrared imaging to monitor the coating process of pharmaceutical tablets. Int J Pharm 370(1–2):8–16

Mery D, Lillo I, Loebel H, Riffo V, Soto A, Cipriano A, Aguilera JM (2011) Automated fish bone detection using X-ray imaging. J Food Eng 105(3):485–492
Mettler-Toledo (2009) The X-ray inspection guide. Royston, UK
Niederquell A, Volker AC, Kuentz M (2012) Introduction of diffusing wave spectroscopy to study self-emulsifying drug delivery systems with respect to liquid filling of capsules. Int J Pharm 426(1–2):144–152
Nielsen MS, Lauridsen T, Christensen LB, Feidenhans'l R (2013) X-ray dark-field imaging for detection of foreign bodies in food. Food Control 30(2):531–535
Pani P, Leva AA, Riva M, Maestrelli A, Torreggiani D (2008) Influence of an osmotic pre-treatment on structure-property relationships of air-dehydrated tomato slices. J Food Eng 86(1):105–112
Parasoglou P, Parrott EPJ, Zeitler JA, Rasburn J, Powell H, Gladden LF (2009) Quantitative moisture content detection in food wafers. In: 34th international conference on infrared, millimeter, and terahertz waves, September 21–25, Busan (Korea)
Parra V, Arrieta AA, Fernandez-Escudero JA, Iniguez M, de Saja JA, Rodriguez-Mendez ML (2006) Monitoring of the ageing of red wines in oak barrels by means of an hybrid electronic tongue. Anal Chim Acta 563(1–2):229–237
Peris M, Escuder-Gilabert L (2009) A 21st century technique for food control: electronic noses. Anal Chim Acta 638(1):1–15
Pine DJ, Weitz DA, Zhu JX, Herbolzheimer E (1990) Diffusing-wave spectroscopy—dynamic light-scattering in the multiple-scattering limit. J Phys 51(18):2101–2127
Pinheiro C, Rodrigues CM, Schafer T, Crespo JG (2002) Monitoring the aroma production during wine-must fermentation with an electronic nose. Biotechnol Bioeng 77(6):632–640
Rudnitskaya A, Schmidtke LM, Delgadillo I, Legin A, Scollary G (2009) Study of the influence of micro-oxygenation and oak chip maceration on wine composition using an electronic tongue and chemical analysis. Anal Chim Acta 642(1–2):235–245
Sarig Y (2000) Potential applications of artificial olfactory sensing for quality evaluation of fresh produce. J Agric Eng Res 77(3):239–258
Schneider V, Wulf J, Geyer M, Schluter O (2008) Fluorimetrical determination as indicators for the properties of pork—Fluorescence spectroscopy as a tool for monitoring pork meat quality along the production chain. Fleischwirtschaft 88(9):127–130
Sharifi M, Young B (2012a) Milk total solids and fat content soft sensing via electrical resistance tomography and temperature measurement. Food Bioprod Process 90(C4):659–666
Sharifi M, Young B (2012b) Qualitative visualization and quantitative analysis of milk flow using electrical resistance tomography. J Food Eng 112(3):227–242
Shen YC (2011) Terahertz pulsed spectroscopy and imaging for pharmaceutical applications: a review. Int J Pharm 417(1–2):48–60
Stetter JR, Penrose WR (2002) Understanding chemical sensors and chemical sensor arrays (electronic noses): past, present, and future. Sens Update 10(1):189–229
Toko K (2000) Taste sensor. J Food Hyg Soc Jpn 41(5):J336–J341
Vlasov Y, Legin A, Rudnitskaya A (2002) Electronic tongues and their analytical application. Anal Bioanal Chem 373(3):136–146
Wadley R, Dawson MK (2005) LIF measurements of blending in static mixers in the turbulent and transitional flow regimes. Chem Eng Sci 60(8–9):2469–2478
Widengren J (2010) Fluorescence-based transient state monitoring for biomolecular spectroscopy and imaging. J Royal Soc Interface 7(49):1135–1144
Winquist F, Bjorklund R, Krantz-Rulcker C, Lundstrom I, Ostergren K, Skoglund T (2005) An electronic tongue in the dairy industry. Sens Actuator B Chem 111:299–304
Wold JP, Mielnik M, Pettersen MK, Aaby K, Baardseth P (2002) Rapid assessment of rancidity in complex meat products by front face fluorescence spectroscopy. J Food Sci 67(6):2397–2404
Worlitschek J, Mazzotti M (2003) Choice of the focal point position using lasentec FBRM. Part Part Syst Charact 20(1):12–17
Yu W, Erickson K (2008) Chord length characterization using focused beam reflectance measurement probe—methodologies and pitfalls. Powder Technol 185(1):24–30

Chapter 12
Food Industry Perspectives on the Implementation of a PAT Strategy

Julie Lundtoft Johnsen

12.1 Introduction

The process analytical technology (PAT) tools and analysers have been known within the food industry for decades. Previously, the focus with these tools was to implement on-/in-line analysers in the production process just to have real-time measurements to monitor the production. Within the past couple of years, the focus has changed from implementation of on-/in-line analysers just for monitoring the product attributes to the use of PAT technologies to understand and control the whole manufacturing process and to consistently ensure a predefined quality at the end of the manufacturing process.

Process variation within a manufacturing process is caused by, for example, uncontrolled disturbances like changes in the milk used in cheese making and the temperature of the surroundings. In an ideal world, there would be no changes to the uncontrolled variables. A given combination of settings for the process set points will always result in a product with a given quality. In the real world, the uncontrolled variables will change and a given combination of settings for the process set points will always result in a product that varies in quality. Variation in the product quality can be reduced if it is possible to determine the optimum settings for the process set point during the manufacturing process.

Prediction of the optimum settings for the process set points requires many years of experience for the production personnel or a control system that can take the changes to the uncontrolled disturbances into account and predict the set points. Such a control system can be developed based on the PAT tools such as on-/in-line analysers and multivariate data analysis. A schematic view of the principle behind a control system is shown in Fig. 12.1. Here, a manufacturing process with two process steps is shown. An advanced control system can be built for the total process

J. L. Johnsen (✉)
Arla Strategic Innovation Centre, Arla Foods amba, Aarhus, Denmark
e-mail: julie.lundtoft.johnsen@arlafoods.com

C. P. O'Donnell et al. (eds.), *Process Analytical Technology for the Food Industry*, Food Engineering Series, DOI 10.1007/978-1-4939-0311-5_12,

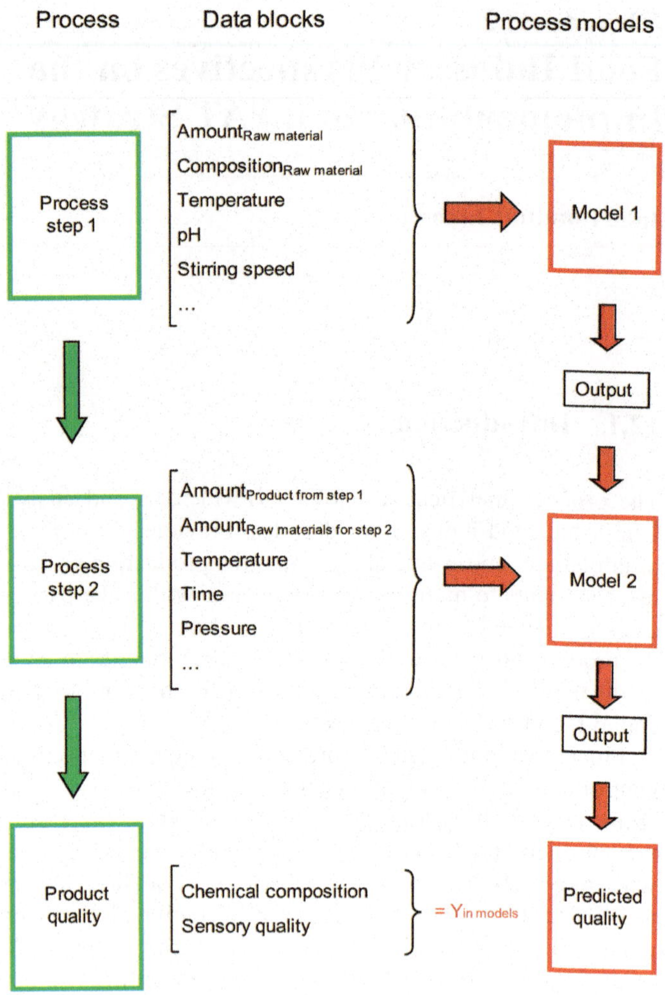

Fig. 12.1 Schematic view of the principle behind a control system

or for each process step. The aims of an advanced control system are to be able to predict the quality of the end product based on knowledge of the raw material and process settings used and to be able to predict the optimum settings for the process set points based on knowledge of the raw material used and specification of the pre-defined quality of the end product. An advanced control system is based on process models, which relates measurements of raw material and process variables to the outcome of a process or process step. The process model for process step 1 relates measurement of the raw material and the different process variables, like temperature, pH, speed and amounts of materials used in process step 1 with measurements of the semimanufactured product from step 1. The process model can be based on multivariate data analysis, first principle models or a combination. After model validation, this process model can be used to predict the optimum process settings for process step 1 based on measurements of the given raw material used and a given setting for the expected quality of the semimanufactured product. A total process model or a combination of individual process models for steps in a production process could be used to optimize a production process.

An advanced control system could be an ultimate solution for understanding and controlling a manufacturing process. Before it is possible to build a control system fully or partly based on multivariate data analysis, the necessary analysers used to measure the critical quality and control attributes need to be implemented and a system for automatic data collection installed. A PAT strategy needs to be developed and implemented.

This chapter focuses on advantages and challenges when implementing PAT and is based on experience gained during 7 years of work within the area. Case studies from dairy processing will demonstrate the potential for improved product consistency and enhanced process control through minimising the variability of critical quality attributes.

12.2 A PAT Strategy

The drivers for an implementation of a PAT strategy are usually to gain a uniform and high product quality, improve yield and reduce production costs. The propagation of PAT within the manufacturing processes has led to focus on process understanding and continuous improvements. This is due to the fact that it has become possible to generate relevant and high-frequency data by use of, for example, spectroscopic analysers, where a measurement is performed within seconds or minutes instead of using reference methods that take hours. The driver for implementation of PAT must be that the new strategy gives value to the business. An economical driver will result in commitment and focus in an organisation.

Within a manufacturing company, it should be discussed if and how a PAT strategy and the implementation of, for example, on-/in-line analysers gives value to the business. The number of possible PAT cases within a production site will depend on the size and complexity of the production.

During the development of a PAT strategy, one of the focus areas should be to identify possible changes to working routines within the organisation. Implementation of a PAT strategy can affect the routines in production and the laboratory. If changes are identified, the management should handle this and make sure that the routines are changed properly. This is an important task to ensure a successful implementation of a PAT strategy and the future work.

The purpose of a given PAT case has to be described in detail. What is the goal? What are the conditions for a successful application? What are the risks? How can the goal be reached—is it by implementation of an analyser, an advanced control system based on multivariate data analysis or both? What are the milestones? What is the business value?

12.2.1 Business Value Calculations

In the calculations of the business value, all the investments and costs needed and the expected economic gains for the implementation of a PAT strategy are listed and estimated.

- Investments and other costs: e.g. analysers, changes to the production equipment and IT systems, costs of samples taken for calibration of an analyser, time used by production technicians, laboratory technicians, engineers and others during implementation, maintenance costs and also costs related to education of personnel and consultant assistance from suppliers.
- Economic gains: e.g. improved yield due to reduced process variation, increased product quality, reduced amount of product outside specifications, reduced product costs and improved process knowledge.

Estimation of the economic gains is based on analysis of historic process and product data. The historic data are analysed to identify the present process and product variation and from that the potential gain is estimated. Estimation of the value of improved process knowledge can be a challenge, but in spite of that, it is an important gain, which needs to be addressed.

Another economic gain not mentioned above is reduced costs in the laboratory since there will be fewer samples to be measured by the reference method after implementation of an at-/on-/in-line analyser. The reason why it is not included in the list is that the calibration, validation and maintenance of an analyser require resources. Even though there will be fewer samples to measure, other tasks need to be taken care of instead, e.g. recalibration of an analyser.

There are different ways to estimate the business value, but a common estimate to use is the net present value (NPV). From the estimated investments, costs and economic gains, an NPV is calculated. The NPV indicates how much value a given project gives to the manufacturing company within a given period (e.g. 5 years).

The NPV is calculated as (12.1)

$$\mathrm{NPV} = C_0 + \frac{C_1}{1+r_1} + \frac{C_2}{(1+r_2)^2} + \cdots + \frac{C_T}{(1+r_T)^T} \quad (12.1)$$

C_0 is the cash flow at date 0. C_0 includes investments and costs of initiating a project and the value will therefore be negative. $C_1, C_2, \ldots, C_T$ are the cash flows from operation at year 1, 2,…, T and $r_1, r_2, \ldots, r_T$ is the discount rate at year 1, 2,…, T (Grinblatt and Titman 2002).

In the screening period before a proposed PAT strategy is accepted and an implementation is initiated, the NPV will be an estimate since the investments and gains can only be estimated. It is important that the NPV is as realistic as possible since it is used to decide whether an implementation of a PAT strategy is initiated. During a project, more details become available and the NPV will be more accurate, because at the time close to the implementation of analysers and other hardware the investments and other costs are known.

A high positive NPV indicates that the business value is high. The economic driver for implementation of a PAT strategy is there and implementation of the proposed PAT strategy may continue. The implementation of the strategy has to be accepted by the supply chain organisation, which is investing the money.

12.2.2 Organisation

Implementation of PAT obviously requires financial resources for investment in new equipment, but it also requires sufficient time resources and the people involved need the right competences. This fact is sometimes neglected, but missing focus and resources results in a high risk that an implementation will fail. Commitment from the supply chain organisation and management at the production site is crucial to a successful implementation.

The implementation of a PAT strategy should be managed by a project group, which is coordinated by a project manager. The project group should at least consist of people with competences in the production process, at-/on-/in-line analysers, laboratory analysis, multivariate data analysis, IT and automation.

Besides the competences within the project group, the following people should be available during an implementation, but they do not necessarily need to be in the project group: people with competences in contract writing and purchase to ensure a good agreement when investing, technicians from production to ensure knowledge sharing related to the manufacturing process and involvement during implementation, a maintenance department to ensure an easy handover, laboratory technicians to ensure the use of right reference methods and maintenance of the analyser and the supplier to ensure support during implementation.

Involvement of the technicians from production and laboratory in an early stage is an advantage, because after implementation, they are the end users and it will be a part of the work to establish their ownership of the new techniques and strategy.

The establishment of ownership and commitment in an organisation is important, and within a project organisation, the steering group has a central role in the work to secure commitment. The steering group consists of representatives from the supply chain organisation and/or management at the production site. The members of the steering group must have the authority to make superior decisions in relation to the project, because it is their responsibility to follow up on the progress of the project and to make superior decisions concerning the project, e.g. if changes are to be made to the purpose or description of a project. It is also the responsibility of the steering group to allocate enough and relevant resources to the project.

12.2.3 *Implementation of an Analyser*

A positive business value for a given project and a project organisation in place are fundamental for initiation of a project. After project initiation, the more technical part starts. The first part of a project is usually implementation of one or more analysers. The implementation of an analyser involves several steps. These steps are described in this part and the focus will be on how to ensure a successful implementation and the challenges that can appear. The steps are:

- Set up requirements for the analyser
- Do a screening of the market and assess one or two relevant techniques
- Approve the investment
- Install and develop method
- Validate the method
- Carry out maintenance

12.2.3.1 Requirements for the Analyser

The investment in a new analyser requires specific knowledge on what the purpose and economic goal are for the implementation of an analyser. Without the knowledge, it can be difficult to find the optimal analyser. Based on the knowledge, it is defined which data the analyser is going to provide to reach the goals and requirements for the analyser should be defined. The requirements can be defined based on questions like the ones listed below:

- What is the purpose of the analyser? For example, is it determination of the content of a quality attribute, like fat, protein and moisture content in cheese, or is it end-point detection, like detection of when a fermentation has ended?
- Should the measurement be at-line, on-line or in-line? What measurement interval is needed? Which type of sample needs to be measured; is it a liquid, a slurry, a powder, a solid?
- Which product characteristics need to be measured? For example, is it physical characteristics, like particle size of a powder; sensory characteristics, like the taste of a cheese; or chemical characteristics, like moisture content in cheese? What ranges or concentrations are the characteristics normally in? What accuracy and precision is needed? What are the accuracy and precision of the reference method?
- Where on the production equipment is the analyser going to be installed; e.g. is it in a pipe or a vat? What is the temperature of the product when measured? Is the product corrosive? Are there any special characteristics of the product or process, e.g. if the measurements are done in-line, what is the pressure in the pipe or vat?

- Is there a risk of segregation or a concentration gradient within the sample? This can affect the decision on if the measurements should be done in reflectance, transflectance or transmission.
- Which data from the analyser should be available for data storage, e.g. calculated concentrations and/or the spectra?
- How are the data going to be used in the production, e.g. for manual or advanced process control? Does the analyser need to be connected to the production server and IT network? If yes, how does it need to connect—through OPC, MODBUS, 4–20 mA, etc.? How should the data be available for the technicians in production, e.g. as a number or time series plot? How do the data need to be stored?
- How much maintenance can be accepted (maximum time to be used)? Should it be possible to control the instrument remotely?

The answers to all of these questions will result in a list of requirements for the analyser. The answers to the questions should focus on what is needed and NOT what would be nice. The list of requirements is then used when screening the market for potential solutions.

12.2.3.2 Screening of Market

A thorough screening of the market before the decision on which analyser to invest in is of high importance. The purpose of the screening is to collect enough information on possible techniques to be able to decide which technique best complies with the requirements. Information is found through a literature search, a search on the Internet, seminars, by talking to people from your network and by having meetings with suppliers.

Be aware that the screening should be wide in the beginning. Do not focus on only one technique or one instrument; for example, the near infrared spectroscopy (NIR) is widely used in many applications within the food industry, like measurements of moisture and protein content in bread, cheese, flour, meat, milk powder and pasta (Osborne et al. 1993). In spite of that, NIR and other spectroscopic techniques are not always the right choice. For some applications, another and sometimes more simple technique is the optimal one, for example, measurements of salt by the use of conductivity measurements. Increased complexity for an analyser can lead to use of more resources for method development and validation. It can be a challenge to find the optimal technique, but the choice can influence whether or not the implementation of the PAT strategy will be a success. Since the analyser is to be implemented into a production environment and people without expert skills in sensor technology and multivariate data analysis are to handle the analyser, it is more important that the analyser is easy to use and robust, does not require a lot of maintenance and that data from the analyser is easy to access instead of choosing an analyser with outstanding accuracy and precision.

For applications where there is no obvious choice of analyser after a screening, it can be a challenge to decide on which technique to invest in. In such a case, it can be an idea to borrow or lease the analyser(s), which seem to be the most relevant one(s), and make screening trials in the laboratory. Trials in the laboratory can give an indication whether the analyser can comply with the requirements.

12.2.3.3 Approval of Investment and Contract Writing

The next step is to get approval of the investment in the analyser decided upon. The investment process is different for the individual companies, but usually it is the project sponsor that needs to approve the investment and the project plan. The investment that needs to be approved should at least cover the analyser, sampling equipment, data collection system, the cost of changes to the production equipment, cost for taking calibration samples and doing reference analysis and cost of training courses in calibration and maintenance of the analyser.

After approval of the investment, a contract between the production site and the supplier should be signed. A detailed contract is normally an advantage for both parties, because the responsibilities during the delivery and installation will be outlined. The contract should cover:

- A description of the aim and functionalities of the analyser, e.g. why is the analyser installed and how is it used?
- A description of the technical solution including installation instructions and timetable for the installation steps with responsibility information. Responsibility is shared between supplier and buyer depending on task.
- A description of the performance requirements, e.g. requirements for measurement accuracy and precision, handling of analyser, data validity, maximum level of unscheduled down time, safety and cleaning. It is also important to be aware if there are any limitations for the analyser.
- General issues like period of guarantee, conditions of guarantee, insurance, payment, service and milestones for the installation including factory acceptance testing.

The contract negotiation can be a challenge, but if the details are not all discussed and agreed upon between the supplier and buyer, there might be bigger challenges during the installation if something deviates from plan and responsibilities are not defined. After signing the contract, the order can be placed.

12.2.3.4 Installation and Calibration

The physical installation of the analyser, the method development and calibration is the next step when the analyser is delivered. The challenges that can appear during the physical installation depend on whether the analyser has to be used at-line or on-/in-line. An at-line installation usually only requires some space in the laboratory or operator room close to the production. On- and in-line installations require changes to be done to the production equipment (vats, pipes, etc.) and are usually more complex. One of the things to be aware of is to ensure that the installation of the analyser comply with the hygienic requirements in the production.

For both at-line and on-/in-line installations, sampling equipment for the calibration and validation samples has to be installed. Sampling needs to be performed in a way where the samples taken are representative. For calibration purposes, the samples taken need to be of the material measured by the analyser at a given time. For example, for flows in pipes and for more heterogeneous material, as seen in the food industry, sampling is a challenge. Sampling errors can be up to 1000 times larger than the analysis error (Gy 1998, 2004), so sampling is an important issue to focus on before implementation.

The physical installation of sampling equipment and the analyser may also require changes to the electrical system. The installation of the analyser and sampling equipment is based on recommendations from the equipment supplier, and it is recommended to follow these instructions closely to avoid performance issues. If any performance issues appear in spite of everything, the supplier should be able to solve the problem, since the instructions from the supplier are based on their experience and knowledge on the given application.

For some on-/in-line installations, it can be a challenge to find the optimal process interfacing (e.g. flow cell, probe head and probe angle) due to a difficult product matrix. In such cases, trials performed in collaboration with the supplier can be an advantage.

The physical installation should be completed by an approval from the supplier to ensure that the recommendations are followed.

Calibration can be initiated when the physical installation is approved, but before calibration samples are taken, calibration procedures should be in place and the responsibility for every step in the procedure should be addressed. The following should be defined and described:

- How often are calibration samples taken?
- How is a representative calibration sample taken?
- Who is responsible for taking samples?
- How is the sample handled after it is taken and until it is analysed in the laboratory? And who is responsible?
- Which reference method needs to be used and what is the exact procedure to follow?
- How is the calibration data reported?

All of the above are described in the calibration procedure and technicians in both the laboratory and production need to be trained in the procedures. If no proper calibration procedure or training is available, there is a risk that samples are taken or handled in a wrong way or that a wrong reference method is used, which can lead to a calibration with unacceptable accuracy and precision or to a long calibration period. The training of persons who have no previous knowledge within analysers, statistics or multivariate data analysis can be a challenge, but it is one of the most important tasks to succeed in. The persons who are to take the samples and do the reference analysis have to understand the importance of the procedures. During preparation of the calibration procedure, one should be aware that the more advanced a procedure is, the more errors or problems it can cause during the calibration period.

Calibration of an analyser is based on results from a reference analysis performed in the laboratory, and to achieve a good calibration it is important that the right reference method is used. The uncertainties of the reference analysis will influence the accuracy and precision of the calibration.

Calibration of an analyser requires samples, which contain varying amounts of the component of interest. Samples used for calibration should be taken in a way where they will cover the area of normal variation within the production. One can be tempted to take many samples from one production run and make a calibration based on these samples, but the calibration will not be robust enough to handle batch-to-batch variation. Calibration samples taken need to cover:

- All the products included in a given calibration
- Process variation for the component of interest
- Batch-to-batch variation, e.g. due to changes in raw materials

Variations in the production need to be included in the calibrations for the at-/on-/in-line analyser.

At the beginning of the calibration step, it can be a good idea to have a limited number of people involved in the calibration procedure. This is due to the fact that with many people involved the risk for mistakes to happen will be higher. When the calibration procedure and routines for one or two technicians in the laboratory and production are successfully worked into the daily routines, more people can be introduced and trained.

12.2.3.5 Method Validation

Validation of the installation and calibration is the final step before the implementation is ended and the analyser is included into the maintenance procedures at the production site. Validation of the installation covers evaluation of whether, based on the experiences from the calibration period, there is anything in the practical handling of the analyser that is not appropriate. It also covers an evaluation of whether the technical requirements are fulfilled.

Validation of a calibration covers the evaluation of whether the required accuracy and precision is reached. This validation is done based on results for validation samples, which are sampled and analysed using the exact same procedures as the calibration samples.

12.2.3.6 Maintenance

Maintenance of the implemented analyser is included into the normal maintenance procedure and is done by the personnel at the production site. The maintenance is

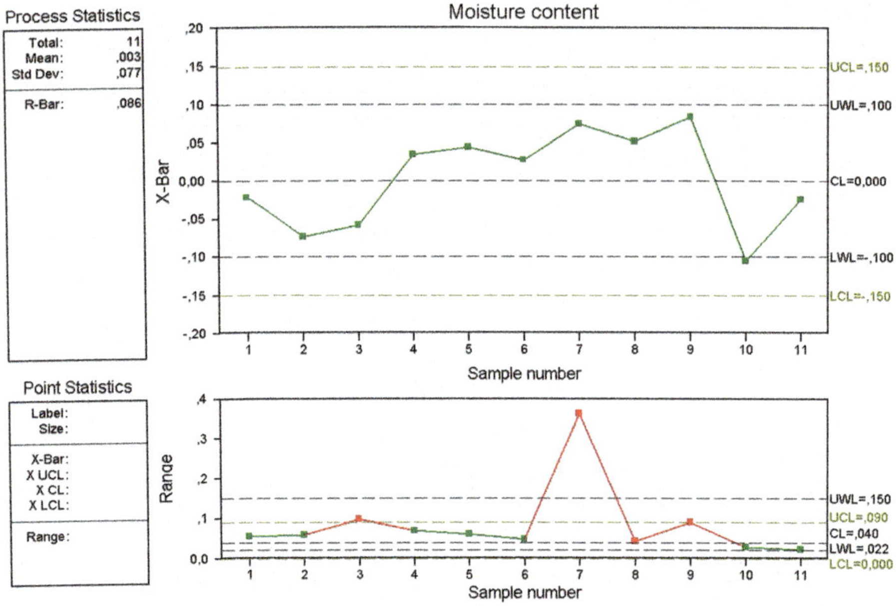

Fig. 12.2 Control chart for monitoring the difference between predicted and reference value

divided into two areas, hardware maintenance and calibration maintenance. Hardware maintenance is done to ensure that the consumable parts, e.g. the light source, are changed in time to ensure that the analyser keeps working. Many analysers have the possibility to set up automatic tests for performance checking of the instruments, and if these tests fail, alarms are activated.

Calibration maintenance is done to monitor and ensure the accuracy and precision of the method. Calibration maintenance includes making control charts for the reference analysis in the laboratory to ensure that the reference analysis is in control and can be used. It also includes making control charts for samples taken on a regular basis to monitor the prediction error (calculated as the difference between the predicted value from the analyser and the reference value from the laboratory). An example of a control chart is shown in Fig. 12.2.

If relevant, the calibrations are recalculated or updated. Recalibration can be necessary due to drift in the analyser signal or changes in the composition of the samples.

Maintenance is important for the continued success of the implementation of a PAT strategy. For the maintenance to be successful, it is important to have focus on resources and competences. The managers at the production site have the responsibility to solve this challenge but it might be an idea to appoint a person, who is responsible for the maintenance and for giving the managers warning if challenges on competences and resources appear.

12.2.4 Control Strategy

Process optimization and process control based on at-/on-/in-line analysers can be more or less automatic. Three different strategies can be used.

- Manual control: The results from the analyser are logged into the data collection system in the production in real time and shown on the production technicians' screen in the control room. The production technicians will take action if necessary, based on these results and their own experience.
- Advisory system: The results from the analyser are logged into the data collection system in the production in real-time together with all the other process variables. Based on the real-time data and a process model calculated from historical data, the advisory system will calculate the optimal settings for the process set point and advise the production technician on which changes to make if necessary.
- Automatic control system: The results from the analyser are logged into the data collection system in the production in real time together with all other process variables. Based on the real-time data and a process model calculated from historical data, the automatic control system will calculate the optimal settings for the process set point, and if necessary, the set points will be changed automatically.

The control strategy to choose will be dependent on the IT structure available, the results of the business value calculation, the size of the production at the production site and on the available resources and competences. The control strategy should be selected based on requirements and not because it is the most simple or advanced strategy.

A manual, advisory or automatic control system, fully or partly based on multivariate data analysis, depends on the relevant process data and data from at-/on-/in-line analysers being logged into the system in real time. Without real-time data collection and data handling systems, a control system is impossible to build into production.

Worldwide, there are many different suppliers of data collection and data handling systems on the market, and which system is the most optimal depends on which other control systems there are on a given production site. Compatibility between the different IT systems on a production site is very important.

12.2.4.1 Data Collection and Data Handling System

Data collection and data handling is critical for whether an advisory or automatic control system will work or not. For the control strategies to succeed, all the relevant process data and data from at-/on-/in-line analysers need to be logged. Process knowledge is important when deciding which data could be relevant and not. All process variables that can influence the final product should be logged. It is better to log all the process data than to log too few data, because the nonsignificant process variables can easily be excluded from the process model during the modelling stage.

Whether the advisory or automatic control system is built for a single process unit or for a total production process, the steps needed to define relevant data are the same. A list of the process variables that are already logged should be made and the different steps in the production process should be analysed, based on this list, to decide if more process variables need to be logged. Afterwards, a plan for how to implement the data collection should be made. In the same way as for a new analyser, a list of requirements should be made for the data collection and data handling system. The other steps like screening of the market, approval of investment and contract writing, installation and validation are just as important for a data collection and handling system as for a new analyser.

Since the process data and data from the at-/on-/in-line analysers are crucial for an automatic control or advisory system, the data always need to be valid and available. In some data collection and data handling systems, it is possible to register if an analyser fails and no data are logged into the system. In such a case, some systems can also give an alarm to raise awareness of missing data, by either sending an email or an SMS. An alarm makes it possible to solve the problem as soon as possible.

Data collection, data handling and IT communication between the different IT systems in a production is crucial to a running PAT system, and focus on IT and IT risk management is important to ensure a well-functioning system.

A benefit from implementation of an extensive data collection and data handling system is that, from the data, there is complete traceability from raw material to final product. All records for a given product will be stored within the data collection and data handling system.

12.2.4.2 Process Modelling

The basis of an advisory or automatic control system is the process model, which can be based on multivariate data analysis, first principle models or a combination. The process model relates measurements of raw material and process variables to the outcome of a process or process step and extracts information on how all the variables are behaving relative to one another. This will contribute to more process knowledge.

Before process modelling is initiated, it should be ensured that the logged data for the process variables and from the analysers are valid and of high quality. Without high-quality data, there is a risk that the process modelling will fail. The task requires resources to validate the process data and it can be a challenge.

In the same way as for the calibration of an analyser, the process data included in the modelling need to cover all the products included in a given process model, batch-to-batch variation and sufficient process variation. A data set with sufficient variation can take months to collect, particularly if variation during normal operating conditions is not big enough to reveal obvious cause–effect relations. In such cases, step tests with planned and supervised changes to the process settings can usefully be made. The challenge is to define how big changes to the process settings can be allowed without the risk of producing product outside specifications.

12.3 Case Studies: Process Optimization Based on Quality Attribute Measurements

Quality attributes for dairy products can be both the chemical composition of a given product like protein, moisture and fat content, and the sensory quality attributes like taste, smell and consistency. The quality attributes of products or semimanufactured products can be measured using PAT tools. Table 12.1 shows some examples of PAT tools employed for various operations within the dairy processing industry. The asterisks in the table indicate how widely the techniques are implemented.

The case studies in this chapter will focus on cheese production and describe the goals, advantages and challenges of implementing a measurement system for moisture in semihard cheese and a sensory quality assurance system for cream cheese.

12.3.1 In-line Moisture Measurement for Semihard Cheese

Measurements of the quality attributes of products are important for process control, and for semihard cheese, one of the quality attributes is moisture content. The variation in and level of moisture in cheeses influences the yield. Control of the moisture content in cheese is therefore also a part of optimization of the yield in cheese production. In-line measurements of the moisture content make it possible to control the cheese vats and pressing system to reach a predefined moisture content. Control based on the measurements results in a reduced batch-to-batch and in-batch variation and gives good and stable quality and a higher yield.

The requirements that were defined for an analyser to measure moisture content in semihard cheese resulted in the selection of a technique based on microwave technology. The advantage of this technique is that some available microwave sensors can measure in transmission through a cheese block, which is important because there is typically a moisture gradient within the block. A cheese block can be up to 140 mm high. The other requirements were, as described in Sect. 12.2.3, related to accuracy and precision of the measurements, data storage, data communication and the process conditions at the measuring point. In the given case, a specific requirement related to cleaning was defined, because the analyser was installed in a place where CIP (clean in place) was performed.

The instrument was delivered, installed into the production and approved by the supplier.

Table 12.1 Examples of PAT tools employed for various operations within the dairy processing industry

Product	Unit operation	Technology					
		NIR	MIR	Microwave	Turbidity	Conductivity	Sensory assessments
Raw milk	Compositional analysis		****				
	Authenticity		*				
Cheese	Compositional analysis	****		***			
	Fermentation		**				
	Sensory quality						****
Butter	Compositional analysis of raw materials	****	****				
	Compositional analysis	****					
	Sensory quality						****
Cream cheese	Compositional analysis of raw materials		****				
	Fermentation		**				
	Separation	***		**		***	
	Compositional analysis	***	***				
	Sensory quality						****
Whey	Compositional analysis		****		*		
	Fractionation		***				
WPC	Compositional analysis	***					
WFC	Compositional analysis	**					
Milk powder	Compositional analysis	***					
Clean in place					****	****	

Asterisks indicate how widely the techniques are implemented
NIR near infrared spectroscopy, *MIR* mid infrared spectroscopy, *WPC* whey protein concentrate, *WFC* whey fat concentrate

12.3.1.1 Changes of Routines in Production and Laboratory

The decision to implement the in-line moisture analyser at a production site resulted in the need for new routines in production and in the laboratory. A routine for how calibration samples are taken during production needed to be introduced and in the laboratory there was a need for new routines concerning reference analysis. The change in routines can be a challenge and can also require new sorts of competences.

Changes in routines can also be caused by the product being measured and sampled. In the case of semihard cheese, one of the challenges in the calibration work was the sampling. The sampling needs to be performed in such a way that the samples taken from a block of cheese of minimum $20 \times 20 \times 9$ cm is representative of the given cheese block and taken without changing the composition of the sample. There is a risk of changing the composition of a sample of fresh semihard cheese because when the sample is taken the cheese is cut, and when a fresh semihard cheese is cut, whey starts to drain from the cheese. The whey drainage results in a decrease in moisture content in the sample taken to the laboratory for reference analysis. To ensure that the calibration for the analyser is as accurate as possible, specific routines were implemented for how to take the calibration samples and how to handle the sample prior to reference analysis. The calibration samples were taken at the production line, and as soon as the sample was cut from the cheese, it was placed into two bags and transported to the laboratory right away. In the laboratory, the sample and the small amount of whey that had drained from the cheese into the bag was mixed prior to the reference analysis to ensure that the measured moisture content was not affected by the whey drainage.

It is not enough just to implement a new routine. Another important part is to have the production and laboratory technicians to understand why it is necessary that they take the sample in the way described and why they must use a given reference method. For production personnel who are not familiar with on-/in-line analysers and do not understand why analysers need to be calibrated, it can be hard to understand why they need to follow routines that sometimes are time consuming and difficult to carry out. Project managers for an on-/in-line implementation need to be aware of this challenge and find a way to handle it. In cases where the new routines are more time consuming during the calibration period, it is also important to get management support to have the necessary resources. If the necessary resources are not made available, there is a risk that the calibration period will be prolonged or that the calibration obtained will not be satisfactory and cannot be used for control purposes.

The calibration work should not be initiated until the routines are in place.

12.3.1.2 Calibration of the In-line Moisture Analyser

Implementation of the in-line moisture analyser required variation in the moisture content of the cheese. As described in Sect. 12.2.3, the calibration samples need to cover all the products included in a given calibration, process variation for the component of interest and batch-to-batch variation.

In a dairy, different sorts of cheeses are produced. The cheeses differ in height, size, moisture, fat and protein content. Based on these variables, the cheeses can be grouped according to similarities. Some types of cheese are sufficiently alike so that they can be included in the same calibration to keep the number of calibrations to a minimum. Less number of calibrations results in less calibration maintenance work.

Calibration of the in-line moisture analyser (using microwave transmission) for each group of cheeses requires that samples are taken for every sort of cheese included in the group and that the samples taken vary in height and moisture content. This seems like a fairly easy task, but within a production environment challenges appear. The planning of when samples are taken can be difficult.

In a cheese plant, the production is planned from week to week and the production is highly flexible. The flexibility and the fact that different sorts of cheese are produced results in a production plan that varies from week to week, and during a day different sorts of cheese are produced. It is rarely the case that one sort of cheese is always produced on a specific day of the week. The flexibility in the production plans is a challenge when the people at the dairy have to schedule the resources for the calibration work.

The production at a dairy typically runs for 24 h a day, and sometimes the product of interest is produced during the night when no laboratory technicians are on site. In other cases, the product of interest is planned for production during the day but a technical problem then arises during production and the production is delayed. Such cases can cause calibration work to be disrupted from time to time.

12.3.1.3 Benefits of Implementation of an In-line Moisture Analyser

Even though implementation of an in-/on-line analyser takes time and requires the right competences and resources, there are important benefits.

The measurements from the in-line moisture analyser resulted in new insight into the cheese-making process. The cheese-making process consists of multiple process steps. Cheese milk is added into a cheese vat and then culture and rennet is added to ensure acidification and coagulation. After the coagulation of the cheese milk, the coagulum is cut and stirred. Whey is removed from the cheese vat before the cheese curd is pumped into a pressing system, where it is pressed and formed into cheese blocks. The cheese blocks are then brined and stored.

The in-line moisture analyser is installed between the pressing and brining step as a measurement of a critical product quality attribute for the fresh produced semi-hard cheese. The measurements of the moisture in the fresh cheeses revealed that the moisture content in most cases decreased within a batch; an example is shown in Fig. 12.3.

The first cheeses within a batch have higher moisture content than the last cheeses within a batch. The measurements confirmed the intuitive knowledge about the process. It had never been possible to document this in detail before because this

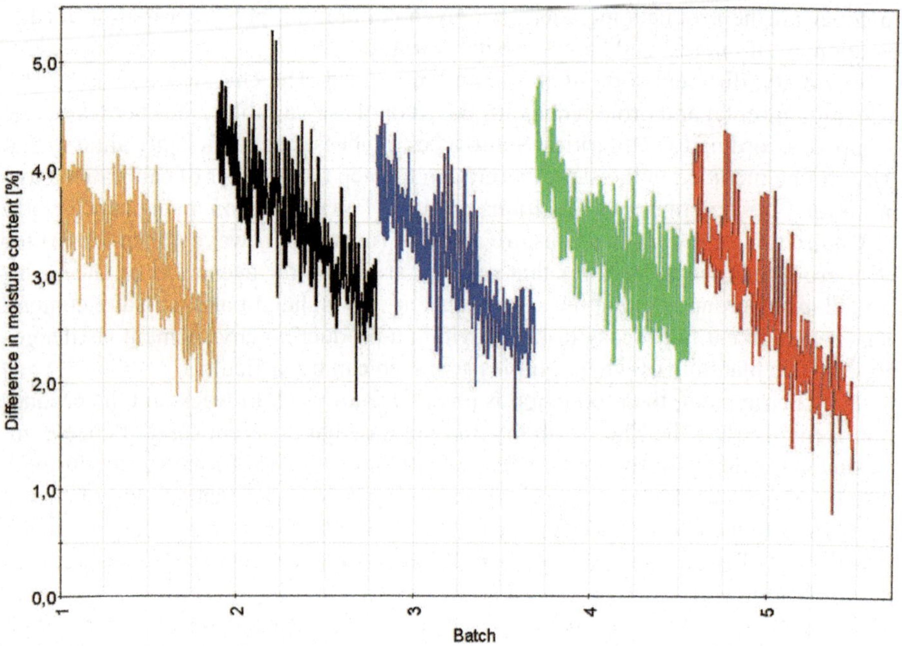

Fig. 12.3 In-line measurements of moisture in cheeses for five different batches

kind of analysis would have required a lot of manual sampling and reference testing in the laboratory, which would have required many resources.

The benefit of the in-line moisture analysis is that the variation within the production can be documented, and based on the measurements it is possible to make changes to the relevant process variables to reduce the systematic decrease in moisture content within a batch of cheese.

An example of in-line moisture measurements for two batches where changes were made to the relevant process variables and three batches without changes is shown in Fig. 12.4.

The in-line measurements for the batches where changes were made to some relevant process variables show that the systematic decrease in moisture content has been reduced compared to the batches without changes. These results show one of the potentials for process optimization based on the in-line moisture measurements.

The process optimization and process control based on the in-line moisture measurements can be done in three ways, manual control based on real-time feedback to the production technicians or control based on advanced process models for an advisory system or automatic control system. The implemented control strategy will vary from dairy to dairy dependent on the result of the business value calculation, the size of the production at the dairy and on the available resources and competences at the given dairy. The in-line measurements can be used for optimization of the internal batch variation by subsidiary optimization for the most relevant

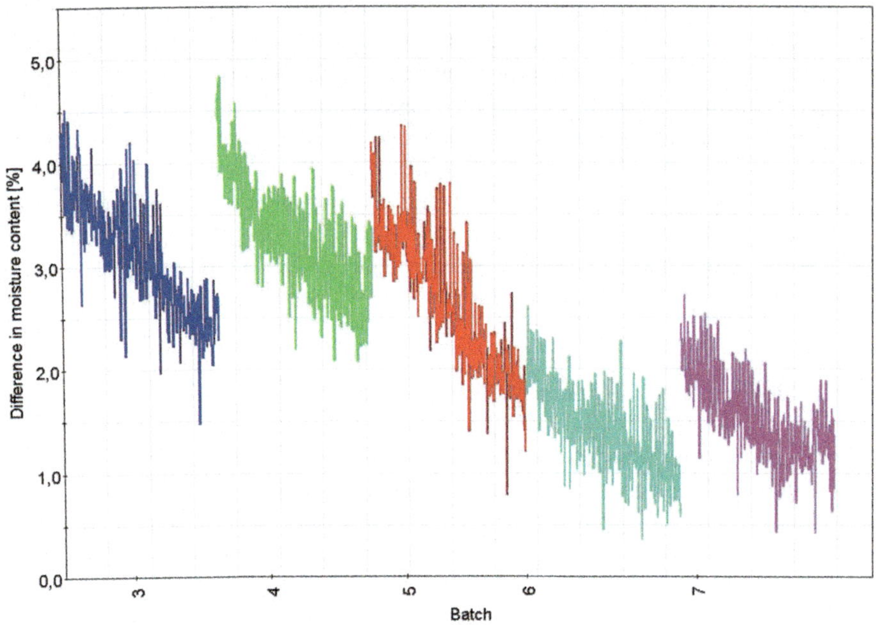

Fig. 12.4 In-line measurements of moisture in cheeses including batches where changes were made to the process variables to reduce the decrease in moisture content. *Light blue* and *purple lines* represent batches with changes. *Blue, green* and *red lines* represent batches without changes

process step, but in the case of the advanced process control the measurements can also be used to build a process model, which is used to predict the optimal settings for the overall process set point to ensure a specified moisture content in the cheese produced.

12.3.2 Sensory Quality Assurance System for Cream Cheese

The key quality attributes for cream cheese are sensory attributes related to the consistency, taste, smell and appearance of the cheese. The sensory attributes need to be measured to be able to optimize the production according to these attributes. Sensory attributes can be difficult to measure with at-/on- and in-line analysers, but even though it is a challenge, there is a potential in optimization of the sensory quality of food products. The economic gains will be a reduction in amount of product outside specifications and thereby in the amount of rejected product.

Instead of using at-/on- or in-line analysers, a sensory quality assurance system was developed with the focus that the assessments could be used in an advanced control system. In the same way as for an instrumental analyser, as described in Sect. 12.2.3, requirements were defined for the sensory quality assurance system both within the performance area and for data collection. The assessments needed to give interval data and to be on a linear scale instead of nominal data and category scale. The assessments also had to focus on possible deviations from product specification for optimum product instead of assessing total characteristics like smell, taste, consistency and appearance. Some possible deviations could be that the cream cheese is too soft or that there are small grains of fat. An assessment of all possible deviations from specification for a product makes it possible to correlate the results to how the production was run, in order to gain process insight, to find the critical control points and to make it possible to control the production to produce a product with a predefined quality and reduce the amount of rejected product.

12.3.2.1 Principle of the Sensory Quality Assurance System

In the sensory quality assurance system, the assessment is divided into two steps. The first step for the assessors after they have tasted the product is to decide whether the product is approved, approved with a remark or rejected, based on the product specifications. "Approved with a remark" means that the product is within product specifications but with small deviations from optimum product. The conclusion from the first step of the sensory assessment is whether the product can be sold.

The second step in the assessment relates to process optimization. Optimization of the process to gain an improved sensory quality is focused on the ability to control the production to reduce the amount of product categorized as "approved with remark" or rejected. To reduce this amount, it is necessary to assess if a given product deviates from specifications for the optimum product, and if it does, then to assess by how much it deviates. Results from the second step in the assessment can then be used to investigate what, in the production process, caused the deviation. For every product type, a list of possible deviations from optimum product is defined—see example in Fig. 12.5.

The degree of deviation is assessed on a line scale, where it is important that the line scale is unstructured. An unstructured line scale has no numbers on the scale, because such numbers can be seen as anchor points on the scale and the assessments will then usually be placed at these points. The only anchor points on the scale are None and Most, for which the assessors have definitions. Assessments on a line scale also give normally distributed data, which is preferred. Normally distributed data would not be achieved if the assessors were just to place a cross on predefined points. Without normally distributed data, it is not possible to make the relevant data analysis, like, for example, ANOVA.

12.3.2.2 Training of the Assessors and the Challenges

Implementation of a sensory quality assurance system includes training of assessors. The assessors need to go through training to be able to manage the assessment system, just like an analyser needs to be calibrated. Training of the assessors is as

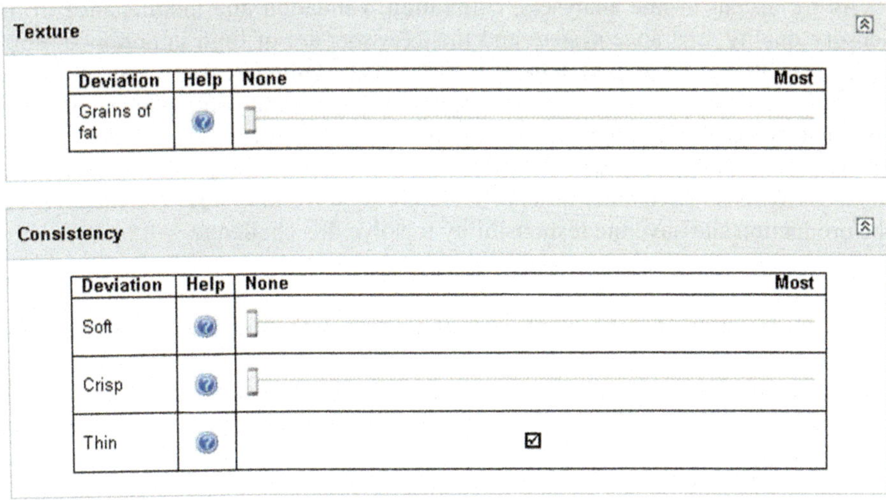

Fig. 12.5 The sensory quality assurance system

important as calibration of at-/on-/in-line analysers and it takes time and effort. The assessors need to get the right competences.

During training, the assessors need to understand the definition of maximum degree of deviations for a given product and have to assess reference samples taken from the production. The reference samples need to have various degrees of deviation. For every reference sample, the assessors need to agree on which area the sample is placed on the line scale, which is done during a discussion between the assessors based on the product specification. The training can be challenging because every assessor needs to agree on the decision. It is necessary to be aware that in a sensory quality assurance system the analyser is human and not a mechanical instrument, which can be reset or recalibrated. In some cases, this requires that work needs to be done to change the mindset of an assessor, who may previously have worked with other assessment systems.

The purpose of the training is to ensure that the assessors can repeat their test results on the same sample over time, that they can separate products with different degrees of deviation from optimum product and that they can rank the products based on degree of deviation.

When this is documented, the system can be taken into use at the production site. To ensure and to document that the assessors can repeat themselves over time, a standard sample or a sample from production, which is used repeatedly, is assessed when products from normal production are being assessed. To document that there is agreement between the assessors, it is important to have replicates, so more assessors need to assess the same products from normal production. This can result in changes in routines in the daily work, because the assessors and their managers need to plan when the assessments are performed.

As for at-/on-/in-line analysers, calibration, validation and maintenance of the sensory quality assurance system and the assessors are of high importance. For a sensory quality assurance system, it requires that training sessions be carried out with regular intervals. For a successful implementation and continued use of a sensory quality assurance system, resources are crucial, but as for calibration and maintenance of an analyser, it can often be a challenge to get enough resources because the assessors also have other tasks to solve during a working day. The managers at the production site have the responsibility to solve this challenge.

12.3.2.3 Benefits of Implementation of a Sensory Quality Assurance System

Implementation of a sensory quality assurance system requires time, training of the assessors and the right competences and resources, but there are very important benefits to gain from the work.

The assessments in the sensory quality assurance system are performed on a computer, and the data are saved into the data collection and data handling system together with relevant process data. The assessment results from the sensory quality assurance system can be analysed by statistics and multivariate data analysis, which makes it possible to correlate the results to how the production was run. This will generate more process knowledge. The assessments could be included into a process model and into a control system to make it possible to control the production to ensure production of a product with a predefined quality and reduce the amount of rejected product.

Besides the economic gains, the results from the assessments can be used as part of a quality management system at a production site. As stated in the previous section about training of the assessors, it should be possible to document that the assessors can repeat themselves over time and to document that there is agreement between the assessors. Since the results from the assessments are automatically saved into a data collection and data handling system, the data will always be available and documentation on the performance of the assessors can be extracted. This documentation can then be used when an audit is performed at the production site.

12.4 Summary

The intent of this chapter was to present some food industry perspectives on the advantages and challenges when implementing a PAT strategy.

The driver behind implementation of a PAT strategy must be an economic gain obtained by improved yield, increased product quality, reduced amount of product outside specifications or reduced product costs. Besides the economic gains, the implementation of PAT results in other benefits like more process knowledge and understanding. As part of the implementation of a PAT strategy, a data collection and data handling system is implemented which can give the advantage of complete traceability in a production site.

Management commitment and support is crucial to ensure a successful implementation of PAT, because the necessary focus, resources and competences have to be available throughout a project and after implementation. It is important to be aware that the investment made during an implementation of a PAT strategy is not only in physical instrumentation (e.g. in an analyser) but also in resources and competences for calibration, validation and maintenance of the analyser and maintenance of a data collection and data handling system. The steps and challenges to overcome when implementing an analyser were described. The list of steps is not necessarily complete, but hopefully it will be an inspiration and basis for further discussion at sites where a PAT strategy is going to be implemented.

References

Grinblatt M, Titman S (2002) Financial markets and corporate strategy, 2nd edn. McGraw-Hill, New York

Gy P (1998) Sampling for analytical purposes. Wiley, West Sussex

Gy P (2004) Sampling of discrete materials—a new introduction to the theory of sampling I. Qualitative approach. Chemom Intell Lab Syst 74(1):7–24

Osborne BG, Fearn T, Hindle PH (1993) Practical NIR spectroscopy with applications in food and beverage analysis, 2nd edn. Longman Scientific & Technical, Essex

Index

C. P. O'Donnell et al. (eds.), *Process Analytical Technology for the Food Industry*, Food Engineering Series, DOI 10.1007/978-1-4939-0311-5,

CPSIA information can be obtained
at www.ICGtesting.com
Printed in the USA
LVOW02*0720041115
461050LV00002B/2/P

9 781493 903108